HAND

HAND SECRETS

Third Edition

Peter J.L. Jebson, MD
Associate Professor and Chief
Division of Elbow, Hand, and Microsurgery
Department of Orthopaedic Surgery
University of Michigan Medical Center
Ann Arbor, Michigan

Morton L. Kasdan, MD, FACS
Clinical Professor
Division of Plastic and Reconstructive Surgery
University of Louisville School of Medicine
Louisville, Kentucky
Clinical Professor
Department of Preventive Medicine and Environmental Health
University of Kentucky Medical Center
Lexington, Kentucky

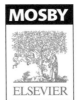

1600 John F. Kennedy Boulevard
Suite 1800
Philadelphia, PA 19103-2899

Hand Secrets ISBN-13: 978-1-56053-623-9
Third Edition ISBN-10: 1-56053-623-3

Copyright © 2006, 2002, 1998 by Elsevier Inc. All rights reserved.

No part of this publication may be reproduced or transmitted in any form or by any means, electronic or mechanical, including photocopying, recording, or any information storage and retrieval system, without permission in writing from the publisher. Permissions may be sought directly from Elsevier's Health Sciences Rights Department in Philadelphia, PA, USA: phone: (+1) 215-239-3804, fax: (+1) 215-239-3805, e-mail: healthpermissions@elsevier.com. You may also complete your request on-line via the Elsevier homepage (http://www.elsevier.com), by selecting "Customer Support" and then "Obtaining Permissions."

NOTICE

Knowledge and best practice in this field are constantly changing. As new research and experience broaden our knowledge, changes in practice, treatment, and drug therapy may become necessary or appropriate. Readers are advised to check the most current information provided (i) on procedures featured or (ii) by the manufacturer of each product to be administered, to verify the recommended dose or formula, the method and duration of administration, and contraindications. It is the responsibility of the practitioner, relying on his or her own experience and knowledge of the patient, to make diagnoses, to determine dosages and the best treatment for each individual patient, and to take all appropriate safety precautions. To the fullest extent of the law, neither the Publisher nor the Editors assume any liability for any injury and/or damage to persons or property arising out or related to any use of the material contained in this book.

Previous editions copyrighted 2002, 1998.
Library of Congress Cataloging-in-Publication Data
Hand secrets/[edited by] Peter J.L. Jebson, Morton L. Kasdan.–3rd ed.
  p.; cm.
  Includes bibliographical references and index.
  ISBN-13: 978-1-56053-623-9
  ISBN-10: 1-56053-623-3
  1. Hand–Wounds and injuries–Examinations, questions, etc. 2. Hand–Diseases–Examinations, questions, etc. I. Jebson, Peter J. L., 1963- II. Kasdan, Morton L.
  [DNLM: 1. Hand Injuries–Examination Questions. 2. Hand–pathology–Examination Questions. WE 18.2 H2362 2006]
RD559.H35993 2006
617.5'750076–dc22

2005057655

Vice President, Medical Education: Linda Belfus
Developmental Editor: Stan Ward
Senior Project Manager: Cecelia Bayruns
Marketing Manager: Kate Rubin

Printed in China.

Last digit is the print number: 9 8 7 6 5 4

CONTENTS

Top 100 Secrets 1

1. Hand and Wrist Anatomy 9
   Nana Mizuguchi, MD, and Terry McCurry, MD

2. Biomechanics of the Hand and Wrist 15
   Lawrence G. Sullivan, MD

3. Physical Examination of the Hand 20
   Terry McCurry, MD, and Morton L. Kasdan, MD

4. Neurologic Evaluation of the Upper Extremity 27
   Hal M. Corwin, MD

5. Anesthesia 32
   Bruce Crider, MD

6. Diagnostic Imaging of the Wrist and Hand 38
   Joel S. Newman, MD, and Arthur H. Newberg, MD

7. Osteoarthritis 50
   Jay T. Bridgeman, MD, DDS, and Sanjiv H. Naidu, MD, PhD

8. Rheumatoid Arthritis 59
   Mark N. Halikis, MD, William M. Weiser, MD, and Julio Taleisnik, MD

9. Other Arthritides 70
   Michael I. Vender, MD, and Prasant Atluri, MD

10. Tendinitis and Tenosynovitis 75
    Peter J.L. Jebson, MD

11. Infections 81
    H. Stephen Maguire, MD

12. Extravasation, Foreign Bodies, and High-Pressure Injection Injuries 85
    Cassandra Robertson, MD, and Douglas P. Hanel, MD

13. Frostbite 94
    Stephen D. Trigg, MD

14. Burns 98
    Vera C. van Aalst-Barker, MD

v

15. Compartment Syndrome 102
    Joseph J. Thoder, MD

16. Dupuytren's Disease 108
    Zvi Margaliot, MD, MS

17. Cerebral Palsy, Stroke, Traumatic Brain Injury, and
    Tetraplegia 114
    Radford J. Hayden, PA-C

18. Complex Regional Pain Syndromes 122
    Dean S. Louis, MD, and Morton L. Kasdan, MD

19. Kienböck's Disease 126
    Asheesh Bedi, MD

20. Wrist Arthroscopy 131
    John V. Hogikyan, MD

21. Distal Radius Injuries 138
    Adam Mirarchi, MD

22. Distal Radioulnar Joint 150
    Michael J. Moskal, MD

23. Carpal Fractures and Instability 158
    Adam Mirarchi, MD

24. Metacarpal and Phalangeal Fractures 175
    Steven Shah, MD

25. Joint Dislocations and Ligament Injuries 185
    Peter J.L. Jebson, MD

26. Arthrodesis and Arthroplasty of the Hand and Wrist 193
    Peter J.L. Jebson, MD, and Edwin Spencer, MD

27. Wrist Pain 200
    Fred M. Hankin, MD

28. Flexor Tendon Injuries 210
    Christopher C. Schmidt, MD, and Samir M. Patel, MD

29. Extensor Tendon Injuries 218
    Mary Lynn Newport, MD

30. Nail Bed and Fingertip Injuries 225
    Jeffrey C. King, MD, and Keith G. Wolter, MD

31. Soft Tissue Coverage of the Hand 235
    Donald P. Condit, MD, and Danielle A. Conaway, MD

32. Skin Flaps of the Upper Extremity 242
    Donald P. Condit, MD, and Danielle A. Conaway, MD

33. Electrodiagnostic Testing 254
    Hal M. Corwin, MD

34. Compression Neuropathies of the Upper Extremity 259
    Michael I. Vender, MD, and Prasant Atluri, MD

35. Peripheral Nerve Injury 266
    Paul S. Cederna, MD

36. Tendon Transfers 274
    Kevin C. Chung, MD, MS

37. Vascular Disorders 280
    Brian M. Braithwaite, MD

38. Principles of Microsurgery 284
    Robert I. Oliver Jr., MD

39. Replantation 289
    Clifford King, MD, PhD, and William M. Kuzon, Jr., MD, PhD

40. Amputation and Prosthetics 296
    Kelly Vanderhave, MD

41. Bone Tumors of the Hand and Wrist 302
    Peter M. Murray, MD, and Theodore W. Parsons, MD, LTC, USAF, MC

42. Soft Tissue Tumors 310
    Edward A. Athanasian, MD

43. Congenital Hand Differences 316
    Brian D. Adams, MD, and Curtis M. Steyers, MD

44. Upper Extremity Occupational Injuries and Illness 328
    V. Jane Derebery, MD

45. Hand Therapy 333
    Nan Boyer OTR, CHT, Cathie Basil, OTR, CHT, Linda Miner, OTR, CHT,
    Denise Justice, OTR, Trisha Mozdzierz, OTR, Carole Dodge, OTR,
    Andrea Eisman, MS, OTR, CHT, and Jeanne Riggs, OTR, CHT

46. Legal Aspects of Hand Surgery 340
    Walter E. Harding, JD

Index 345

CONTRIBUTORS

Brian D. Adams, MD
Professor, Department of Orthopedic Surgery, University of Iowa Hospital and Clinics, Iowa City, Iowa

Edward A. Athanasian, MD
Assistant Professor, Department of Orthopedics, Weil Cornell Medical College; Memorial Sloan-Kettering Cancer Center; Hospital for Special Surgery, New York, New York

Prasant Atluri, MD
Hand Surgery Associates, SC, Arlington Heights, Illinois

Cathie Basil, OTR, CHT
Occupational Therapy Division, Department of Physical Medicine and Rehabilitation, University of Michigan Medical Center, Ann Arbor, Michigan

Asheesh Bedi, MD
Department of Orthopaedic Surgery, University of Michigan Medical Center, Ann Arbor, Michigan

Nan Boyer, OTR, CHT
Occupational Therapy Division, Department of Physical Medicine and Rehabilitation, University of Michigan Medical Center, Ann Arbor, Michigan

Brian M. Braithwaite, MD
Chief Resident, Division of Plastic and Reconstructive Surgery, University of Louisville School of Medicine, Louisville, Kentucky

Jay T. Bridgeman, MD, DDS
Penn State Orthopaedics, Hershey, Pennsylvania

Paul S. Cederna, MD
Associate Professor, Section of Plastic and Reconstructive Surgery, Department of Surgery, University of Michigan Medical Center; Attending Surgeon, Veterans Affairs Medical Center, Ann Arbor, Michigan

Kevin C. Chung, MD, MS
Associate Professor, Section of Plastic and Reconstructive Surgery, Department of Surgery, University of Michigan Medical Center, Ann Arbor, Michigan

Danielle A. Conaway, MD
Grand Rapids Medical Education and Research Center, Orthopaedic Surgery Registry, Grand Rapids, Michigan

Donald P. Condit, MD
Clinical Associate Professor of Surgery, Michigan State University, Grand Rapids, Michigan

Hal M. Corwin, MD
Assistant Clinical Professor of Neurology, University of Louisville School of Medicine; Neuroscience Associates, PFC, Louisville, Kentucky

CONTRIBUTORS

Bruce Crider, MD
Assistant Professor, Department of Anesthesiology, University of Michigan Medical Center, Ann Arbor, Michigan

V. Jane Derebery, MD, FACOEM
Concentra Health Services, San Antonio, Texas

Carole Dodge, OTR
Occupational Therapy Division, Department of Physical Medicine and Rehabilitation, University of Michigan Medical Center, Ann Arbor, Michigan

Andrea Eisman, MS, OTR, CHT
Occupational Therapy Division, Department of Physical Medicine and Rehabilitation, University of Michigan Medical Center, Ann Arbor, Michigan

Mark N. Halikis, MD
Associate Clinical Professor, Department of Orthopaedic Surgery, University of California at Irvine, Irvine, California; Orthopedic Specialty Institute, Orange, California; Chief of Hand Surgery Service, Veterans Affairs Medical Center, Long Beach, California

Douglas P. Hanel, MD
Professor, Departments of Orthopedic Surgery and Sports Medicine, University of Washington School of Medicine; Department of Orthopedics, Harborview Medical Center, Seattle, Washington

Fred M. Hankin, MD
Community Orthopedic Surgery, Ypsilanti, Michigan

Walter E. Harding, JD
Partner, Specialist in Workers' Compensation, Boehl, Stopher, and Graves, LLP, Louisville, Kentucky

Radford J. Hayden, PA-C
Department of Orthopedic Surgery, University of Michigan Medical Center, Ann Arbor, Michigan

John V. Hogikyan, MD
Department of Orthopedic Surgery, University of Michigan Medical Center, Ann Arbor, Michigan; Staff Surgeon, St. Joseph's Mercy Hospital, Superior Township, Michigan; Community Orthopedic Surgery, Ypsilanti, Michigan

Peter J.L. Jebson, MD
Associate Professor and Chief, Division of Elbow, Hand, and Microsurgery, Department of Orthopaedic Surgery, University of Michigan Medical Center, Ann Arbor, Michigan

Denise Justice, OTR
Occupational Therapy Division, Department of Physical Medicine and Rehabilitation, University of Michigan Medical Center, Ann Arbor, Michigan

Morton L. Kasdan, MD, FACS
Clinical Professor, Division of Plastic and Reconstructive Surgery, University of Louisville School of Medicine, Louisville, Kentucky; Clinical Professor, Department of Preventive Medicine and Environmental Health, University of Kentucky Medical Center, Lexington, Kentucky

Clifford King, MD, PhD
Dean Medical Center, Madison, Wisconsin

Jeffrey C. King, MD
Clinical Assistant Professor, College of Human Medicine, Michigan State University, East Lansing; Healthcare Midwest Hand and Elbow Surgery, Kalamazoo, Michigan

CONTRIBUTORS

William M. Kuzon, Jr., MD, PhD
Professor and Section Head, Section of Plastic and Reconstructive Surgery, Department of Surgery, University of Michigan Medical Center, Ann Arbor, Michigan

Dean S. Louis, MD
Professor, Division of Elbow, Hand, and Microsurgery, Department of Orthopaedic Surgery, University of Michigan Medical Center, Ann Arbor, Michigan

Zvi Margaliot, MD, MS, FRCSC
Trillium Health Center, Mississauga, Ontario, Canada

Terry McCurry, MD
Department of Orthopedic Surgery, University of Louisville School of Medicine, Louisville, Kentucky

H. Stephen McGuire, MD
University of Louisville School of Medicine, Louisville, Kentucky

Linda Miner, OTR, CHT
Occupational Therapy Division, Department of Physical Medicine and Rehabilitation, University of Michigan Medical Center, Ann Arbor, Michigan

Adam Mirarchi, MD
Department of Orthopedics, Case Western Reserve University Hospitals and Clinics, Cleveland, Ohio

Nana Mizuguchi, MD
Department of Orthopedic Surgery, University of Louisville School of Medicine, Louisville, Kentucky

Michael J. Moskal, MD
Clinical Instructor, Department of Orthopedic Surgery and Sports Medicine, University of Washington School of Medicine, Seattle, Washington; Clinical Instructor, Department of Orthopedic Surgery, University of Louisville School of Medicine, Louisville, Kentucky

Trisha Mozdzierz, OTR
Occupational Therapy Division, Department of Physical Medicine and Rehabilitation, University of Michigan Medical Center, Ann Arbor, Michigan

Peter M. Murray, MD
Adjunct Associate Professor, Department of Bioengineering, Clemson University, Clemson, South Carolina; Senior Consultant, Division of Hand and Microsurgery, Department of Orthopedic Surgery, Mayo Clinic, Jacksonville, Florida

Sanjiv H. Naidu, MD, PhD
Department of Orthopedics and Rehabilitation, Pennsylvania State University College of Medicine, Hershey, Pennsylvania

Arthur H. Newberg, MD
Professor, Departments of Radiology and Orthopaedics, Tufts University School of Medicine; Chief, Musculoskeletal Imaging, Department of Radiology, New England Baptist Hospital, Boston, Massachusetts

Joel S. Newman, MD
Associate Clinical Professor of Radiology, Tufts University School of Medicine; Associate Chair, Department of Radiology, New England Baptist Hospital, Boston, Massachusetts

Mary Lynn Newport, MD
Associate Professor, Department of Orthopedic Surgery, University of Connecticut School of Medicine, Farmington, Connecticut

Robert I. Oliver, Jr., MD
Chief Resident, Division of Plastic and Reconstructive Surgery, University of Louisville School of Medicine, Louisville, Kentucky

Theodore W. Parsons, MD, LTC, USAF, MC
Chairman, Department of Orthopedic Surgery, Wilford Hall Medical Center, Lackland Air Force Base;
Department of Surgery, Uniformed Services School of the Health Sciences, San Antonio, Texas

Samir M. Patel, MD
Department of Orthopedic Surgery, Allegheny General Hospital, Pittsburgh, Pennsylvania

Jeanne Riggs, OTR, CHT
Occupational Therapy Division, Department of Physical Medicine and Rehabilitation,
University of Michigan Medical Center, Ann Arbor, Michigan

Cassandra Robertson, MD
Fellow, Hand Surgery, University of Washington School of Medicine, Seattle, Washington

Christopher C. Schmidt, MD
Director, Upper Extremity Microvascular Reconstruction, Department of Orthopedic Surgery,
Allegheny General Hospital, Pittsburgh, Pennsylvania

Steven Shah, MD
Department of Orthopaedic Surgery, University of Michigan Medical Center, Ann Arbor, Michigan

Edwin Spencer, MD
Knoxville Orthopaedic Clinic, Knoxville, Tennessee

Curtis M. Steyers, MD
Professor, Department of Orthopedic Surgery, University of Iowa Hospital and Clinics, Iowa City, Iowa

Lawrence G. Sullivan, MD
Fellow, Hand Surgery, Loma Linda Medical Center, Loma Linda, California

Julio Taleisnik, MD
Clinical Professor, Department of Orthopaedic Surgery, University of California at Irvine School of Medicine,
Irvine, California

Joseph J. Thoder, MD
Professor and Chairman, Department of Orthopaedic Surgery, Temple University School of Medicine,
Philadelphia, Pennsylvania

Stephen D. Trigg, MD
Department of Orthopedic Surgery, Mayo Clinic, Jacksonville, Florida

Vera C. van Aalst-Barker, MD
Clinical Instructor, Division of Plastic Surgery, University of Louisville School of Medicine,
Louisville, Kentucky

Kelly Vanderhave, MD
Assistant Professor, Department of Orthopedics and Rehabilitation, Pennsylvania State University
Milton S. Hershey Medical Center, Hershey, Pennsylvania

Michael I. Vender, MD
Hand Surgery Associates, SC, Arlington Heights, Illinois

William M. Weiser, MD
Department of Orthopaedic Surgery, Kaiser Permanente, Bellflower Hospital, Bellflower, California

Keith G. Wolter, MD
Section of Plastic and Reconstructive Surgery, Department of Surgery, University of Michigan Medical
Center, Ann Arbor, Michigan

PREFACE

Never cease trying to be the best you can become.
—Joshua Hugh Wooden

Success is peace of mind which is a direct result of self-satisfaction in knowing you made the effort to become the best of which you are capable.
—John Wooden

The 3rd edition of *Hand Secrets* is intended to be a reader-friendly study guide to be used by medical students, residents, practicing physicians, and occupational therapists as they care for patients within the discipline of hand surgery. Dr. Morton Kasdan and I have assembled a diverse but dedicated group of professionals who have contributed chapters on a wide range of topics. We are indebted to each of them for giving freely of their time and their commitment. We hope that you enjoy the new format with the addition of Key Points and Top 100 Secrets and the many new illustrations and images. We hope that this book will assist you as you prepare for the In-Training, Board Certification, and Certificate of Added Qualification in Hand Surgery (CAQ) examinations. As always, we welcome your feedback and suggestions.

Thank you to my mentors, Drs. Bill Engber, Curtis Steyers, Bill Blair, Brian Adams, and Dean Louis for fueling my passion for the greatest discipline in medicine: hand surgery.

Peter J.L. Jebson, MD

WEBSITES

The following websites are valuable resources for the topics covered in this book:

1. www.e-hand.com/ (hand anatomy)
2. www.wheelessonline.com/ortho (hand anatomy)
3. www.fpnotebook.com/ORT74.htm (hand anatomy)
4. www.ptcentral.com/radiology/Wrist/Wrist.html (radiology of the hand/wrist)
5. www.aafp.org/afp/20030215/745.html (injection techniques in the hand/wrist)
6. www.nlm.nih.gov/medlineplus/handinjuriesanddisorders.html (National Institutes of Health)
7. www.assh.org (Hand Society)
8. www.aaos.org (Academy of Orthopaedic Surgery)
9. www.handsurgery.org (American Association for Hand Surgery)

\+ ulnar fovea sign = disruption of distal radioulnar
(thumb base of   lig & ulnotriquetral lig injuries
pisiform)

TOP 100 SECRETS

These secrets are 100 of the top board alerts. They summarize the concepts, principles, and most salient details of hand pathology, diagnosis, and therapy.

1. The metacarpophalangeal (MCP) joint has the greatest joint contact area in the hand.

2. Wrist extension occurs primarily through the radiocarpal joint, while flexion occurs primarily through the midcarpal joint.

3. The longitudinal axis of forearm rotation occurs through the center of the radial head and the foveal region of the ulnar head.

4. An ultra-quick examination of the hand is summarized in the following list.

	Intrinsic Muscle	Extrinsic Muscle	Sensory Territory
Ulnar nerve	First dorsal interosseus	FDP to little finger	Volar pad to little finger
Median nerve	Thumb opposition	FDP to index finger	Volar pad to index finger
Radial nerve	None	Wrist extension	Dorsal side of first web space

FDP = flexor digitorum profundus.

5. Ulnar neuropathy at the elbow may involve forearm muscles. Sensation in the forearm is normal.

6. Peripheral neuropathy may result from many illnesses other than diabetes mellitus and deserves a full neurologic evaluation.

7. Patients with severe respiratory compromise may not be good candidates for brachial plexus blockade due to the risk of phrenic nerve paralysis. This may be further exacerbated by contralateral phrenic nerve damage caused by previous surgery, such as carotid surgery or neck dissection.

8. A local anesthetic solution containing epinephrine should be used cautiously for digital blockade because of the risk of ischemic injury to the digit.

9. The following features noted on plain radiographs are suggestive of an aggressive or malignant tumor in the hand: cortical destruction, periosteal reaction, indistinct demarcation, or an associated soft tissue mass.

10. When evaluating a patient with osteoarthritis (OA) of the hand, assess for such associated conditions as carpal tunnel syndrome, de Quervain's tenosynovitis, trigger finger, and flexor carpi radialis tendinitis.

11. The most common type of degenerative arthritis in the wrist is the scapholunate advanced collapse (SLAC) pattern. The preferred surgical approach for treating a symptomatic SLAC pattern of wrist arthritis (proximal row carpectomy versus a "four corner" arthrodesis) is determined by the status of the articular surfaces of the proximal capitate and radiolunate articulations.

12. When a patient presents with an ulnar deviation deformity of the MCP joints, think of inflammatory arthritis, the most likely being rheumatoid arthritis. If you see exuberant tenosynovitis, especially involving the finger and wrist extensors, think of rheumatoid arthritis.

13. When a patient with rheumatoid arthritis presents with the sudden loss of finger extension, differentiate tendon rupture from subluxation at the MCP joints by passively extending the joints and asking the patient to "maintain" extension. Differentiate tendon rupture from posterior interosseous nerve palsy by testing the tenodesis effect of intact finger extensors by passively flexing the wrist.

14. Avoid soft tissue procedures for deformities associated with systemic lupus erythematosus.

15. Nail pitting may represent psoriatic arthritis or a fungal infection of the nail (onychomycosis).

16. Septic arthritis is a surgical emergency.

17. The most common complications following a corticosteroid injection in the hand or wrist are fat necrosis and skin depigmentation, infection, tendon rupture, and nerve injury. Diabetic patients who receive a corticosteroid injection in the hand or wrist should be advised that their glucose level will increase temporarily after injection.

18. A corticosteroid injection is usually unsuccessful in treating a trigger finger if there are multiple trigger fingers, if the symptoms have been present for more than 6 months, if the trigger finger is of the diffuse type, or if the patient is diabetic.

19. Allen Buchner Kanavel (1874–1938), an American surgeon, described the four classic findings of pyogenic flexor tenosynovitis: (1) fusiform swelling of the digit, (2) flexed posture of the digit, (3) pain with passive extension of the digit, and (4) tenderness along the flexor tendon sheath.

20. Paronychia, an infection of the tissue surrounding the eponychial fold, is the most common infection of the hand. *Staphylococcus aureus* is the most common infecting organism. Of all the areas of the nail bed, the hyponychium is the most resistant to infection.

21. The different types of deep subfascial space infections of the hand are (1) subaponeurotic, (2) midpalmar, (3), thenar, (4) interdigital (web space), and (5) Parona's space.

22. Strategies to prevent extravasation and secondary tissue injury should be considered whenever a vesicant substance is administered intravenously.

23. Real-time high-resolution ultrasound is both highly sensitive (94%) and specific (99%) for evaluating the presence of a foreign body in the hand.

24. Injuries due to high-pressure injection of a toxic substance into the hand may initially appear relatively innocuous. Aggressive surgical management, however, is often necessary to treat the extensive internal injury and reduce the risk of digit amputation.

25. Rapid rewarming of a frost-bitten extremity can be extremely painful. The patient should receive adequate analgesia and sedation.

26. The most common angular deformity seen after a frostbite injury in a child is radial deviation of the small finger distal interphalangeal (DIP) joint. Frostbite in children can result in shortened digits due to physeal closure 6–12 months after the injury. Late sequelae in adults following a frostbite injury include cold intolerance, hyperhidrosis, trophic changes, and degenerative arthritis.

27. Significant hand burns require the patient to be transferred to a burn center. Burn wounds to the hand that are not expected to heal within 2–3 weeks should undergo early debridement and skin grafting. Thrombosed veins within the burned area are pathognomonic of a deep, full-thickness injury.

28. Silver sulfadiazine cream (Silvadene) is the most commonly used topical agent for a thermal hand burn. The main side effect is neutropenia.

29. If a patient does not experience significant pain relief following the splitting of a tight cast/splint or dressing, a compartment syndrome should be suspected.

30. Prophylactic fasciotomy should be performed following limb reperfusion with an ischemic time of greater than 4 hours.

31. The indications for surgical treatment in Dupuytren's disease are functional limitations, any proximal interphalangeal (PIP) joint contracture, an MCP joint contracture greater than 30 degrees, or the inability to place the hand flat on a table (Hueston's tabletop test). Injection of clostridial collagenase has proved safe and efficacious in the patient with Dupuytren's contracture involving the MCP and PIP joints.

32. Dupuytren's diathesis is associated with a strong family history, aggressive disease, involvement of the feet and genitalia, and early recurrence following surgery. Dermofasciectomy is the preferred surgical technique for recurrent disease in the patient with Dupuytren's diathesis. The spiral cord in Dupuytren's disease arises from four fascial bands: the pretendinous band, spiral band, lateral digital band, and Grayson's ligament.

33. The spastic clenched fist in the patient who has undergone a stroke and a traumatic brain injury results from unmasking of the primitive grasp reflex.

34. Botulinum toxin is useful for the treatment of spasticity because it inhibits the release of acetylcholine in the neuromuscular junction by attaching to the presynaptic nerve terminal.

35. About 95% of tetraplegic patients consider improved function of their hands as the key to an improved overall quality of life.

36. Symptoms of burning pain, swelling, digital stiffness, and discoloration following a distal radius fracture should alert the physician that a complex regional pain syndrome (CRPS) might be developing. If these symptoms follow a known peripheral nerve distribution, the physician should order valid electrodiagnostic studies for the patient. The diagnosis of CRPS is a cinical one and should be confirmed with valid imaging studies (x-rays and/or triple-phase bone scan).

37. The lunate most at risk for developing Kienböck's disease is associated with ulnar minus variance and has a single extraosseous nutrient vessel and poor intraosseous vascular anastomoses. A reliable method for differentiating stage 3A from stage 3B Kienböck's disease is a radioscaphoid angle >60 degrees (Lichtman stage 3B).

38. Radial shortening is the preferred method of joint leveling for patients with Kienböck's disease and ulnar minus variance because of the morbidity associated with bone graft harvesting and the risk of nonunion associated with ulnar lengthening.

39. Kienböck's disease in children under 12 years of age should be treated conservatively with a trial of cast immobilization; children older than 12 years of age should be treated similarly to an adult.

40. The most common technical difficulty encountered with arthroscopic excision of a dorsal wrist ganglion is poor visualization of the stalk.

41. The most common indications for wrist arthroscopy are chronic wrist pain, an intercarpal ligament tear, and a tear or avulsion of the triangular fibrocartilage (TFC).

42. Arthroscopic-assisted reduction and fixation (AARF) of a distal radius fracture is performed 5–7 days after injury to minimize bleeding from the fracture site and to avoid difficulties in visualizing the articular surfaces and fracture fragments.

43. Partial or complete premature closure of the epiphyseal plate is common following a Salter-Harris type II fracture of the distal radius.

44. Radiographic parameters considered acceptable following a closed reduction of a distal radius fracture include radial length within 5 mm of the uninjured wrist, radial inclination >15 degrees, sagittal tilt <10 degrees dorsal and up to 20 degrees volar, and <1 mm of articular incongruity. The median nerve is commonly injured in association with a fracture of the distal radius.

45. Arthrosis and/or instability of the distal radioulnar joint (DRUJ) is a common cause of prolonged morbidity and a poor outcome following a distal radius fracture.

46. The DRUJ is a weight-bearing joint. Stability of the DRUJ is primarily due to the soft tissue constraints, including the triangular fibrocartilage complex (TFCC) and dorsal and volar radioulnar ligaments. A durable repair of peripheral (ulnar) TFCC tears is possible due to the robust vascularity at this location.

47. The natural history of the neglected symptomatic scaphoid nonunion involves a predictable pattern of progressive posttraumatic arthritis of the wrist. Percutaneous screw fixation has been advocated for nondisplaced scaphoid waist fractures because the morbidity of prolonged cast immobilization is avoided, a return to full activity occurs sooner, and the time to union is shorter.

48. The naviculocapitate syndrome consists of a fracture of the scaphoid waist and capitate neck regions with rotation of the proximal capitate fragment such that the articular surface is directed distally.

49. During open reduction internal fixation (ORIF) of a displaced proximal phalanx condyle fracture, the collateral ligament attachment should be preserved to avoid osteonecrosis or nonunion.

50. The indications for ORIF of a metacarpal shaft fracture include failed closed treatment, open fracture, multiple fractures, segmental bone loss, and associated soft tissue loss.

51. The presence of skin dimpling or a sesamoid within the metacarpophalangeal (MP) joint on plain radiographs is considered pathognomonic for a complex MP joint dislocation.

52. Treatment of a thumb carpometacarpal (CMC) dislocation is controversial. Recent evidence suggests that volar beak ligament reconstruction using the flexor carpi radialis tendon results in a better clinical outcome with a lower incidence of persistent instability.

53. Longitudinal traction and/or hyperextension should be avoided during the reduction of a simple MCP joint dislocation because you may convert the injury into a complex dislocation requiring surgical treatment.

54. The most common complication following excision of the distal ulna (Darrach procedure) is instability of the proximal ulnar stump.

55. When performing a scaphoid excision and "four corner" intercarpal arthrodesis in the patient with the SLAC pattern of wrist arthritis, the extended posture of the lunate must be corrected, or painful impingement against the dorsal radius will occur during wrist extension.

56. It is far more likely for a child to sustain a distal radius growth plate fracture than a wrist sprain following a traumatic injury.

57. Pseudogout or calcium pyrophosphate dihydrate crystal deposition disease (CPPD) is a common but frequently unrecognized cause of acute wrist pain, particularly in elderly patients.

58. Multiple fractures in various stages of healing are suggestive of child abuse.

59. Persistent dorsoradial wrist pain, tenderness in the anatomic "snuffbox" region, and normal plain radiographs following a fall on the outstretched hand must be treated as an occult scaphoid fracture with appropriate immobilization and a repeat clinical and radiographic examination 10–14 days after injury.

60. The A2 and A4 pulleys are the most critical for preventing tendon bowstringing.

61. The strength of a flexor tendon repair is proportional to the number of core strands crossing the repair site. A flexor tendon repair is weakest at 10–12 days postoperatively. Epitendinous suture is advocated during repair of a flexor tendon laceration to tidy the repair site, to avoid adhesion formation or triggering, and to augment the strength of the repair.

62. Zone 2 injuries have the worst results following flexor tendon repair. A poor outcome following extensor tendon repair is noted in patients with an associated fracture, soft tissue loss, or distal injury.

63. An acute boutonnière injury requires central slip disruption and attenuation of the triangular ligament with volar displacement of the lateral bands below to the rotation axis of the PIP joint.

64. Surgical treatment options for a neglected mallet finger injury with a secondary swan-neck deformity include oblique retinacular ligament reconstruction, flexor digitorum superficialis (FDS) tenodesis, or central slip tenotomy (Fowler's procedure).

65. The most common complaints following a fingertip amputation, regardless of the treatment approach, are cold intolerance, diminished sensibility, and hypersensitivity with use.

66. A nail grows at a rate of 0.1 mm/day, with complete growth taking between 70 and 160 days.

67. The goals of treatment for a fingertip amputation are stable skin coverage, good sensibility, satisfactory function, acceptable cosmesis, and a short period of convalescence.

68. The indications for using a thenar flap include a transverse or volar oblique fingertip amputation of the index or long fingers in a young person with no preexisting arthritis or

stiffness in the hand. Split-thickness skin grafting is not used in the management of a fingertip amputation because of breakdown and a lack of durability.

69. Full-thickness skin grafts are preferred for palmar hand wounds, while split-thickness skin grafts are preferred for dorsal wounds.

70. The most common causes of skin graft failure are hematoma formation and inadequate postoperative immobilization. A bacterial count of 10^5 organisms per gram of tissue is associated with a high incidence of graft failure.

71. Survival of a skin graft depends on the ingrowth of blood vessels into existing vascular channels (inosculation) and new vessel formation (neovascularization).

72. Soft tissue coverage in the hand should involve tissue that is thin, pliable, and durable and should permit tendon gliding while avoiding joint and web space contracture formation.

73. A false-positive nerve conduction study may be reported if the limb was cool during the test. An abnormal nerve conduction study should always be compared to the opposite limb to avoid the misdiagnosis of a systemic disease of the nervous system.

74. When you are suspicious of an upper extremity neuropathy, order a bilateral upper extremity electromyography (EMG). Physical exam alone is unreliable for diagnosing peripheral neuropathies of the upper extremity.

75. In "double crush" neuropathies, all sites of compression must be addressed to reliably eliminate symptoms.

76. Internal topography of a nerve refers to the precise location and organization of sensory and motor fascicles within a peripheral nerve.

77. Nerve endings may be matched during repair by visualizing fascicular groups and/or vascular markings in the epineurium. A nerve gap may be overcome by transposing the nerve or shortening the extremity prior to repair or using a nerve graft or conduit. Maximal axonal regeneration in humans is 1–2 mm/day.

78. The indications for tendon transfer are to replace absent motor function, augment motor power, and balance motor function.

79. The ulnar artery is usually the main contributor to the superficial palmar arch, and the radial artery is the main contributor to the deep palmar arch.

80. A true aneurysm contains elements of all three layers of the arterial wall.

81. Unrecognized tension during free-flap inset rivals technical errors of the anastamosis as a cause of vessel thrombosis and flap necrosis.

82. Persistent failure of a free flap or replantation despite bedside maneuvers, such as limb elevation and dressing removal, is a relative indication for urgent reexploration to assess vessel patency.

83. The ischemic tolerance of an amputated part is inversely related to the amount of muscle in the amputated part.

84. Remember the principle of *spare parts surgery* before discarding an amputated part; the amputated part can be a valuable source of nerve, artery, vein, tendon, and/or skin graft.

85. Reperfusion syndrome is the constellation of clinical findings that occurs following the replantation of a major limb. These findings include acidosis, myoglobinuria, cardiogenic shock, renal failure, multisystem organ failure, and occasional death.

86. Arterial insufficiency following replantation is characterized by a cool, pale extremity with absent capillary refill and poor tissue turgor. Venous insufficiency following replantation is characterized by a congested, ruborous part with rapid capillary refill and increased tissue turgor.

87. The motion required to operate the terminal device in a patient with a below-elbow amputation is glenohumeral flexion and scapular abduction.

88. Prosthetic fitting in the child with an acquired or congenital below-elbow amputation should be initiated at 6–9 months of age when the child is sitting upright and beginning bimanual activities.

89. In addition to plain radiographs of the lesion, the patient with a suspected bone malignancy should also have a chest radiograph, chest and abdominal computed tomography (CT) scan, and a three-phase bone scan performed. The biopsy of a bone tumor can be performed using fine needle aspiration, a core needle technique, or an open biopsy (with the open biopsy having the highest accuracy rate).

90. Giant cell tumor of bone is a benign aggressive lesion with a predilection to recurrence and the capability of metastasizing to the lungs.

91. A metastasis to the hand (acrometastasis) is a poor prognostic sign; most patients with this condition live less than 6 months from time of diagnosis.

92. The factors associated with recurrence following excision of a giant cell tumor of the tendon sheath are DIP joint involvement, more distal location, prior recurrence, associated bone erosion, and radiographic evidence of OA.

93. The most important prognostic factor in the patient with a subungual melanoma is the depth of invasion.

94. The prognosis of a soft tissue sarcoma is affected by the tumor grade, size, location, and presence of metastases. The preferred treatment for a soft tissue sarcoma in the hand/upper extremity is wide surgical excision or amputation.

95. Unilateral syndactyly with ipsilateral deficiency of the sternocostal head of the pectoralis major muscle is known as Poland's syndrome.

96. The most common thumb duplication includes both the proximal and distal phalanges (Wassel type IV). The most common congenital amputation in the upper extremity is a short below-elbow amputation.

97. The three signaling centers that control limb development are the apical ectodermal ridge (proximal to distal limb development and interdigital necrosis), the zone of polarizing activity (radioulnar limb formation), and the Wnt pathway of the dorsal ectoderm (dorsoventral limb development).

98. The mnemonic **BICEPS** can help the clinician remember the basics for the management of workers' compensation injuries or illness:
    - **B = Brevity:** Treatment should be brief, and the patient should be advised early on that a rapid recovery is anticipated. A time frame of expected treatment should be discussed.
    - **I = Immediacy:** Recognize early in the course of treatment when a patient is not recovering as expected or when psychogenicity is evident.
    - **C = Centrality:** Patients should be treated in a central area by clinicians and therapists familiar with workers' compensation issues and dynamics.
    - **E = Expectancy:** Patients should be given both verbal and nonverbal messages that their injury or illness is not disabling and that they will improve rapidly.
    - **P = Proximity:** Patients should be kept at work if at all possible, with restrictions, if necessary. This reduces a crucial issue of secondary gain.
    - **S = Simplicity:** Disability, not impairment, is frequently a maladaptive response to stress or a difficult life situation (e.g., financial problems, family strife, poor job performance). Intensive psychotherapy is not appropriate. Treatment is best kept simple, focusing on teaching more adaptive ways to handle stress (e.g., exercise, relaxation techniques).

99. The tensile strength of an immobilized tendon decreases in the first 3–5 days postoperatively. The tensile strength and excursion of a tendon that is mobilized following repair are superior to that of an immobilized or delayed mobilized tendon.

100. Excessively tight compression can exacerbate persistent edema due to lymphatic system inhibition. Light compression and low stretch wraps with massage and range of motion exercises are more effective in managing persistent edema.

HAND AND WRIST ANATOMY
Nana Mizuguchi, MD, and Terry McCurry, MD

1. **How often is the palmaris longus tendon present?**
   The palmaris longus tendon is present in the forearm of approximately 85% of individuals. It is best found by opposing the thumb to the little finger and slightly flexing the wrist.

2. **The flexor digitorum superficialis (FDS) of the small finger is absent in what percentage of the population?**
   Around 30%. The presence of the FDS in the little finger varies greatly among individuals.

3. **What is carpal height?**
   Carpal height is the distance from the distal articular surface of the radius to the distal articular surface of the capitate. Shortening of the carpal height may be caused by malrotation of the proximal carpal row as seen in rheumatoid arthritis, scapholunate dissociation (SLAC wrist), and scaphoid nonunion.

4. **What are the boundaries and contents of the carpal tunnel?**
   The transverse carpal ligament forms the volar border, and the carpal bones form the dorsal border. The median nerve, four flexor digitorum profundus (FDP) tendons, four FDS tendons, and the flexor pollicis longus traverse the carpal tunnel.

5. **What is the most radial structure within the carpal tunnel?**
   The flexor pollicis longus.

6. **Name the flexor pulleys in the fingers.**
   There are five annular (A) and three cruciform (C) pulleys. The odd-numbered annular pulleys (A1, A3, and A5) are located over the metacarpophalangeal (MCP), proximal interphalangeal (PIP), and distal interphalangeal (DIP) joints, respectively. The A2 pulley is positioned over the proximal half of the proximal phalanx, and the A4 pulley is located over the center of the middle phalanx. The C1 pulley is situated between the A2 and A3 pulleys; C2 is between A3 and A4; and C3 is between A4 and A5.

7. **Name the pulleys in the thumb.**
   The thumb has two annular pulleys that are located over the metacarpophalangeal and interphalangeal joints, respectively, and an oblique pulley lies over the proximal phalanx.

8. **What is the function of the flexor tendon sheath?**
   The flexor tendon sheath, a synovial-lined fibro-osseous tunnel, has three important functions:
   - Providing a smooth gliding plane for the tendon
   - Maintaining the flexor tendons close to the volar surface to prevent bowstringing
   - Providing the area for tendon nutrition

9. **In zone 2 reconstructions of the flexor tendons, which pulley system should be preserved to prevent bowstringing?**
   The A2 and A4 pulleys in the fingers and the oblique pulley in the thumb are the most essential pulleys to prevent bowstringing. The A2 and A4 pulleys are broader and stronger and do not overlie the joints.

10. **What is the typical distal attachment of the interosseous muscles? What is the exception?**
The interosseous muscle has two typical points of distal attachment: one to the proximal phalanx of the particular digit and another to the extensor aponeurosis of the same digit. The exception is the first dorsal interosseous muscle. Its entire insertion is to the base of the proximal phalanx of the index finger.

11. **Describe the actions of the interosseous muscles.**
The interosseous muscles are ulnar and radial deviators of the fingers as well as flexors of the MCP joints and extensors of the interphalangeal joints. The dorsal interossei abduct the fingers (Fig. 1-1), while the palmar interossei adduct the fingers (Fig. 1-2). All interossei act as prime flexors of the MCP joints because they pass palmar to the joint axis.

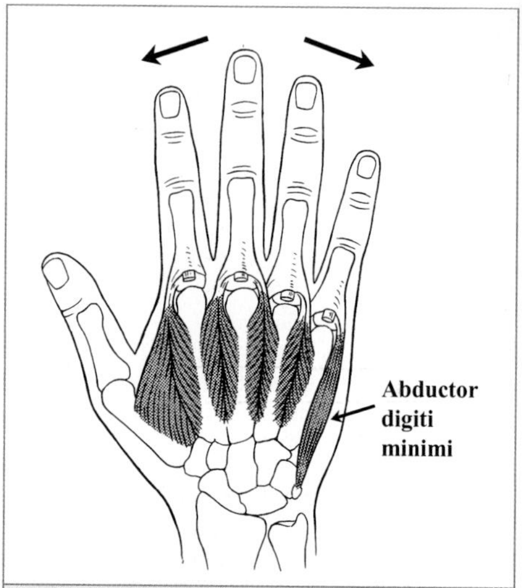

12. **When present, do the extensor indicis tendon and extensor digiti minimi tendon lie ulnar or radial to the extensor digitorum communis tendon?**
Ulnar.

Figure 1-1. The dorsal interossei function to abduct (direction of arrows) the fingers away from the axis of the hand (the longer finger ray). The abductor digiti minimi muscle abducts the small finger. (From Concannon MJ: Common Hand Problems in Primary Care. Philadelphia, Hanley & Belfus, 1999.)

13. **Describe the insertion and action of the abductor pollicis longus (APL), extensor pollicis longus (EPL), and extensor pollicis brevis (EPB).**
   - The **APL** inserts on the base of the thumb metacarpal on the radial side of the extensor surface. Its primary action is to radially abduct the metacarpal, and it also aids in radial deviation of the wrist.
   - The **EPL** inserts on the distal phalanx and is the primary extensor of the interphalangeal joint.
   - The **EPB** inserts on the proximal phalanx and acts as an extensor of the MCP joint.

14. **What separates the plane of the lumbricals from the interossei?**
The deep transverse metacarpal ligament connecting the metacarpal heads.

15. **Which tendon is most likely to rupture in association with a fracture of the distal radius?**
The EPL passes over the distal third of the radius and can rupture because of mechanical factors or disruption of the delicate blood supply to the tendon.

16. **What are the contents of Guyon's canal?**
Guyon's canal is ulnar to the carpal tunnel, and is located between the hook of the hamate and the pisiform bone. The canal contains the ulnar artery and nerve. Compression in this area is called *ulnar tunnel syndrome*. The most common cause is a ganglion.

17. Which intrinsic muscles in the hand are innervated by the median nerve? By the ulnar nerve?
    - The **median nerve** innervates the two radial lumbricals, the opponens pollicis, the abductor pollicis brevis, and the superficial head of the flexor pollicis brevis. The mnemonic is **LOAF**.
    - The **ulnar nerve** innervates the remainder of the intrinsic muscles of the hand: the adductor pollicis, the deep head of the flexor pollicis brevis, the palmaris brevis, the abductor digiti minimi, the flexor digiti minimi, the opponens digiti minimi, and all interossei.

18. What is the most common source of innervations of the extensor carpi radialis brevis (ECRB) muscle?
    The most common innervation source of the ECRB is the deep branch of the radial nerve, also called the *posterior interosseous nerve*. Compression of the nerve by the ECRB can result in the inability to extend the MCP joints of the affected hand (referred to as *posterior interosseous syndrome*).

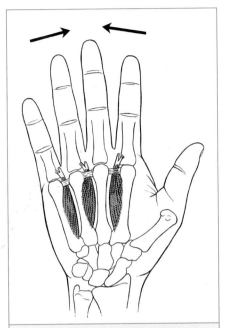

Figure 1-2. The palmar interossei function to adduct (direction of arrows) the fingers. There are no palmar interossei attached to the long finger because it is the axis. (From Concannon MJ: Common Hand Problems in Primary Care. Philadelphia, Hanley & Belfus, 1999.)

19. Why is the brachioradialis muscle unique in regard to its innervation?
    Because it is a flexor of the elbow but is innervated by the radial nerve.

20. Which area of the nail is most resistant to the development of infection?
    The hyponychium is particularly resistant to infection.

21. Fingernail production occurs in which areas?
    The germinal matrix produces 90% of the nail volume. The remainder of the volume is provided by the germinal layer in the dorsal roof of the proximal nail fold and by the sterile matrix.

22. Where do the extensor carpi radialis longus and extensor carpi radialis brevis insert?
    They insert at the base of the index and long metacarpals, respectively.

23. Name the tendons in each of the six dorsal wrist compartments, from radial to ulnar.
    1. Extensor pollicis brevis (EPB), abductor pollicis longus (APL)
    2. Extensor carpi radialis brevis (ECRB), extensor carpi radialis longus (ECRL)
    3. Extensor pollicis longus (EPL)
    4. Extensor indicis proprius (EIP), extensor digitorum communis (EDC)
    5. Extensor digiti minimi (EDM)
    6. Extensor carpi ulnaris (ECU)

24. **What separates the second and third compartments of the wrist?**
    Lister's tubercle on the dorsal distal radius.

25. **Describe the anatomic snuffbox.**
    The snuffbox is a hollow area on the dorsal radial aspect of the hand. It is defined by an area between the first and third dorsal wrist compartments, the radial styloid proximally, and the base of the thumb metacarpal distally. The radial artery courses through this area. Pain in the snuffbox indicates a possible scaphoid fracture.

26. **Which muscles arise from a flexor tendon and insert onto an extensor tendon?**
    Lumbricals.

27. **What is a spiral cord?**
    A spiral cord is composed of a pretendinous band, a spiral band, a lateral digital sheet, and the Grayson's ligament and is a frequent cause of PIP joint flexion contracture in the patient with Dupuytren's contracture.

28. **Which palmar arch provides the predominant blood supply to the thumb?**
    The dorsal branch of the radial artery.

29. **What is the order of ossification of the carpal bones?**
    - **Capitate:** first year
    - **Hamate:** first year
    - **Triquetrum:** third year
    - **Lunate:** fifth year
    - **Trapezium:** fifth year
    - **Scaphoid:** sixth year
    - **Trapezoid:** eighth year
    - **Pisiform:** twelfth year

    Figure 1-3 depicts the eight carpal bones.

30. **Where do the median, ulnar, and radial sensory nerves have the least amount of crossover?**
    - **Median nerve:** tip of index finger
    - **Ulnar nerve:** tip of little finger
    - **Radial nerve:** dorsal surface of first web space.

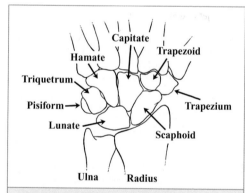

Figure 1-3. The carpal bones. (From Concannon MJ: Common Hand Problems in Primary Care. Philadelphia, Hanley & Belfus, 1999.)

31. **Where is the maximal crossover sensory innervation on the hand located?**
    On the dorsal surface of the middle phalanx of the ring finger.

32. **In harvesting the radial forearm flap, the distal radial artery is found between which two muscles or tendons?**
    The brachioradialis and flexor carpi radialis.

33. **What is the fixed unit of the hand?**
    The fixed unit of the hand consists of the second and third metacarpals and the distal carpal row. This stable unit provides the foundation for the other three mobile units.

KEY POINTS: HAND AND WRIST ANATOMY

1. Carpal height is the distance from the distal articular surface of the radius to the distal articular surface of the capitate.
2. There are five annular (A) and three cruciform (C) pulleys in the fingers.
3. The thumb has two annular pulleys located over the MCP and interphalangeal joints. The oblique pulley lies over the proximal phalanx.
4. Guyon's canal is ulnar to the carpal tunnel and is between the hook of the hamate and the pisiform bones. The canal contains the ulnar artery and nerve.
5. The fixed unit of the hand consists of the second and third metacarpals and the distal carpal row. This stable unit provides the foundation for the other three mobile units.

34. **In a Bennett's fracture-dislocation, which ligament maintains the first metacarpal base fragment in its anatomic position?**
    The anterior oblique carpometacarpal ligament. This ligament extends from the first metacarpal to the anterior crest of the trapezium.

35. **What is the maximum flexion arc of the finger joints?**
    Finger flexion is provided by the combination of intrinsic, FDS, and FDP motor function. The average amount of flexion at the MCP joint is 90 degrees (intrinsic); the average flexion of the PIP joint is 120 degrees (FDS), and flexion at the DIP joint is 80 degrees. The combined effort is the total flexion arc.

36. **What is the checkrein ligament of the interphalangeal (IP) joint?**
    The checkrein ligament is the thickened lateral portion of the volar plate composed of a part of the dorsal flexor sheath, the horizontal reflection of the accessory collateral ligament, and the lateral margin of the volar plate. This ligament protects the transverse branch of the digital artery as it traverses into the joint and also prevents PIP joint hyperextension.

37. **What is the only carpal bone to have a tendon insert on it?**
    The flexor carpi ulnaris inserts on the pisiform.

38. **What is the space of Poirier?**
    An area of weakness in the volar wrist capsule.

39. **What is the "pseudo-Terry Thomas" sign?**
    The Terry Thomas sign (named after the British actor with a gap between his two front teeth) is the gap between the scaphoid and lunate, which may indicate a scapholunate ligament disruption. In children, ossification of the scaphoid bone occurs from distal to proximal. During this period, there is an apparent gap between the ossified scaphoid and lunate. This is referred to as the *pseudo-Terry Thomas sign* and is part of normal development.

40. **In the normal wrist, what is the angle between the longitudinal axis of the scaphoid and the longitudinal axis of the hand?**
    Between 30 and 60 degrees.

41. **Describe the location of the epiphyseal plates in a child's hand.**
    The epiphyseal plates are located at the distal ends of the second through the fifth metacarpals. The remaining epiphyseal plates are located at the proximal end (thumb metacarpal and proximal and distal phalanges).

42. **Why is the scaphoid bone susceptible to avascular necrosis?**
    The blood supply to the scaphoid enters distally and at the waist region and then proceeds proximally. The proximal segment, therefore, is at high risk for developing avascular necrosis with displaced fractures.

43. **What is the "mobile wad"?**
    The mobile wad includes the muscles innervated by the radial nerve prior to dividing into the PIN and superficial radial branches. The muscles are the brachioradialis, ECRL, and ECRB.

BIBLIOGRAPHY

1. Agur A: Grant's Atlas of Anatomy, 9th ed. Baltimore, Williams & Williams, 1991.
2. Green DP, Hotchkiss RN, Peterson WC (eds): Green's Operative Hand Surgery, 4th ed. New York, Churchill Livingstone, 1999.
3. Idler R: The Hand: Examination and Diagnosis, 3rd ed. New York, Churchill Livingstone, 1990.
4. Lister G: The Hand: Diagnosis and Indications, 3rd ed. London, Churchill Livingstone, 1993.
5. Tubiana R, Thomine J, Mackin E: Examination of the Hand and Wrist, 2nd ed. London, Mosby, 1996.

BIOMECHANICS OF THE HAND AND WRIST

Lawrence G. Sullivan, MD

1. **What are the two primary functions of the hand?**
   - Prehension (holding an object between any two surfaces of the hand)
   - Touch

2. **How much wrist motion is required for most activities of daily living (ADLs)?**
   Palmer et al. (1985) concluded that 5 degrees of flexion, 30 degrees of extension, 10 degrees of radial deviation, and 15 degrees of ulnar deviation are necessary to achieve most ADLs. In contrast, Ryu et al. (1991) determined that most ADLs require 40 degrees of flexion, 40 degrees of extension, 10 degrees of radial deviation, and 30 degrees of ulnar deviation.

3. **How much stability of the distal radioulnar joint (DRUJ) is provided by the sigmoid notch and ulnar head articulation?**
   It has been shown that approximately 30% of DRUJ stability is conferred by the sigmoid notch and the ulnar head articular relationship.

4. **Which component of the lunotriquetral (LT) ligament is the strongest?**
   The volar component of the LT ligament is the strongest.

5. **How many degrees of freedom does the wrist joint have?**
   Three: flexion/extension, radial/ulnar deviation, and axial rotation (prosupination).

6. **How much wrist motion is lost after a radiocarpal or intercarpal arthrodesis?**
   Radiocarpal fusion, fusion between both carpal rows, and fusion within a carpal row result in the loss of approximately 55%, 25%, and 10% of wrist motion, respectively.

7. **Most functional activities are performed with the wrist in what position?**
   Extension and ulnar deviation.

8. **Where is the center of rotation for the wrist located?**
   The wrist's center of rotation during flexion/extension (Fig. 2-1) is in the proximal capitate. During radioulnar deviation, the center of rotation is located slightly to the ulnar side of the longitudinal axis at a point one fourth of the total length of the capitate distal to its proximal end (Fig. 2-2). It has been determined, however, that the wrist's center of rotation is not a fixed point but rather changes during planar motions. The instantaneous center of rotation (ICR) translates along a curvilinear path between the proximal capitate and midscaphoid regions (Fig. 2-3).

9. **Which intercarpal fusion is associated with the least loss of wrist flexion/extension?**
   LT or capitohamate.

10. **Describe the motion of the proximal carpal row during wrist deviation.**
    During radial to ulnar deviation of the wrist, the proximal carpal row moves from a flexed to an extended position.

11. **How much motion occurs within the proximal carpal row during wrist flexion/extension?**
    As the wrist moves from flexion to extension, 24–34 degrees of motion occurs between the scaphoid and lunate and 12–18 degrees between the lunate and triquetrum.

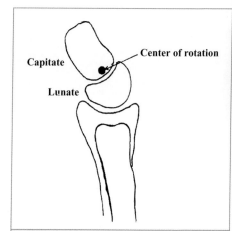

Figure 2-1. Flexion/extension of the wrist. During flexion/extension, the wrist has a fixed center of rotation that is located in the proximal part of the capitate. (Modified from Peimer CA [ed]: Surgery of the Hand and Upper Extremity. New York, McGraw-Hill, 1996.)

12. **What kinematic changes occur in the proximal carpal row after sectioning of the scapholunate (SL) ligament?**
    Scaphoid flexion and supination in conjunction with lunate extension.

13. **Which part of the scapholunate interosseous ligament (SLIL) is the most important in preventing motion between the scaphoid and lunate?**
    The dorsal SLIL.

14. **How does force distribution through the distal radius and ulna change with wrist and forearm position?**
    In the neutral position of forearm rotation, approximately 80% of force is transmitted through the radius and 20% through the distal ulna. With gripping, ulnar deviation of the wrist—or forearm pronation— increased force is transmitted through the distal ulna.

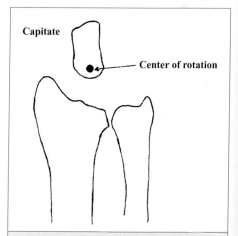

Figure 2-2. Radioulnar deviation of the wrist. A fixed center of rotation, found during radioulnar deviation of the wrist, is located in the head of the capitate. (Modified from Peimer CA [ed]: Surgery of the Hand and Upper Extremity. New York, McGraw-Hill, 1996.)

15. **How much stability of the DRUJ is provided by the sigmoid notch and ulnar head articulation?**
    Approximately 30%.

16. **How is force transmitted across each articular fossa of the distal radius?**
    Approximately 50% of the total load occurs at the scaphoid fossa and 30% at the lunate fossa.

17. **How does motion vary between the carpometacarpal (CMC) joints of the fingers?**
    The index and long finger CMC joints are relatively immobile. Between 5 and 10 degrees of motion is possible at the ring CMC joint, and between 15 and 20 degrees is possible at the small CMC joint.

18. **Describe the center of rotation for the trapeziometacarpal joint.**
    The center of rotation for the trapeziometacarpal joint varies with thumb position. It is located at the center of the distal trapezium joint surface in the resting position, within the body of trapezium during flexion/extension, and within the first metacarpal base during thumb abduction/adduction.

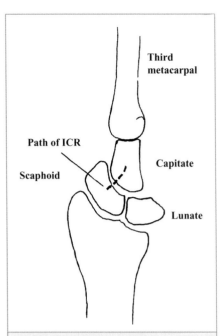

Figure 2-3. Wrist's center of rotation during planar motions. The wrist does not have a fixed ICR as it moves through a 35-degree motion arc from ulnar to radial deviation. (Modified from Peimer CA [ed]: Surgery of the Hand and Upper Extremity. New York, McGraw-Hill, 1996.)

19. **Which tendon has the greatest excursion at the metacarpophalangeal (MCP) joint during digital flexion/extension?**
    The flexor digitorum superficialis.

20. **Which finger joint has the greatest joint contact area?**
    The MCP joint.

21. **What wrist position permits maximal power grip?**
    Extension (dorsiflexion) combined with ulnar deviation.

22. **What are the relative contributions of the radiocarpal and midcarpal joints to wrist flexion/extension?**
    - **Extension:** 66% radiocarpal, 34% midcarpal
    - **Flexion:** 40% radiocarpal, 60% midcarpal

23. **True or false: Forearm rotation occurs entirely through the radioulnar joints.**
    False. About 90% occurs through the radioulnar joints, while 10% occurs within the radiocarpal and intercarpal joints.

24. **How many degrees of freedom do the interphalangeal joints have?**
    Only 1 degree of freedom; in flexion/extension.

25. **The longitudinal axis of rotation for the forearm passes through which anatomic sites?**
    The center of the radial head and the foveal region of the distal ulnar head.

> ## KEY POINTS: BIOMECHANICS OF THE HAND AND WRIST
> 1. Most activities of daily living are performed with the wrist in extension (dorsiflexion) and ulnar deviation.
> 2. The center of wrist motion is located in the proximal capitate.
> 3. The dorsal component is the strongest of the three identified regions of the SL ligament, while the volar component is the strongest of the three identified regions of the LT ligament.
> 4. In the ulnar neutral wrist, 80% of load is transmitted across the radiocarpal articulation, while 20% occurs across the ulnocarpal articulation.

26. **Describe the distribution of forces across the midcarpal joint when the wrist is in the neutral position.**
    - **Scaphotrapeziotrapezoidal (STT) joint:** 31%
    - **Scaphocapitate:** 19%
    - **Lunocapitate:** 29%
    - **Triquetrohamate:** 21%

27. **Which changes in load occur with dorsal angulation of the distal radius articular surface?**
    A progressive increase in load at the radioscaphoid and radiolunate articulations occurs as dorsal angulation increases beyond 30 degrees.

28. **How much of the total available articular surface between the scaphoid and lunate and the distal radius and triangular fibrocartilage complex (TFCC) is in actual contact?**
    A total of 20%, regardless of wrist position.

29. **Which intercarpal ligaments are the strongest?**
    The SL and LT.

30. **What effect does a 2.5 mm ulnar shortening osteotomy have on load transmission?**
    About 80% of load is transmitted across the radius and 20% across the ulna in the ulna neutral wrist. With a 2.5 mm ulnar shortening ostoeotomy, ulnocarpal load is reduced to only 4%.

31. **Describe the contributions of the STT, scaphocapitate, lunocapitate, and triquetrohamate articulations to total midcarpal joint contact.**
    Less than 40% of the available total articular surface of the midcarpal joint is in actual contact at any one time. The lunocapitate joint contributes 29%, the scaphocapitate joint provides 28%, the STT joint provides 23%, and the triquetrohamate joint contributes 20%.

32. **How much force is generated in the digital flexor tendons during passive range of motion (ROM)?**
    Between 2 and 4 N.

33. **How much force is generated during thumb to index fingertip pinch?**
    A total of 120 N.

34. **What percentage of longitudinal stiffness is provided by the central band of the interosseous membrane following a radial head excision?**
    About 71%.

35. **What is the biomechanical function of the flexor tendon sheath?**
    To maintain the flexor tendons close to the phalanges and the axes of joint rotation, thereby maximizing joint flexion as a function of tendon excursion.

36. **What is the function of the radioscapholunate (RSL) ligament (ligament of Testut)?**
    The RSL ligament functions primarily as a neurovascular conduit.

BIBLIOGRAPHY

1. An KN, Berger RA, Cooney WP (eds): Biomechanics of the Wrist Joint. New York, Springer, 1991.
2. Berger RA, Imaeda T, Berglund LJ, Linscheid RL: Anatomy and material properties of the scapholunate ligament. Trans Orthop Res Soc 18:317, 1993.
3. Crosby CA, Wehbe MA, Mawr B: Hand strength: Normative values. J Hand Surg 19A:665, 1994.
4. Imaeda T, Niebur G, Cooney WP, et al: Kinematics of the normal trapeziometacarpal joint. J Orthop Res 12:197, 1994.
5. Minami A, An KN, Cooney WP, et al: Ligamentous structures of the metacarpophalangeal joint: A quantitative anatomic study. J Orthop Res 1:361, 1984.
6. Minami A, An KN, Cooney WP, et al: Ligament stability of the metacarpophalangeal joint: A biomechanical study. J Hand Surg 10A:255, 1985.
7. Palmer AK, Werner FW: Biomechanics of the distal radioulnar joint. Clin Orthop 187:26, 1984.
8. Palmer AK, Werner FW, Murphy D, Glisson R: Functional wrist motion: A biomechanical study. J Hand Surg 10A:39, 1985.
9. Ryu J, Cooney WP, Askew LJ, et al: Functional ranges of motion of the wrist joint. J Hand Surg 16A:409, 1991.
10. Short WH, Werner FH, Fortino MD, Palmer AK: Distribution of pressures and forces on the wrist after simulated intercarpal fusion and Kienböck's disease. J Hand Surg 17A:443, 1992.
11. Short WH, Werner FH, Fortino MD, et al: Dynamic kinematics of the scaphoid and lunate before and after scapholunate interosseous ligament sectioning. Trans Orthop Res Soc 18:115, 1993.
12. Stuart PR, Berger RA, Linscheid RL, An K: The dorsopalmar stability of the distal radioulnar joint. J Hand Surg 25A:689–699, 2000.
13. Watanabe K, Nakamura R, Horii E, Miura T: Biomechanical analysis of radial wedge osteotomy for the treatment of Kienböck's disease. J Hand Surg 18A:686–689, 1993.
14. Youm Y, McMurtry RY, Flatt AE, Gillespie TE: Kinematics of the wrist. J Bone Joint Surg 60A:423, 1978.

CHAPTER 3
PHYSICAL EXAMINATION OF THE HAND
Terry McCurry, MD, and Morton L. Kasdan, MD

1. **List the priorities in managing a patient with a hand injury.**
   - Resuscitation
   - History
   - Physical examination
   - Blood supply
   - Debridement
   - Skeletal stabilization
   - Repair of nerve and tendon
   - Soft tissue coverage
   - Dressings
   - Tetanus/antibiotics
   - Evaluation of psyche and motivation for rehabilitation
   - Secondary reconstruction

2. **What pertinent points concerning the history of hand trauma should the examiner gather?**
   - **Time:** When did the injury occur? How much time has elapsed?
   - **Place:** Where did the injury occur? At home, work, or play? How clean was the environment?
   - **Mechanism:** What exactly happened? Crush, avulsion, or clean amputation?
   - **Previous treatment:** Has any intervention already occurred? What was it?

3. **What pertinent points concerning the history of hand pain should the examiner gather?**
   - **Time:** What was the first abnormality? When did it happen? Has it progressed? Have there been any previous episodes of pain or disability?
   - **Impairment/motivation:** What limitation is the patient experiencing? Why is the patient seeking treatment now?
   - **Location:** Where exactly is the pain? Does it radiate? Are there any other similar pains in other parts of the body or in the opposite hand?
   - **Relieving/exacerbating factors:** What, if anything, improves the pain? Does the pain level vary between night and day? What, if anything, aggravates the pain?

4. **List several risk factors for developing a *Clostridium tetani* infection.**
   - Wound >6 hours old
   - Depth >1 cm
   - Missile injury, crush, burn, frostbite
   - Stellate pattern, avulsion, abrasion
   - Presence of devitalized tissue
   - Contamination with dirt, soil, feces, or saliva

5. **How does the American Society for Surgery of the Hand (ASSH) name the digits of the hand?**
   Thumb, index finger, middle finger, ring finger, and little finger. This system prevents confusion because it allows the use of the first letter of each name as an abbreviation (e.g., RLF = right little finger, LMF = left middle finger).

6. **Name the topographical areas of the hand and the joints of each digit.**
   The hand is composed of the thenar region at the base of the thumb, the hypothenar region at the base of the little finger, and the midpalmar space. Each finger has three joints: the metacarpophangeal (MCP), the proximal interphalangeal (PIP), and the distal interphalangeal (DIP). The thumb has two joints: the interphalangeal (IP) and the MCP.

7. **What is the initial step in the assessment of the injured hand?**
   Inspection. The manner in which the patient holds the hand, the loss of normal finger cascade, the gross malposition of the joints, and the location and size of tissue deficits aid the examiner in initially defining the injury.

8. **In addition to the answers in question 7, what else should be assessed when examining an injured hand?**
   Skin, vascular supply, nerves, bones, joints, muscles, tendons, and ligaments.

9. **A patient is to have a radial forearm free flap raised to fill a pharyngeal defect. What is the name of the bedside screening test that should be used to assess the hand? What does the test tell you?**
   Allen's test is used to assess the patency of the superficial palmar arch. This is the communication between the radial artery and ulnar artery through the hand. As long as the arch is patent, the entire hand can be perfused by *either* the radial *or* the ulnar artery. Therefore, sacrifice of the radial artery with the free flap will not jeopardize the blood supply to the fingers.

10. **How do you perform the Allen's test?**
    Manually occlude both the radial and the ulnar arteries. Have the patient repeatedly make and release a fist to exsanguinate the hand. Then have the patient tightly clench, hold, and then open the digits. Release the radial artery. The hand and all the digits should blush. This indicates good collateral flow and is referred to as a negative test. Repeat the procedure, this time releasing the ulnar artery first. A positive test occurs when some of the digits do not blush (Fig. 3-1).

11. **Is the Allen's test the most accurate method for assessing the blood supply to the hand?**
    It is an excellent screening test but may be abnormal in a small number of patients with adequate collateral circulation. Laser Doppler or digital systolic pressure measurements are more accurate, but the arteriogram is the gold standard.

12. **What test is used to determine the presence of a nerve injury? What common item can be used to perform this test?**
    Two-point discrimination measures the ability of the patient to detect the difference in static pressure and to differentiate between one point and two points of pressure in the same area of the finger. A simple paper clip can be bent into a U shape and the ends adjusted to various distances apart for this test. The examiner presses the two points into the volar pad and then gradually increases the distance between the points until the patient discerns two distinct points of pressure.

13. **What is a "normal" value for two-point discrimination testing?**
    This question is tricky. The patient should first watch a test to know what to expect. Then serial testing should be done in each zone, and the points should lie in an axis parallel to that of the finger tested. The flow of the exam should go from distal to proximal. The examinee should get two out of three correct to receive credit. This distance varies with the anatomic area tested. It can range from as low as 3–5 mm at the tip of the finger to as much as 7–10 mm at the base of the palm and wrist. An abnormal finding should always be compared with the opposite hand.

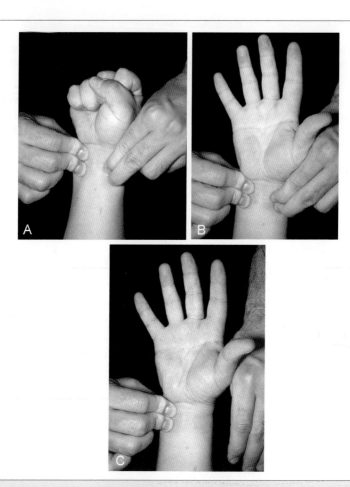

Figure 3-1. Allen's test for assessment of the collateral circulation in the hand. *A*, The examiner applies direct pressure over the radial and ulnar arteries at the wrist. Then the patient tightly squeezes and opens the fist several times to force the blood out. *B*, While pressure over the radial and ulnar arteries is maintained, the patient holds the fingers in extension. The palmar skin is visibly pale. *C*, The radial artery is released while occlusion is maintained over the ulnar artery. In this example, particular attention is paid to the ulnar aspect of the hand to ensure that good perfusion occurs via an intact radial artery. (From Concannon MJ: Common Hand Problems in Primary Care. Philadelphia, Hanley & Belfus, 1999.)

14. As the on-call hand surgeon, you are asked to evaluate the hand of a patient with a severe closed head injury. He is unconscious and ventilated. What four tests or maneuvers can you use to evaluate the hand?
    - Plain x-ray series of the hand
    - Assessment of capillary refill for the hand and digits
    - Prune test
    - Check of passive tenodesis

15. **What is the prune test?**
    An uninjured hand immersed in water for 5 to 10 minutes will exhibit "pruning" or wrinkling of the glabrous skin. Areas of denervated skin, as noted with a digital nerve injury, will not show this characteristic.

16. **What is passive tenodesis?**
    The uninjured hand maintains a delicate balance between the flexor and extensor tendon systems. On passive extension of the wrist, the fingers start to flex. On passive flexion of the wrist, the fingers start to extend. Any obvious disruption in this process or the natural cascade of the fingers suggests tendon imbalance, such as from a laceration or avulsion.

17. **What is the difference between the intrinsic and extrinsic muscles of the hand?**
    Extrinsic muscles, such as the long finger flexors (FDS, FDP), originate in the forearm and insert within the hand. Intrinsic muscles, such as the interossei, have *both* their origin *and* insertion within the hand.

18. **What are the three main peripheral nerves that supply the hand? Which one is responsible for sensation alone?**
    The ulnar, median, and radial nerves innervate the skin of the hand and intrinsic muscles. Only the ulnar and median nerves supply the intrinsic muscles. Therefore, the radial nerve supplies sensation only to the hand itself.

19. **What sensory disturbance is typically present in carpal tunnel syndrome?**
    Paresthesias progressing to pain can be present in any of the digits innervated by the median nerve.

20. **Is the palm of the hand affected in carpal tunnel syndrome?**
    Not typically. The superficial palmar branch arises proximal to and travels over the flexor retinaculum to innervate the palm and base of the thumb. In pronator syndrome, however, the location of nerve entrapment is proximal to the superficial palmar branch; therefore, the palm of the hand is not spared.

21. **What is Phalen's test?**
    Phalen's test is passive flexion of the wrist with the elbows extended. Anatomic studies show a decrease in volume of the carpal canal by about one third with the test. This maneuver can aggravate compression and therefore elicit symptoms. However, it is a purely subjective test.

22. **Describe the one clinical finding that helps distinguish ulnar nerve compression in the cubital tunnel from entrapment in Guyon's canal.**
    Neurosensory changes in the dorsoulnar aspect of the hand are typically present with cubital tunnel syndrome and absent with compression in Guyon's canal. This occurs because the dorsal sensory branch that innervates the dorsoulnar part of the hand arises distal to the cubital tunnel but proximal to Guyon's canal.

23. **Describe Wartenberg's sign and Froment's sign.**
    Both can occur with an ulnar nerve injury. Weakness in the intrinsic hypothenar muscles and palmar interosseous is overcome by the radial-innervated extensor digiti minimi, causing abduction and slight hyperextension of the little finger. This is **Wartenberg's sign**. Patients typically complain that they "catch" the little finger when placing the hand into a pocket. Hyperflexion of the thumb IP joint compensates for weakness in the adductor pollicis when the patient attempts to hold objects between the thumb and radial side of the index finger. This is **Froment's sign**. An abnormal test should always be compared with the opposite side when feasible.

24. **Your patient cannot make the "okay" sign. His hand appears flattened because he can press together the volar pads but not the actual tips of the thumb and index finger. What is wrong?**
    The patient appears to have weakness in the flexor pollicis longus (FPL) and flexor digitorum profundus (FDP) to the index finger. This finding can be caused by injury to the anterior interosseus nerve, a purely motor branch of the median nerve.

25. **An intoxicated college student punched a wall and presents to the emergency department 2 days later with a painful, swollen hand, especially on the ulnar side near the knuckle. What is wrong?**
    This patient probably has a fracture at the neck of the fifth metacarpal. This is also known as a boxer's fracture. Routine x-rays are diagnostic.

26. **How do you properly assess the aforementioned patient for a rotational deformity?**
    With the fingers extended, a significant deformity can be underestimated. To assess properly for rotational deformity, have the patient make a fist as much as possible. Any malrotation will become more obvious.

27. **A football player accidentally punched his teammate in the mouth during a game. He presents to the emergency department 2 days later with a painful, swollen hand. What is wrong?**
    This could be another boxer's fracture, but the examiner should carefully inspect the skin over the knuckle. Any sign of foreign body penetration may have come from a tooth and herald infection of the soft tissues and possibly the MCP joint itself. This is known as a "fight bite." An x-ray should always be taken to look for any foreign bodies as well as fractures.

28. **A carpenter presents to the emergency department 2 days after accidentally driving a finish nail into the volar side of his left index finger over the proximal phalanx. The x-rays show no fracture or foreign body. The index finger is held in slight flexion, swollen in a fusiform fashion, and tender along its length. What additional maneuver will you use to confirm the diagnosis?**
    You should suspect suppurative flexor tenosynovitis or an infection of the flexor tendon sheath. The fourth sign is extreme pain with passive extension of the digit. This maneuver produces the most sensitive physical finding. The four findings are known as Kanavel's cardinal signs.

KEY POINTS: KANAVEL'S CARDINAL SIGNS OF SUPPURATIVE FLEXOR TENOSYNOVITIS

1. Fusiform swelling of the digit
2. Flexed resting posture
3. Tenderness all along the volar side of the digit
4. Exquisite pain with passive extension

29. **Describe the boutonnière deformity.**
    The boutonnière deformity consists of PIP flexion and DIP hyperextension.

30. **Describe the swan-neck deformity.**
    The swan-neck deformity consists of PIP joint hyperextension and DIP flexion.

31. **A high school football player makes a difficult tackle but then is unable to extend the DIP joint of his right index finger. What is wrong?**
    This patient probably has disruption of the terminal slip of the extensor mechanism, causing a mallet finger deformity.

32. **How do you test for intrinsic tightness?**
    The patient's MCP joints are passively extended and stabilized. Note the degree of passive PIP flexion. Then the MCP joints are stabilized in flexion. Now note the degree of passive PIP motion. If the amount of PIP flexion is greater with the MCP joints flexed compared to when they are extended, intrinsic tightness is present.

33. **Define the lumbrical-plus deformity.**
    Prior disruption of the FDP tendon, as with distal phalanx amputation, causes the proximal tendon to retract. This increases the resting tension in the lumbrical muscle of that digit. With attempted finger flexion, the tension on the lumbrical increases and overcomes the power of the flexor digitorum superficialis (FDS), causing a paradoxical *extension* of the affected digit. Division of the lumbrical tendon usually corrects this problem. Anchoring the FDP tendon to the middle phalanx can prevent this problem.

34. **Define the quadriga effect.**
    Injury resulting in a shortened or adhered FDP tendon impairs the motion of the FDP tendons in the other fingers since they arise from a common muscle belly. This effect commonly occurs when the FDP tendon is sutured to the extensor tendon during closure of an amputation. With attempted grasp, all the force is used in the injured tendon; therefore, the other FDP tendons cannot achieve full excursion, resulting in a weak grip. The term *quadriga* is derived from the days when Roman chariots were used. These chariots were typically pulled by four horses, and the weakest of the four horses would slow down or hold back the others.

35. **A 20-year-old man complains of pain on the radial side of the wrist after falling 3 days ago. Based on the history, you suspect a scaphoid fracture. How do you know where to find the scaphoid bone on examination?**
    Palpate over the anatomic snuffbox, which lies distal to the radial tuberosity between the extensor pollicis longus (EPL) and the combination of the abductor pollicis longus (APL) and extensor pollicis brevis (EPB). Tenderness here suggests a fracture of the scaphoid. If initial x-rays are negative, immobilize the wrist and then reexamine the wrist in 10–14 days. At this point, x-rays should again be taken. If the x-rays remain negative, a bone scan, CT scan, or MRI should be ordered.

36. **How do you find the EPL tendon?**
    Have the patient place the hand palm down on a table. Ask him or her to lift the thumb off the table. The tight bowstring-like structure is the EPL tendon.

37. **A 60-year-old seamstress complains of worsening pain over the base of her thumb. You suspect basilar joint arthritis. How do you assess the thumb during physical examination?**
    Perform the grind test. Apply axial compression of the thumb metacarpal against the trapezium while gently rotating it back and forth. In the presence of basilar joint or carpometacarpal (CMC) arthritis, this maneuver reproduces the pain and crepitus.

38. **A 30-year-old machine shop worker complains of pain over the radial aspect of her wrist. A grind test is negative. What other test is important to document?**
    Finkelstein's test, which is performed by passively adducting the thumb while simultaneously deviating the wrist to the ulnar side. A positive test reproduces the pain and suggests inflammation of the tendons in the first dorsal compartment (de Quervain's tenosynovitis).

39. **A 75-year-old veteran complains of the inability to play his guitar because his right ring and little fingers are drawn into the palm of his hand. What is the likely diagnosis? Are his children at risk?**
   He probably has Dupuytren's disease, a hereditary disorder with variable penetrance. Fair-skinned people of northern European ancestry and those with a history of diabetes are at a higher risk of developing this condition.

40. **A patient with joint pain that is present in the morning and gradually resolves during the course of the day probably has what kind of arthritis?**
   Rheumatoid arthritis. This finding is classic. Osteoarthritis tends to improve with rest and becomes worse with repeated usage.

BIBLIOGRAPHY

1. Concannon MJ: Common Hand Problems in Primary Care. Philadelphia, Hanley & Belfus, 1999.
2. Green DP, Hotchkiss R, Pederson W (eds): Green's Operative Hand Surgery, 4th ed. New York, Churchill Livingstone, 1999.
3. Levinsohn DG, Gordon L, Sessler DL: The Allen's test: Analysis of four methods. J Hand Surg 16A:279–282, 1991.
4. Omer GE Jr, Spinner M, Van Beek AL (eds): Management of Peripheral Nerve Problems, 2nd ed. Philadelphia, W.B. Saunders, 1998.
5. Seiler JG III: Essentials of Hand Surgery. Philadelphia, Lippincott Williams & Wilkins, 2002.
6. Tubiana R, Leclerq C, Hurst LC, et al (eds): Dupuytren's Disease. London, Martin Durity, 2000.

NEUROLOGIC EVALUATION OF THE UPPER EXTREMITY

Hal M. Corwin, MD

1. **What is carpal tunnel syndrome?**
   Carpal tunnel syndrome is a complex of symptoms consisting of pain, numbness, tingling, or a burning sensation in the median nerve distribution caused by compression of the median nerve in the carpal tunnel (Fig. 4-1).

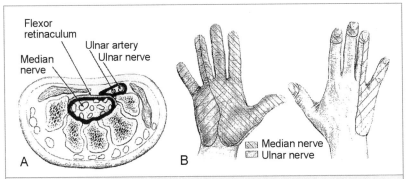

Figure 4-1. Wrist anatomy. *A*, The median nerve through the carpal tunnel in close proximity to Guyon's canal, where the ulnar nerve resides. *B*, Median and ulnar nerve sensory distribution. (From West SG [ed]: Rheumatology Secrets. Philadelphia, Hanley & Belfus, 1997.)

2. **Define cubital tunnel syndrome.**
   The cubital tunnel is located 3–5 cm distal to the medial epicondyle and is formed by the humeroulnar aponeurotic arcade that joins the humeral and ulnar heads of the flexor carpi ulnaris origin. Compression of the ulnar nerve at this site is called cubital tunnel syndrome. Symptoms include intermittent hyperesthesia in the ulnar nerve distribution that is associated with elbow flexion and often relieved by elbow extension. Weakness and atrophy of the ulnar-innervated intrinsic hand muscles are other nonspecific signs.

3. **What are the three sites of ulnar nerve compression at the elbow?**
   The major points of compression are at the humeroulnar aponeurotic arcade, the deep flexor pronator aponeurosis, and the retroepicondylar groove. The most common mechanism for ulnar neuropathy at the retroepicondylar groove is external compression.

4. **What is hand dystonia?**
   Hand dystonia is a movement disorder causing a distortion of the normal posture of the hand. It is due to slow and sustained contractions of the muscles. The muscular tone is increased and may be painful. Contractures and deformities may result in chronic cases.

5. **How is botulinum toxin used to treat focal hand dystonia?**
   Intramuscular injections of botulinum toxin type A (Botox) have proved to be the most effective treatment for focal hand dystonia. The neurotoxin prevents release of acetylcholine from the peripheral nerve terminal by working at the presynaptic membrane. This results in muscle paralysis. The procedure is often performed with electromyographic guidance to target the actively contracting muscle. Botox can be used to treat torticollis, blepharospasm, and tremor as well as focal hand dystonia.

6. **What is muscle wasting due to a cervical rib?**
   Also called neurogenic thoracic outlet syndrome, this condition consists of wasting of the hand muscles, occasionally accompanied by lesser degrees of weakness band wasting of forearm muscles, associated with an elongated C7 transverse process. A fibrous band extending from the C7 transverse process or from a rudimentary cervical rib to the region of the first rib causes compression of the lower trunk of the brachial plexus or the C8 to T1 nerve roots. Associated vascular changes may or may not be present.

7. **What are the presenting upper extremity symptoms of amyotrophic lateral sclerosis (ALS)?**
   ALS is a type of motor neuron disease characterized by atrophy of the skeletal muscles of the body associated with degeneration of the motor neurons. Patients with ALS have weakness, atrophy, and fasciculations of the muscles. The weakness does not respect spinal root or peripheral nerve anatomic distribution and is often asymmetric. A patient may complain of muscle twitching, pain, or paresthesias but loss of sensation is not usually present.

8. **How often does ALS present with symptoms involving the upper extremity?**
   Approximately one third of patients with ALS present with the initial symptoms in one or both upper extremities.

9. **What is the main cause of cervical radiculopathy?**
   Spondylosis. This condition includes degeneration of the cervical disc; formation of osteophytes; hypertrophy of facet surfaces; and nerve root impingement.

10. **What are the common presenting symptoms and signs of cervical radiculopathy?**
    Pain may be exclusively extracervical in nature. However, it is often lateralized in the neck with radiation to the scapula, occipital region, or arm and hand. Paresthesias may also occur in a radicular distribution. The pain and sensory symptoms may be precipitated by coughing or neck movement. Atrophy of muscles corresponding to the affected nerve root may occur in advanced cases with an associated loss of deep tendon reflexes.

11. **Which diagnostic studies are useful in evaluating the patient with suspected cervical radiculopathy?**
    Plain radiographs of the cervical spine can determine loss of disc space, vertebral subluxation, and facet hypertrophy with encroachment on neural foramina. Magnetic resonance imaging (MRI) and myelography with axial computerized tomography (CT) sections are superior to plain x-ray films for assessing this disease. Electromyography (EMG) provides physiologic information concerning the spinal roots.

12. **What upper extremity symptoms are associated with Parkinson's disease?**
    The onset of the illness is often noted by the appearance of tremor in one limb, with the arm more likely than the leg. There is a tendency for the opposite limb to become involved within 1 year, followed by generalized slowing of movements and gait. Patients with Parkinson's disease may complain of difficulties with dressing, using eating utensils, or handwriting.

They may perform these tasks more slowly or with great difficulty. Reduced finger tapping, slow clenching and unclenching of the fists, and reduced arm swinging while ambulating are all common symptoms.

13. **At what age do the symptoms of Parkinson's disease typically begin?**
    The mean age of onset is in the mid-50s with about 30% of patients reporting symptoms before the age of 50.

14. **Will C7 radiculopathy cause weakness of the intrinsic hand muscles?**
    No. The intrinsic hand muscles are innervated by the C8 and T1 motor fibers.

15. **Name the central nervous system diseases that may mimic carpal tunnel syndrome.**
    Vascular, neoplastic, infectious, and degenerative disorders of the brain or spinal cord, including stroke, multiple sclerosis, cervical myelopathy from degenerative disc disease, and other such conditions, may mimic carpal tunnel syndrome or other upper extremity peripheral nerve entrapment.

16. **What is Lhermitte's sign?**
    Lhermitte's sign is the presence of a sudden painful or electrical sensation that occurs with flexion of the neck and spreads to the lower cervical or lumbosacral spine or into the limbs. This sign may be noted in patients with multiple sclerosis, cervical myelopathy, or recent head injury.

KEY POINTS: NEUROLOGIC EVALUATION OF THE UPPER EXTREMITY

1. Peripheral nerve entrapment disorders are diagnosed clinically with support from electrodiagnostic testing.
2. Central nervous system illness may often present with upper extremity symptoms of motor deficit, hyperreflexia, movement disorder, and vague sensory complaints.
3. Insufficient neurologic examination is a common source of misdiagnosis.

17. **Why is it important to evaluate the function of the medial antebrachial cutaneous (MABC) nerve in the patient with hand numbness?**
    MABC function is impaired with lesions of the lower trunk of the brachial plexus but spared in ulnar neuropathy. Both diseases may cause overlapping symptoms of numbness in the medial hand, but only lower trunk lesions cause numbness in the distribution of the MABC.

18. **Name a common nontraumatic cause of brachial plexopathy.**
    Parsonage-Turner syndrome, or idiopathic brachial plexus neuritis, is a common cause of brachial plexopathy. The condition may be postinflammatory but also can occur after periods of stress or hospitalization.

19. **How do the sensory pathways for pain and temperature differ from the sensory pathways for vibratory sensation, two-point discrimination, and limb proprioception?**
    The sensory pathway for pain and temperature sensation travels through the lateral columns of the spinal cord to the thalamus. The sensory pathways for vibratory sense, two-point discrimination, and proprioception travel through the posterior columns of the spinal cord to the

thalamus. Therefore, lesions of the posterior columns of the spinal cord will not diminish pain or temperature sensation.

20. **Name the most common causes of peripheral neuropathy.**
    - Diabetes mellitus
    - Hypothyroidism
    - Idiopathic disease
    - Monoclonal gammopathy
    - Vitamin deficiency
    - Familial disorder

21. **How are upper motor neuron lesions differentiated from lower motor neuron lesions in a clinical examination?**
    If weakness is combined with atrophy, the lesion is localized to the motor unit. In the absence of atrophy, examination of muscle stretch reflexes and limb tone will determine whether the diagnosis is one of upper or lower motor neuron dysfunction. Lesions of the upper motor neuron result in hyperreflexia, clonus, a Babinski sign, weakness without evidence of atrophy, and spasticity. Lower motor neuron lesions result in hyporeflexia, hypotonia, or weakness with atrophy in the distribution of a peripheral nerve or spinal root.

22. **What is Froment's sign? What does it imply?**
    The primary muscle that enables key pinch between the radial border of the index finger and the thumb is the adductor pollicis (ADP). When the ADP is weak due to ulnar neuropathy, key pinch strength is enhanced by flexing the thumb at the interphalangeal (IP) joint using the median nerve innervated flexor pollicis longus. Froment's sign is the clinical observation that the patient hyperflexes the thumb at the IP joint during key pinch.

23. **Name two forearm muscles innervated by the ulnar nerve.**
    Flexor carpi ulnaris and flexor digitorum profundus supplying the ring and small fingers.

24. **What are the peripheral nerves and spinal roots corresponding to upper extremity reflexes?**
    See Table 4-1.

TABLE 4-1. PERIPHERAL NERVES AND SPINAL ROOTS CORRESPONDING TO UPPER EXTREMITY REFLEXES

Reflexes	Root	Nerve
Brachioradialis	C5-C6	Radial
Biceps	C5-C6	Musculocutaneous
Triceps	C6-C8	Radial
Finger flexion	C6-T1	Median and ulnar

25. **True or false: Sensory nerve conduction studies are normal in cervical radiculopathy.**
    True. The peripheral sensory fibers originating in the dorsal root ganglion are not injured; therefore, the sensory nerve conduction studies are normal.

26. **A patient with multiple sclerosis may present with numbness in the hand or arm. The MRI abnormalities are located in which anatomic region?**
    The lesion may appear in the white matter of the brain, brainstem, or cervical spinal cord. An MRI of the brain and cervical spinal cord should be obtained if multiple sclerosis is suspected. Spinal fluid studies may also support this diagnosis.

BIBLIOGRAPHY

1. Campbell WW: Ulnar neuropathy at the elbow: Anatomy, causes, rational management, and surgical strategies. AAEM Course D:7–12, 1991.
2. Caroscio JT, Mulvihill M, Sterling R, et al: Amyotrophic lateral sclerosis, its natural history. Neurol Clin 5:1–8, 1987.
3. Chung KW: Gross Anatomy, 3rd ed. Baltimore, Williams & Wilkins, 1995.
4. DeLong RN: The Neurologic Examination, 5th ed. Philadelphia, Lippincott-Raven, 1979.
5. DeMyer W: Technique of the Neurologic Examination, 4th ed. New York, McGraw-Hill, 1994.
6. Duvoisin RC: A brief history of Parkinsonism. Neurol Clin 10:301–317, 1992.
7. Feindel W, Stratford J: The role of the cubital tunnel in tardy ulnar palsy. Can J Surg 1:287–382, 1958.
8. Ferguson RJL, Caplan LR: Cervical spondylitis myelopathy. Neurol Clin 3:373–382, 1985.
9. Fix JD: Neuroanatomy, 2nd ed. Baltimore, Williams & Wilkins, 1995.
10. Gilliatt RW, LeQuesne PM, Logue V, et al: Wasting of the hand associated with a cervical rib or band. J Neurol Neurosurg Psychiatry 33:615–624, 1970.
11. Goetz CG, Pappert EJ: Textbook of Clinical Neurology. Philadelphia, W.B. Saunders, 1999.
12. Hoehn MM: The natural history of Parkinson's disease in the pre-levodopa and post-levodopa eras. Neurol Clin 10:331, 1992.
13. Kimora J: Electrodiagnosis in Diseases of Nerve and Muscle: Principles and Practice, 2nd ed. Philadelphia, F.A. Davis, 1989.
14. Wilbourn AJ: The brachial plexus: Anatomy, pathophysiology, and electrodiagnostic assessment. AAEM Course C:7–30, 1994.

CHAPTER 5
ANESTHESIA
Bruce Crider, MD

1. **What is conscious sedation anesthesia?**
   Conscious sedation is a pharmacologically induced state of depressed consciousness that preserves the patient's ability to maintain his or her own airway, spontaneously ventilate, and respond to both physical and verbal stimuli.

2. **When do "anesthesia standards" apply if a patient is to receive conscious sedation?**
   The Joint Commission on Accreditation of Healthcare Organizations (JCAHO) has stated that anesthesia standards apply when patients are to receive general anesthesia, regional anesthesia, or sedation—analgesia for the purposes of undergoing a procedure during which there is a risk of the patient losing protective airway reflexes.

3. **Which patient assessments are necessary *before* proceeding with a procedure done under conscious sedation?**
   - *Non per os* (nothing by mouth, NPO) status
   - Airway risk assessment (e.g., Mallampati airway classification)
   - Preexisting level of consciousness/sedation
   - Anticipated level of sedation required for the procedure
   - Overall health of the patient
   - Status of comorbidities (i.e., American Society of Anesthesiologists [ASA] classification)

4. **What are the minimum monitoring requirements for conscious sedation?**
   - Level of sedation
   - Heart rate
   - Respiratory rate
   - Blood pressure
   - Oxygen saturation
   - Possible electrocardiogram (ECG) for patients with cardiac comorbidities

> ### KEY POINTS: MAJOR CONSIDERATIONS FOR MANAGING CONSCIOUS SEDATION
>
> 1. Aspiration risk (NPO status)
> 2. Airway assessment (Mallampati class)
> 3. Comorbidity assessment (ASA classification)
> 4. Required level of sedation for the procedure
> 5. Monitoring requirements

5. **What are the major advantages of regional anesthesia?**
   Regional anesthesia causes less disturbance of normal body physiology than any other type of anesthesia and is the only anesthetic technique that prevents afferent impulses from reaching the central nervous system.

6. **What are the minimal requirements for monitoring patients during regional anesthesia?**
   With the exception of a digital or simple nerve block, which involves injecting a small amount of local anesthetic, basic monitoring includes periodic assessment of blood pressure, cardiac rhythm, and pulse oximetry. Basic resuscitation equipment should be readily available. The patient should receive oxygen via a nasal cannula because the inadvertent systemic introduction of a large amount of local anesthetic may result in systemic toxicity.

7. **List the different regional anesthetic techniques used for surgery of the upper extremity.**
   - Brachial plexus block (at different levels)
   - Peripheral nerve blocks at the elbow or wrist
   - Digital block
   - Intravenous regional anesthesia (Bier's block)

8. **What regional anesthetic techniques are particularly useful in the treatment of hand injuries?**
   - Local infiltration anesthesia
   - Wrist block
   - Digital block
   - Bier's block

9. **Do local anesthetics cause allergic reactions?**
   An allergic reaction to a local anesthetic agent, especially an amide type, is extremely rare. If the patient has a history of allergic reaction to a local anesthetic agent, it is usually in response to an ester-type agent. Cross-reactivity between ester- and amide-type local anesthetic agents does *not* occur.

10. **How do local anesthetic agents become systemically toxic?**
    Toxicity occurs after injection of an excessively large volume of anesthetic agent or when the agent is inadvertently injected intravascularly. Toxicity may occur at lower doses in elderly and critically ill patients.

11. **What clinical findings suggest systemic toxicity with a local anesthetic?**
    - Metallic taste
    - Perioral numbness
    - Ringing in the ears
    - Twitching of the face
    - Convulsions

12. **How are patients with local anesthetic toxicity treated?**
    Convulsions are usually transient. An airway should be established for the patient, and he or she should be ventilated. This strategy is usually sufficient for treating the patient because the blood level of the local anesthetic agent rapidly decreases. If convulsions persist, a small dose of intravenous thiopental (Pentothal), diazepam (Valium), or midazolam (Versed) is given. Full cardiovascular resuscitation is required if cardiovascular toxicity (e.g., arrhythmias, cardiac arrest) occurs. Cardiovascular compromise most commonly occurs with toxicity due to high-concentration bupivacaine.

13. **How do the volume and concentration of a local anesthetic agent affect a nerve block?**
    - Volume determines the extent of analgesia.
    - Concentration determines the density of the sensory and motor block.

14. **What is the major advantage of using a low-concentration local anesthetic agent in a regional block?**
    An increased volume can be used, which is important for achieving a successful block of the brachial plexus.

15. **Why is a nerve block technique preferred over local infiltration of an anesthetic agent?**
    The nerve block technique provides complete, longer-lasting anesthesia and does not distort the soft tissues in the operative field.

16. **What type and volume of local anesthetic agents are commonly used with brachial plexus block anesthesia?**
    The agents used are 1% lidocaine and 0.25–0.5% bupivacaine. An interscalene block requires 20–30 ml; a supraclavicular block requires 20–30 ml, and an axillary block requires 35–50 ml of the local anesthetic agent.

KEY POINTS: MAJOR CONSIDERATIONS FOR CONDUCTING UPPER EXTREMITY REGIONAL ANESTHESIA

1. Patient's mental status
2. Preoperative and operative nerve involvement
3. Potential block-associated complications or consequences (including pneumothorax, phrenic nerve block, Horner's syndrome, local anesthetic toxicity, and masking of postoperative pain, such as compartment syndrome)
4. Monitoring requirements
5. Accessibility of resuscitation equipment

17. **What is the benefit of combining a local anesthetic agent with epinephrine?**
    Although it is controversial, epinephrine, usually used in a 1:200,000 concentration, produces vasoconstriction and thus minimizes blood loss. It also delays vascular uptake, thereby lowering the peak blood level of the local anesthetic agent, reducing the potential for toxicity, and increasing the length of block time.

18. **When should you avoid using a local anesthetic and epinephrine combination?**
    Epinephrine should not be used in digital block anesthesia because of the resulting vasoconstriction and possible ischemia with secondary digital loss. The combination should also be avoided in patients with ischemic heart disease.

19. **Which brachial plexus block technique should be used to anesthetize the upper medial aspect of the arm?**
    This area cannot be anesthetized by any brachial plexus block technique because it is innervated by the intercostobrachial nerve (T2). Therefore, the upper medial aspect of the arm can be anesthetized only by subcutaneous infiltration of a local anesthetic agent across the axillary artery just below the axilla.

20. **Which brachial plexus block technique is most frequently used to provide anesthesia for hand surgery?**
    An axillary block. The technique is simple (because of the location of the artery and nerves), and major complications seen with other techniques are minimal.

21. **What are the advantages of a supraclavicular block?**
    This technique results in the rapid onset of reliable anesthesia yet requires a low volume of local anesthetic. It may be used for most surgical procedures of the upper extremity and has minimal side effects and complications. A supraclavicular block can be performed with the arm at the side, thus avoiding movement of the injured or painful extremity.

22. **What are the major anatomic landmarks for the supraclavicular technique of a brachial plexus block?**
    Subclavian pulse (immediately posterior and superior to the midpoint of the clavicle), clavicle, sternocleidomastoid muscle, and interscalene groove.

23. **What complications are associated with a supraclavicular block?**
    Pneumothorax, phrenic nerve block, and Horner's syndrome. With the exception of a pneumothorax, the effects last for the duration of anesthesia. A pneumothorax can occur up to 12 hours after a supraclavicular block. Treatment is indicated if more than 20% of the lung volume is involved. A typical scenario is the delayed onset of chest pain and discomfort during breathing 10–12 hours after a supraclavicular block.

24. **Which nerve is most frequently *not* anesthetized with a supraclavicular block?**
    The ulnar nerve.

25. **What are the indications for an interscalene, supraclavicular, and axillary block anesthetic?**
    - The **interscalene block** is ideal for shoulder and upper arm surgeries. Increasing the volume of the local anesthetic agent to 30–40 ml anesthetizes the whole arm but misses the ulnar nerve distribution in approximately 50% of patients.
    - The **supraclavicular block** is the most consistently effective approach to anesthetize the brachial plexus. The entire arm, with the exception of the skin over the shoulder, is anesthetized with a supraclavicular block.
    - The **axillary block** reliably anesthetizes the medial aspect of the arm, forearm, and hand but may fail to block the musculocutaneous and radial nerves in approximately 25% of cases.

26. **What are the major landmarks for a brachial plexus block performed with an interscalene technique?**
    - Cricoid cartilage (C6 level)
    - Sternocleidomastoid muscle
    - Interscalene groove

27. **What are the major complications of an interscalene brachial plexus block?**
    - Phrenic nerve block (which can be problematic in patients with significant lung disease)
    - Inadvertent intravascular, subarachnoid, or epidural injection
    - Recurrent laryngeal nerve block
    - Cervical sympathetic chain block with a resultant Horner's syndrome

28. **What is the safest time period to supplement a failed brachial plexus block?**
    Studies have shown that additional dosing with 3.5 mg/kg of body weight of mepivacaine with epinephrine 90 minutes after the initial dose may be successful. The maximal initial dose is 7 mg/kg of body weight. The Sanofi-Winthrop mepivacaine package insert for dosage

recommendations reads as follows: "The total dose for any 24-hour period should not exceed 1000 mg because of a slow accumulation of the anesthetic or its derivatives or slower than normal metabolic degradation or detoxification with repeat administration."

29. **When is it safe to release the tourniquet after establishing intravenous regional anesthesia?**
To avoid a potential toxic reaction, the minimal time between introduction of a regional anesthetic and deflation of the tourniquet should be no less than 20 minutes. The tourniquet should be deflated and reinflated over a period of 2–3 minutes before being finally discontinued.

30. **What is the best location for insertion of an intravenous catheter for the introduction of a Bier's block?**
Ideally, the catheter should be inserted as far distally as possible (i.e., dorsum of the hand). However, the cannula may be inserted anywhere distal to the tourniquet cuff if the hand cannot be used.

31. **Why are nerve blocks at the elbow not recommended for hand or wrist surgery?**
Anesthetizing the whole hand subjects the patient to multiple injections to block each individual nerve. In addition, the technique is unreliable because of the overlap and variation of anatomic distribution of the nerves.

32. **What are the landmarks for a median nerve block at the wrist?**
The median nerve lies between the palmaris longus and flexor carpi radialis (FCR) tendon. If the palmaris longus is absent, the needle should be inserted on the ulnar side of the FCR tendon.

33. **What are the two local anesthetic agents of choice in a pregnant woman who requires regional anesthesia for emergency hand surgery?**
Chloroprocaine (Nesacaine) and bupivacaine (Marcaine). Chloroprocaine is rapidly hydrolyzed, resulting in a low blood level that prevents placental transfer. Most of the bupivacaine is protein bound, which subsequently minimizes the amount of free drug that is available to cross the placenta.

34. **A patient is scheduled for elective hand surgery and requests a regional anesthetic. Preoperative examination reveals an ipsilateral frozen shoulder (adhesive capsulitis). Which technique cannot be used?**
An axillary block should not be used because introduction of the local anesthetic agent requires the arm to be positioned in 90 degrees abduction with slight external rotation.

35. **A patient with a history of a complex regional pain syndrome of the arm following a distal radius fracture malunion is scheduled for a corrective osteotomy. Which anesthetic technique is recommended?**
The major concern in this clinical scenario is that the patient may experience a recurrence or aggravation of the pain syndrome after surgery. It is recommended that the patient undergo a brachial plexus block to provide both intraoperative anesthesia and sympathetic blockade. Ideally, a catheter should be placed several hours before the procedure. The catheter should remain indwelling and be dosed to provide adequate intraoperative anesthesia. It should remain in place for 24–48 hours after surgery.

36. **What are the anesthetic concerns for patients with insulin-dependent diabetes and peripheral neuropathy who are scheduled for hand surgery?**
The peripheral nerves of patients with a neuropathy are more sensitive to local anesthetic exposure and may have prolonged motor and sensory deficits after a peripheral nerve or brachial plexus block. Any preexisting neurologic deficit should be defined so as not to confuse the effects of a block with preexisting disease.

37. A patient is scheduled for elective wrist disarticulation after a nonsalvageable, severe crush injury to the hand. The patient has undergone multiple previous procedures, but the hand remains stiff, painful, insensate, and nonfunctional. The patient asks about an anesthetic technique that prevents phantom pain. How should you respond?

   Phantom pain after the loss of a digit or limb is extremely difficult to treat. Ideally, prevention of phantom limb pain is preferable. Although a small series of clinical trials have produced conflicting data, it has been suggested that using a brachial plexus block both before and after surgery diminishes the incidence and severity of phantom pain following an upper extremity amputation. The brachial plexus block should be placed several hours before surgery and continue for 24–48 hours postoperatively.

38. A 19-year-old college student sustains a dorsal interphalangeal fracture dislocation of the thumb. Closed reduction is necessary. What anesthetic technique should be used to anesthetize the thumb?

   The thumb is innervated by branches of the superficial radial and lateral antebrachial cutaneous nerves and the digital branches of the median nerve. A complete sensory block of the thumb is achieved with a median nerve block at the wrist or a digital block at the base of the thumb. Circumferential infiltration of a local anesthetic at the dorsal base of the thumb to anesthetize the branches of the superficial radial nerve and lateral antebrachial cutaneous nerves is also necessary.

BIBLIOGRAPHY

1. Brown DM: Regional Anesthesia and Analgesia. Philadelphia, W.B. Saunders, 1996.
2. Pearce H, Lindsay D, Leslie K: Axillary brachial plexus block in two hundred consecutive patients. Anaesth Intens Care 24:453–458, 1996.
3. Porter JM, Inglefield CJ: An audit of peripheral nerve blocks for hand surgery. Ann R Coll Surg Engl 75:325–329, 1993.

CHAPTER 6

DIAGNOSTIC IMAGING OF THE WRIST AND HAND

Joel S. Newman, MD, and Arthur H. Newberg, MD

1. **Which radiographic views should be obtained in the evaluation of every patient with a digital, hand, or wrist injury?**
   - Posteroanterior (PA)
   - Lateral
   - Oblique

2. **Describe the normal carpal arcs on a PA radiograph of the wrist, as described by Gilula.**
   In the PA view of the wrist in a neutral position, three smooth arcs can be identified along the proximal articular surfaces of the scaphoid, lunate, and triquetrum; along the distal articular surfaces of the proximal carpal row; and along the proximal margins of the capitate and hamate (Fig. 6-1). Disruption of any of these arcs or carpal overlap is suggestive of carpal subluxation or dislocation.

3. **How is carpal alignment assessed on a true lateral view of the wrist?**
   The lunate, radius, and capitate should appear colinear or within 10 degrees of each other (Fig. 6-2).

4. **What common carpal injuries are detected on a carpal tunnel view of the wrist?**
   - Occult fracture of the volar aspect of the trapezium
   - Pisiform fracture
   - Hook of the hamate fracture

5. **What is ulnar variance?**
   The relationship between the distal articular surfaces of the radius and ulna as seen on a PA view of the wrist (Fig. 6-3). Three types of variance exist: ulnar neutral, ulnar minus, and ulnar plus.

6. **Which radiographic view is used to determine ulnar variance?**
   A neutral rotation PA view of the wrist. The patient is positioned with the shoulder at 90 degrees of abduction; the elbow is flexed at 90 degrees, and the forearm is in neutral rotation with the palm of the hand placed flat on the x-ray cassette.

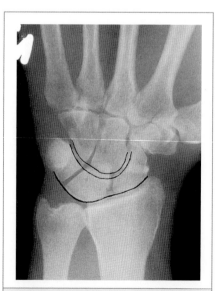

Figure 6-1. The three carpal arcs as described by Gilula.

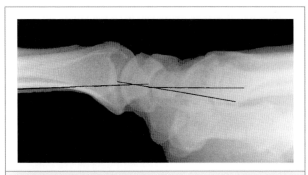

Figure 6-2. Lateral view of the wrist demonstrating colinear alignment of the radius, lunate, and capitate. Lines are perpendicular to radiolunate and lunatocapitate articulations, respectively.

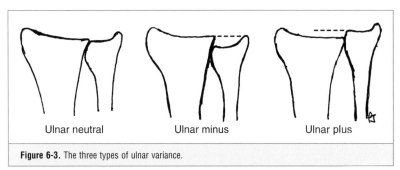

Figure 6-3. The three types of ulnar variance.

7. **What are the three methods for measuring ulnar variance?**
   - The **project-a-line technique** involves drawing a transverse line from the ulnar side of the distal radial articular surface toward the ulna. Variance is the distance between the distal ulnar articular surface and the line.
   - The **concentric circle method** involves placing a concentric circle template on the distal cortical margin of the radius. The distance between the articular surface of the distal ulna and the template curve that most closely approximates the distal radial cortical margin is measured to determine ulnar variance.
   - The **method of perpendiculars** involves drawing a line along the longitudinal axis of the radius. A second line is drawn through the distal ulnar aspect of the radius and perpendicular to the longitudinal axis. The distance between this line and the articular surface of the distal ulna is the ulnar variance (Fig. 6-4).

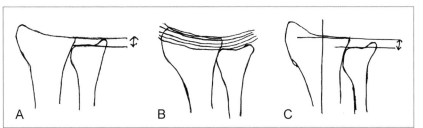

Figure 6-4. Techniques for measuring ulnar variance. ***A***, Project-a-line technique. ***B***, Concentric circle method. ***C***, Method of perpendiculars.

8. **Which radiographic view best demonstrates a fracture of the triquetrum?**
   A lateral view of the hand. Most fractures involve the dorsal cortical surface of the triquetrum.

9. **A 22-year-old cyclist sustained a nondisplaced scaphoid waist fracture after a fall. About 5 months later, the fracture is not united, and collapse and fragmentation of the proximal pole of the scaphoid are noted. What has occurred?**
   Scaphoid nonunion with avascular necrosis (AVN) of the proximal pole.

10. **Describe the radiographic features of Madelung's deformity.**
    - Abnormal palmar and volar ulnar tilting of the distal radial articular surface may exist because of growth disturbance of the palmar ulnar aspect of the distal radial physis.
    - Carpal bones may appear wedged between the v-shaped, deformed radius and ulna.
    - The ulna may be subluxed dorsally.

11. **A college football player is unable to flex the distal interphalangeal (DIP) joint of the ring finger after he tackles an opponent. What should you specifically look for on radiographic images?**
    The distal phalanx and entire volar aspect of the digit on the lateral radiograph should be scrutinized for an avulsion fracture of the proximal palmar aspect of the distal phalanx. This finding would suggest a flexor digitorum profundus tendon avulsion.

12. **What radiographic features are helpful in differentiating rheumatoid from psoriatic arthritis of the hand?**
    Psoriatic arthritis (Fig. 6-5) may be differentiated from rheumatoid arthritis by DIP joint involvement, characteristic periosteal new bone proliferation, and frequent joint ankylosis.

13. **What is acro-osteolysis?**
    Erosion of the terminal tufts of the fingers. This condition is seen in connective tissue diseases with vasculitis (e.g., scleroderma, Raynaud's syndrome), hyperparathyroidism, trauma (e.g., frostbite, burn injury), polyvinyl chloride intoxication, and psoriatic arthritis.

14. **Describe the characteristic findings on radiographs of the hand in a patient with hyperparathyroidism.**
    Subperiosteal bone resorption along the radial aspects of the phalanges and resorption of the terminal tufts. In hyperparathyroidism secondary to renal disease, the bones of the hand may appear dense with trabecular coarsening.

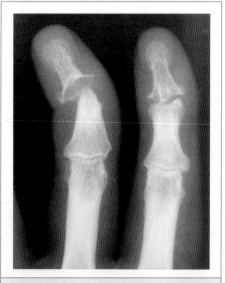

Figure 6-5. Psoriatic arthritis with involvement of the DIP joints of the finger.

15. **Describe the characteristic radiographic features of osteoarthritis.**
    - Joint space narrowing
    - Bony outgrowths or osteophytes
    - Subchondral sclerosis or eburnation
    - Subchondral cyst formation

16. **Chondrocalcinosis or calcification of the triangular fibrocartilage complex (TFCC) occurs in what disease process?**
    Pseudogout or calcium pyrophosphate dihydrate crystal deposition disease (CPPD).

17. **How is carpal height measured?**
    Carpal height is the distance between the distal radial articular surface and the proximal articular surface of the third metacarpal (Fig. 6-6). Carpal height ratio (CHR) is defined as the carpal height divided by the length of the third metacarpal. The normal value of CHR is 0.54 ± 0.03.

18. **Describe the radiographic features of silicone particulate synovitis of the wrist or hand.**
    - Implant deformation
    - Fragmentation and/or fracture of the silicone prosthesis
    - Soft tissue swelling (synovitis)
    - Erosions and/or cysts involving the carpus

19. **Summarize the Salter-Harris classification of epiphyseal fractures.**
    - **Type I:** Epiphyseal plate separation with the fracture line through the physis
    - **Type II:** Metaphyseal fragment associated with an epiphyseal fracture
    - **Type III:** Fracture through the growth plate and epiphysis
    - **Type IV:** Fracture through the growth plate into and including a metaphyseal fragment
    - **Type V:** Crush injury of the epiphyseal plate with initial radiographs possibly normal

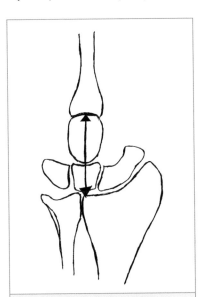

Figure 6-6. Carpal height *(arrow)* is the distance between the distal radial articular surface and the base of the third metacarpal.

20. **What radiographic features suggest a previous injury (fracture or crush) of the epiphyseal plate?**
    Shortening or angulation of the involved bone with premature epiphyseal closure.

21. **Describe the radiographic features of the scapholunate advanced collapse (SLAC) pattern of wrist arthritis.**
    - Widening of the scapholunate interval (>3 mm on a PA view of the wrist)
    - Foreshortening and palmar flexion of the scaphoid
    - Increased scapholunate angle with a cortical ring sign representing the volar cortex of the distal scaphoid pole
    - Degenerative arthritis involving the radioscaphoid articulation (Fig. 6-7)
    - Subchondral sclerosis of the scaphoid facet of the distal radial articular surface
    - Carpal collapse
    - Scaphocapitate and capitolunate arthrosis

22. **The wrist instability series, as described by Gilula, consists of which radiographic views?**
    - PA view in neutral, radial, and ulnar deviation
    - Clenched-fist anteroposterior (AP) view
    - Lateral view in neutral, dorsiflexion, and palmar flexion
    - Oblique view
    - Lateral view in 30 degrees of supination

23. **What intercarpal angles are measured in the assessment of carpal instability?**
    Capitolunate, scapholunate, and radiolunate angles are measured on a *true* lateral view of the wrist.

24. **What are the normal values for each of the angles described in the answer to question 23?**
    - Capitolunate angle of 0 degrees
    - Scapholunate angle varying between 30 and 60 degrees
    - Radiolunate angle varying between 0 and 15 degrees

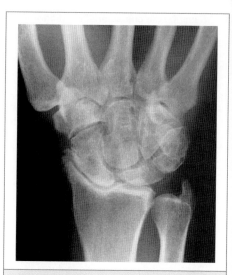

Figure 6-7. Narrowing of the radioscaphoid articulation and erosion of the scaphoid facet of the distal radius as seen in advanced stages of the SLAC pattern of wrist arthritis.

25. **What is a scaphoid view radiograph?**
    A radiographic technique used to improve visualization of the scaphoid. To obtain this projection, the patient is seated with the forearm and hand resting palm down on the radiographic table, the elbow flexed at 90 degrees, and the wrist in maximal ulnar deviation. The x-ray beam is directed toward the carpus.

26. **Which radiographic view profiles the pisotriquetral joint?**
    A supinated oblique view of the wrist.

27. **How is the hand positioned for a carpal tunnel view of the wrist?**
    The wrist is maximally dorsiflexed (extended) with the forearm pronated and resting on the film cassette. The central beam is directed approximately 15 degrees toward the palm of the hand.

28. **Describe the radiographic features of dorsiflexed intercalated segmental instability (DISI).**
    The DISI pattern of carpal instability is seen on a lateral view of the wrist (Fig. 6-8). The lunate is dorsiflexed and the scaphoid palmarflexed perpendicular to the longitudinal axis of the radius, resulting in an increased scapholunate angle (usually greater than 60 degrees).

29. **What standard parameters are used to assess radiographs of the wrist after reduction of a distal radius fracture?**
    - Radial length (normal value of 11–13 mm)
    - Radial inclination (normal value of 15–25 degrees)
    - Amount of gap or incongruity between articular fragments
    - Palmar inclination of the distal radial articular surface (normally 10 degrees of palmar tilt)
    - Amount of dorsal or palmar translation
    - Extent and amount of metaphyseal comminution (expressed as the percent of the sagittal width of the radius)

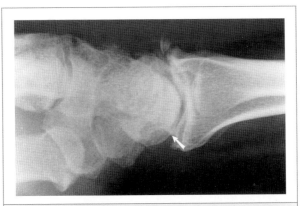

Figure 6-8. The DISI deformity. In this deformity, the lunate *(arrow)* extends and the scaphoid flexes, resulting in an increased scapholunate angle.

30. **What is the "spilled tea cup" sign?**
    Volar rotation of the lunate as detected on a lateral view of the wrist in the patient with a lunate dislocation (Fig. 6-9).

31. **Describe the characteristic radiographic features for rheumatoid arthritis of the wrist and hand.**
    - Generalized osteopenia
    - Marginal erosions
    - Diffuse joint space narrowing
    - Periarticular soft tissue swelling
    - Joint subluxation and/or dislocation
    - Radial deviation of the wrist
    - Subluxation or dislocation of the metacarpophalangeal (MCP) and proximal interphalangeal (PIP) joints
    - Ulnar drift (deviation) of the fingers
    - Erosion and destruction of the MCP and PIP joints
    - Boutonnière and swan-neck deformities
    - Telescoping of the fingers
    - Possible soft tissue nodules (Fig. 6-10)

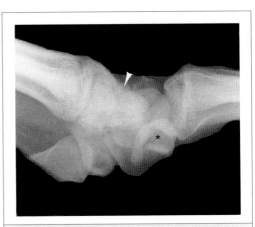

Figure 6-9. The "spilled tea cup" sign. Note the malrotated and dislocated lunate *(asterisk)*. The capitate *(arrowhead)* is subluxed dorsally on the lunate.

32. **Which radiographic series should be obtained in all patients with rheumatoid arthritis who are scheduled for surgery?**
    A cervical spine series (AP, lateral, and odontoid views) to rule out potential asymptomatic cranial settling or atlantoaxial instability. Failure to recognize cervical spine instability may result in significant injury to the brain or spinal cord during the intubation for general anesthesia.

33. **Define volarflexed intercalated segmental instability (VISI).**
    VISI refers to the pattern of carpal instability in which the lunate is palmar flexed, the scapholunate angle is decreased, and the capitolunate angle is increased.

34. **Which radiographic features suggest a slow growing or benign tumor of the hand?**
    - Preservation of cortical margins
    - Well-demarcated lesion
    - Sclerotic margins
    - Contiguous and solid periosteal reaction

35. **What radiographic features suggest an aggressive or a malignant bone tumor of the hand?**
    - Cortical destruction and/or erosion
    - Associated soft tissue mass
    - Irregular periosteal reaction
    - Indistinct demarcation

36. **What are the radiographic features of septic arthritis of the hand?**
    - **Acute:** Soft tissue swelling and/or a joint effusion is present.
    - **Subacute:** Juxta-articular osteoporosis may occur. If concomitant juxta-articular osteomyelitis is present, erosive cortical lesions and "punched-out" irregular lucencies in the cancellous bone with no marginal reaction are seen.
    - **End-stage:** Cartilage erosion, joint space narrowing, and destruction with bony ankylosis and deformity are features.

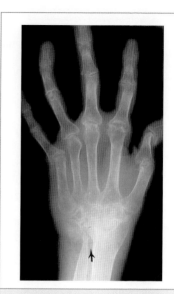

Figure 6-10. Advanced rheumatoid arthritis of the hand. Note the eroded distal ulna *(arrow)*, loss of carpal height, erosions of the distal radius, loss of joint space, and periarticular erosions involving the MCP joints.

37. **Describe the radiographic features of acute osteomyelitis.**
    The findings vary with the disease stage. The earliest finding may be soft tissue edema and loss of normal soft tissue fat planes. Soft tissue changes are followed by permeative intramedullary destruction with irregular lucencies. Subsequent changes include erosive cortical destruction, endosteal scalloping, and a reactive periosteal response.

38. **How long after the onset of acute osteomyelitis are radiographic changes first noted?**
    Approximately 10–14 days.

39. **What are the most common indications for a computed tomography (CT) scan of the wrist?**
    - Detecting an occult carpal fracture
    - Delineating a distal radius fracture pattern
    - Assessing fracture healing, specifically a fracture of the scaphoid

40. **What is the significance of sclerosis of the proximal scaphoid pole after a waist fracture?**
    A CT scan and plain radiographs commonly demonstrate sclerosis of the proximal scaphoid pole in the intervening months after a scaphoid fracture, particularly those fractures involving the proximal third and waist regions. This finding is believed to indicate relative hypoperfusion of the proximal pole. Many, but not all, fractures with sclerosis of the proximal pole achieve union. If fragmentation and collapse of the proximal pole occurs, a radiographic diagnosis of AVN is made.

41. **What is the preferred imaging method for assessing vascularity of the proximal scaphoid in an established nonunion?**
    Magnetic resonance imaging (MRI) with intravenous gadolinium.

42. **What is the preferred imaging technique for confirming a suspected fracture of the hook of the hamate?**
    Multiplanar thin-cut CT scan (Fig. 6-11).

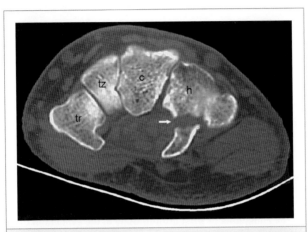

Figure 6-11. A hook of the hamate fracture. The fracture *(arrow)* is best detected by CT. tr = trapezium, tz = trapezoid, c = capacitate, h = hamate.

43. **Which imaging modality is used to assess for distal radioulnar joint (DRUJ) subluxation or dislocation?**
    A CT scan of *both* wrists, side by side, with the forearms in neutral, full pronation, and full supination positions.

44. **What is wrist arthrography?**
    The injection of a radiopaque contrast material into the distal radioulnar, radiocarpal, and/or midcarpal joints. Wrist arthrography is used frequently to evaluate patients with persistent wrist pain and is particularly helpful in identifying a triangular fibrocartilage (TFC) and/or intercarpal ligament tear.

45. **What is a triple-injection, or three-compartment, wrist arthrogram?**
    The sequential injection of all three joints (radiocarpal, DRUJ, and midcarpal) that increases the likelihood of detecting an abnormality.

46. **Describe the technique and indications for a CT scan of the scaphoid.**
    A CT scan of the scaphoid is performed in the plane sagittal to the long axis of the scaphoid itself using thin (1–1.5 mm) sections or a spiral technique with reformatted images. A CT scan of the scaphoid is most commonly obtained to assess for collapse of a scaphoid nonunion or to assess scaphoid fracture healing (Fig. 6-12).

47. **How is a triple-injection wrist arthrogram performed?**
    The radiocarpal joint is injected with 2–3 ml of contrast. Plain radiographs and/or videofluoroscopy images are obtained immediately after injection and dye distribution. The wrist is manipulated, and imaging is repeated. Approximately 1–2 hours later, as the contrast from the radiocarpal injection begins to fade, injection of the DRUJ is performed, and imaging of the wrist

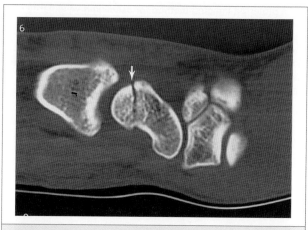

Figure 6-12. A CT scan of a nondisplaced fracture *(arrow)* of the proximal pole of the scaphoid.

is repeated. Finally, the midcarpal joint is injected, and the manipulation and imaging sequence is repeated.

48. **What are the limitations of wrist arthrography in the evaluation of a patient with wrist pain?**
    There is a high incidence of asymptomatic age-related attritional tears of the TFCC, scapholunate, and lunotriquetral ligaments that may not represent the source of pain. Thus, performing arthrograms of both wrists has been advocated, but a significant percentage of patients have comparable findings in the asymptomatic opposite wrist. *Thus, abnormal arthrographic findings should be interpreted cautiously and correlated with the patient's complaints and physical findings.*

49. **What is a bone scan?**
    A nuclear medicine imaging technique that involves the intravenous injection of a radiolabeled phosphorus-based compound marker to detect abnormalities of bone metabolism.

KEY POINTS: DIAGNOSTIC IMAGING OF THE WRIST AND HAND

1. Disruption of any of the three carpal arcs on a PA view of the wrist suggests a carpal subluxation or dislocation.
2. Ultrasound is the best diagnostic imaging technique to assess for a suspected foreign body in the hand.
3. Calcification of the TFCC on plain radiographs is suggestive of calcium pyrophosphate dihydrate or crystal deposition disease (CPPD).
4. It may take up to 10–14 days for plain radiographic changes to appear in the patient with acute osteomyelitis.
5. Gadolinium-enhanced MR imaging is a useful imaging method for investigating the vascularity of the proximal pole in the patient with a scaphoid nonunion.

50. **What is a three-phase bone scan?**
    A three-phase bone scan consists of a radionuclide angiogram of the area of interest obtained during injection of the radiolabeled marker, images of blood pooling (soft tissues) at the anatomic area of interest after injection, and delayed images obtained between 2 and 3 hours after injection.

51. **Which clinical disorders of the hand are assessed with a bone scan?**
    - Infection
    - Occult fracture
    - Tumors (e.g., osteoid osteoma)
    - Suspected reflex sympathetic dystrophy (RSD)

52. **What is the next diagnostic step in the patient with focal increased carpal bone uptake on a bone scan?**
    Despite its high sensitivity, a finding of increased uptake on a bone scan is nonspecific. Therefore, such a finding must be correlated with plain radiographs. If a diagnosis still cannot be made, a thin-section CT scan of the involved portion of the carpus or an MRI may be the next appropriate diagnostic step.

53. **What are the findings on a three-phase bone scan in the patient with osteomyelitis of the hand?**
    - Focally increased blood flow on the radionuclide angiogram
    - Increased activity noted on blood pool and delayed images

54. **What are the bone scan findings in the patient with RSD?**
    - Increased blood flow with increased activity on blood pool images
    - Increased periarticular tracer uptake on delayed images

55. **Which MRI findings are noted in a patient with tenosynovitis?**
    - **On the T1-weighted image,** an abnormal low signal surrounds the involved tendons (Fig. 6-13).
    - **On the T2-weighted image,** a thick band of high signal intensity surrounds the tendons.

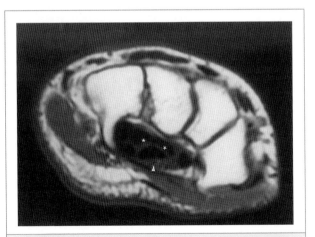

Figure 6-13. A T1-weighted MRI. The tenosynovium *(arrow)* surrounds the flexor tendons *(asterisks)*.

DIAGNOSTIC IMAGING OF THE WRIST AND HAND

56. **A suspected soft tissue tumor of the hand is best evaluated with what imaging modality?**
    An MRI.

57. **An MRI is useful for evaluating what other conditions of the hand?**
    - Osteonecrosis of the lunate (Kienböck's disease) or scaphoid (Preiser's disease)
    - Occult fracture
    - Soft tissue neoplasm
    - Infection
    - Tendon disruption
    - Tear of the TFCC or intrinsic intercarpal (interosseous) ligaments
    - Proliferative synovitis

58. **What are the indications for ultrasound examination of the hand and wrist?**
    Ultrasound is best used as a focused examination to answer a specific clinical question or to evaluate a single anatomic site. Indications include the evaluation of tendon pathology or integrity, fluid collections, foreign body presence and localization, and soft tissue masses.

59. **Fluoroscopy is helpful for assessing which wrist disorders?**
    Disorders related to altered wrist kinematics and carpal instability.

60. **Which nuclear medicine tests may be performed in the patient with osteomyelitis and an equivocal bone scan?**
    A gallium–67 citrate-, technetium-, or indium–111-labeled leukocyte scan. The indium–111 leukocyte scan has a lower sensitivity than the gallium scan for detecting chronic osteomyelitis.

61. **What is magnetic resonance (MR) arthrography?**
    A combination of an MRI with distension of the imaged joint via the intra-articular injection of a contrast agent.

62. **How is MR arthrography of the wrist performed?**
    A radiocarpal arthrogram is performed with the injection of a saline solution combined with gadolinium (an MRI contrast agent) and conventional radiopaque contrast. MRI is performed immediately after the injection.

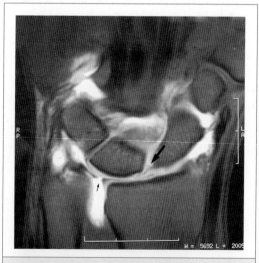

Figure 6-14. MR arthrogram of the wrist. Coronal, fat-suppressed, T1-weighted scan after radiocarpal contrast injection demonstrates gadolinium contrast in all three wrist compartments. A defect in the TFC is evident *(short arrow)*. The abnormal scapholunate ligament is also indicated *(large arrow)*.

63. **What are the indications for MR arthrography of the wrist?**
    MR arthrography appears to provide additional benefit over conventional MRI in

detecting communicating defects of the TFC, scapholunate, and lunotriquetral interosseous ligaments (Fig. 6-14).

64. A 38-year-old woman presents with a 2-month history of index finger pain. Radiographs demonstrate a slightly expansile, lobulated, well-marginated lucent lesion in the proximal phalanx with endosteal erosion and subtle mineralization. What is the diagnosis?
Enchondroma.

BIBLIOGRAPHY

1. Brown RR, Fliszar E, Cotten A, et al: Extrinsic and intrinsic ligaments of the wrist: Normal and pathologic anatomy at MR arthrography with three-compartment enhancement. RadioGraphics 18:667–674, 1998.
2. Cerezal L, del Pinal F, Abascal F, et al: Imaging findings in ulnar sided wrist impaction syndromes. RadioGraphics 22:105–121, 2002.
3. Clavero JA, Alomar X, Monill JM, et al: MR imaging of ligament and tendon injuries of the fingers. RadioGraphics 22:237–256, 2002.
4. Dalinka MK: MR imaging of the wrist. AJR 164:1–9, 1995.
5. DeFlaviis L, Musso MG: Hand wrist. In Fornage BD (ed): Musculoskeletal Ultrasound. New York, Churchill Livingstone, 1995, pp 151–178.
6. Gilula LA, Yin Y: Imaging of the Wrist and Hand. Philadelphia, W.B. Saunders, 1996.
7. Garcia-Elias M: Carpal dislocations instabilities. In Green DP, Hotchkiss R, Pederson W (eds): Green's Operative Hand Surgery, 4th ed. New York, Churchill Livingstone, 1999, pp 865–928.
8. Haims AH, Schweitzer ME, Morrison WB, et al: Internal derangement of the Wrist: Indirect MR arthrography versus unenhanced MR imaging. Radiology 227:701–707, 2003.
9. Haims AH, Schweitzer ME, Morrison WB, et al: Limitations of MR imaging in the diagnosis of peripheral tears of the triangular fibrocartilage of the wrist. AJR 178:419–422, 2002.
10. Jacobson JA, Powell A, Craig JG, et al: Wooden foreign bodies in soft tissue: Detection at US. Radiology 206:45–48, 1998.
11. Levinsohn EM, Rosen ID, Palmer AK: Wrist arthrography: Value of the three-compartment injection method. Radiology 179:231–239, 1991.
12. Manaster BJ: The clinical efficiency of triple-injection wrist arthrography. Radiology 178:267–270, 1991.
13. Mann FA, Wilson AJ, Gilula LA: Radiographic evaluation of the wrist: What does the hand surgeon want to know? Radiology 185:15–24, 1992.
14. Resnick D: Diagnosis of Bone and Joint Disorders, 3rd ed. Philadelphia, W.B. Saunders, 1995.
15. Rogers LF: Radiology of Skeletal Trauma, 2nd ed. New York, Churchill Livingstone, 1992.
16. Stewart MR, Gilula LA: CT of the wrist: A tailored approach. Radiology 183:13–20, 1992.

OSTEOARTHRITIS

Jay T. Bridgeman, MD, DDS, and Sanjiv H. Naidu, MD, PhD

1. **What is the most common joint disease in the upper extremity?**
   Osteoarthritis (OA) or degenerative joint disease (DJD).

2. **Define OA.**
   OA is a degenerative condition of diarthrodial joints characterized by the deterioration of articular cartilage and formation of new bone at the articular margins and joint surface.

3. **How prevalent is OA in the hand?**
   - About 88% of men and 94% of women aged 70 years or more have radiographic evidence of osteoarthritis.
   - Of those with radiographic osteoarthritis in the hand, approximately 9% of men and 17% of women have clinical symptoms.

4. **British authors refer to osteoarthritis as *osteoarthrosis* because of the belief that it is a noninflammatory process. Is this belief correct?**
   No. Synovially derived inflammatory mediators, such as interleukins, have been identified as important mediators in the pathophysiology of OA.

5. **Name and describe the two main variants of OA in the hand.**
   1. **Primary generalized OA:**
      - Involves the distal interphalangeal (DIP) and trapeziometacarpal (TMC) joints
      - Affects the TMC joint as the most symptomatic joint
   2. **Erosive OA:**
      - Involves the proximal interphalangeal (PIP) joints
      - Rarely involves the TMC joint
      - Associated with a greater incidence of bony erosions and fewer mucous cysts
      - Linked to a 15% incidence of seropositivity and subsequent indolent, benign course of rheumatoid arthritis (RA)

6. **Describe the clinical features of OA.**
   - Pain with joint use and motion
   - Deformity
   - Joint enlargement
   - Crepitation
   - Loss of joint motion

7. **What functional limitations are most affected by symptomatic osteoarthritis of the hand?**
   - Precise pincer grasp
   - Power grasp.

8. **Which joints are most commonly involved in generalized OA of the hand?**
   - DIP joints
   - TMC joint (Fig. 7-1)

9. **Which osteoarthritic hand joint most frequently undergoes surgical treatment?**
   TMC joint of the thumb (basilar joint).

10. **What is the proposed mode of inheritance for OA?**
    The familial predisposition for development of primary generalized OA in women and their female offspring is poorly understood. The predisposition is believed to be multifactorial, with age, sex, and heredity the most influential factors. DIP joint involvement in primary generalized OA is thought to be inherited in an autosomal dominant pattern in women and autosomal recessive pattern in men.

11. **What specific genetic markers have been identified in OA?**
    In female twins with primary generalized OA, a single mutation that substitutes cysteine for arginine in position 519 of the type II procollagen gene (*COL2A1*) has been identified.

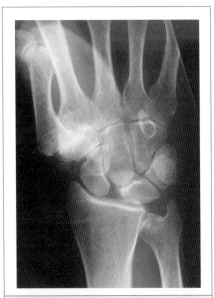

Figure 7-1. Osteoarthritis of the trapeziometacarpal (TMC) joint. Note the adducted first metacarpal, compensatory hyperextension at the MCP joint, and subchondral sclerosis at the metacarpal base and trapezium.

12. **What other factors are associated with the development of OA?**
    - Age and sex
    - Heredity and environment
    - Ethnicity
    - Infection and systemic disease
    - Repetitive stresses
    - Biomechanical features of the involved joints

13. **What are Heberden's nodes?**
    Bony spurs on the dorsal aspect of the DIP joints of the fingers, first described by William Heberden in 1802. They are usually small, but because of their subcutaneous location, they give the DIP joint a knobby appearance. Heberden's nodes are *the* most common clinical manifestation of OA.

14. **Which gender has a higher incidence of Heberden's nodes?**
    Heberden's nodes are 10–20 times more common in women than men.

15. **What are Bouchard's nodes?**
    PIP joint osteophytes. The finger has a distinctive sausage-like appearance, particularly in patients with erosive OA.

16. **What is a mucous cyst?**
    A subdermal cyst of variable size that is filled with a viscous, mucoid fluid. Most mucous cysts are located on the dorsum of the DIP or PIP joints. They arise from the joint capsule and frequently occur in association with Heberden's nodes.

17. **What are the indications for excision of a mucous cyst?**
    - Progressive enlargement
    - Rupture and drainage
    - Vulnerability to local trauma
    - Infection
    - Pain
    - Secondary nail involvement

18. **How is a mucous cyst treated?**
    - Aspiration with a high risk of recurrence
    - Surgery consisting of cyst excision and DIP joint osteophyte debridement

19. **List the treatment goals for OA of the hand.**
    - Pain relief
    - Maintenance of function
    - Prevention of associated deformities
    - Patient education

20. **What nonoperative treatments are used for OA of the hand?**
    - Rest and activity restriction
    - Splint immobilization
    - Nonsteroidal anti-inflammatory drugs (NSAIDs)
    - Therapy program to preserve and maintain motion and strength

21. **What types of deformities occur with OA of the DIP joint?**
    - DIP joint flexion with or without radial or ulnar deviation
    - Mucous cyst formation
    - Mallet finger deformity due to attenuation of the terminal slip of the extensor mechanism by a dorsal osteophyte

22. **List the indications for surgical treatment of a symptomatic osteoarthritic DIP joint.**
    - Intractable pain
    - Deformity with functional limitation
    - Painful instability
    - Problematic mucous cyst

    Surgery performed primarily for cosmetic reasons should be avoided.

23. **What is the preferred surgical procedure for OA of the DIP joint?**
    Arthrodesis.

24. **List the indications for surgical treatment of a symptomatic osteoarthritic PIP joint.**
    - Intractable pain
    - Deformity
    - Painful instability
    - Restricted motion interfering with functional activities

    Surgery performed primarily for cosmetic reasons should be avoided.

25. **What are the primary goals of performing surgery on an osteoarthritic PIP joint?**
    To provide painless range of motion and lateral joint stability so that functional pinch and power grip may be restored and/or preserved.

26. **What are the most common surgical procedures for treating OA of the PIP joint?**
    - Arthroplasty
    - Arthrodesis

27. **When is arthrodesis the preferred procedure?**
    Arthrodesis is recommended for OA of the index finger PIP joint because it provides lateral stability and a stable base for pinch. Arthrodesis is also preferred for treating severe PIP joint instability or treating young active people or manual laborers who have experienced a failed joint implant, patients with loss of bone stock, and patients with preoperative PIP joint motion <30 degrees.

KEY POINTS: OSTEOARTHRITIS

1. Radiographic evidence of OA is much more prevalent than clinical symptoms.
2. Surgery for symptomatic OA in the hand is most commonly performed on the trapeziometacarpal joint.
3. The treatment goals for OA in the hand are to provide pain relief, maintain function, and prevent deformity.
4. The deformities resulting from OA of the interphalangeal joints are referred to as Heberden's nodes (DIP joint) and Bouchard's nodes (PIP joint).
5. Arthrodesis is the preferred surgical treatment for symptomatic OA of the index finger PIP joint.

28. **What type of fixation technique provides the best union rate for PIP arthrodesis?**
    Intramedullary screw fixation.

29. **When is arthroplasty preferred for treating OA of the PIP joint?**
    Arthroplasty is preferred for treating OA of the ring and small finger PIP joints so that motion critical for the development of a power grip is preserved.

30. **What is the most common complication of a PIP joint implant arthroplasty?**
    Instability.

31. **Describe OA of the TMC joint of the thumb.**
    - The condition typically occurs in women older than 50 years of age.
    - Patient complaints include pain at the basilar region of the thumb, loss of thumb motion, and weakness with grip and pinch.

32. **Degeneration of which structure is believed to result in OA of the thumb TMC joint?**
    The volar beak ligament, an intracapsular ligament that originates on the trapezium and inserts on the volar-ulnar aspect of the thumb metacarpal base.

33. **Describe the pathophysiology of OA in the basal joint.**
    Degeneration of the beak ligament, which is the primary stabilizer of the TMC joint, permits excessive dorsopalmar translation of the first metacarpal on the trapezium during lateral pinch. This pathologic translation generates shear forces with resultant wear of the volar articular surfaces of the trapezium and first metacarpal.

34. **What is the grind test?**
    A provocative test to assess for symptomatic OA of the TMC joint. The test involves simultaneous rotation and axial loading of the TMC joint by the examiner.

35. **What is the crank test?**
    Another provocative test to assess for OA of the TMC joint. The test involves axial loading of the TMC joint of the thumb with simultaneous passive flexion and extension of the metacarpal base.

36. **What clinical information is gleaned from the crank and grind tests?**
    Both tests are provocative maneuvers used to elicit pain and crepitance within the thumb TMC joint. The findings must be correlated carefully with the patient's complaints and radiographic findings.

37. **What are the surgical indications for OA of the TMC joint?**
    - Incapacitating pain that limits functional activities
    - Failed nonoperative treatment
    - Deformity due to metacarpophalangeal (MCP) joint instability
    - Contracted first web space that interferes with function

38. **What are the surgical options for osteoarthritis of the thumb TMC joint?**
    Arthrodesis is preferred for treatment of highly active young patients. Most patients require excision of the trapezium and a soft tissue interpositional arthroplasty with or without reconstruction of the volar beak ligament. Resection of the distal half of the trapezium may be performed if the TMC joint alone is involved. The entire trapezium is typically removed, particularly in the presence of concomitant scaphotrapezial arthritis. Resection of the entire trapezium with soft tissue interpositional arthroplasty alone may be performed; however, because of potential proximal migration of the first metacarpal with secondary impingement against the distal pole of the scaphoid, most surgeons perform concomitant reconstruction of the volar beak ligament using the flexor carpi radialis, the abductor pollicis longus, or one of the wrist extensor tendons (extensor carpi radialis brevis or longus).

39. **Why does a swan-neck deformity of the thumb develop in patients with OA of the TMC joint?**
    The primary deformity is radial subluxation of the first metacarpal base, which leads to a thumb adduction contracture. Compensatory MCP joint hyperextension to permit adequate positioning of the thumb during grasp occurs.

40. **Describe the synovial fluid analysis in patients with OA.**
    Findings include thick and viscous fluid with a normal glucose level, low total cell count, and a minimal increase in white blood cells. Monosodium urate and calcium pyrophosphate crystals are absent.

41. **What clinical conditions are commonly associated with OA of the hand?**
    - Carpal tunnel syndrome
    - de Quervain's tenosynovitis
    - Flexor carpi radialis tendinitis
    - Stenosing tendovaginitis of the finger (trigger finger)

42. **List the most common disease processes/conditions that result in secondary OA of the hand.**
    - Gout
    - Calcium pyrophosphate deposition (CPPD) disease
    - Wilson's disease

- Hemochromatosis
- Trauma
- Avascular necrosis of the lunate (Kienböck's disease)
- Avascular necrosis of the scaphoid (Preiser's disease)
- Scaphoid nonunion
- Carpal instability
- Infection
- Hyperparathyroidism
- Acromegaly
- Ehlers-Danlos syndrome
- Ochronosis
- Hemophilia
- Frostbite

43. **What is the differential diagnosis of the patient who has clinical signs and symptoms suggestive of OA?**
    - Seronegative rheumatoid arthritis
    - Gouty arthritis
    - CPPD arthropathy
    - Osteonecrosis
    - Seronegative spondyloarthropathy, including psoriatic arthritis and Reiter's syndrome

44. **Which condition is associated with isolated MCP joint OA?**
    Hemochromatosis.

45. **What activities are painful for the patient with OA of the thumb TMC joint?**
    Any activity that results in increased loading of the TMC joint, such as key or chuck pinch activities, opening the lid of a jar, turning a key in a door, or brushing/combing hair.

46. **What patterns of degenerative arthritis occur in the wrist?**
    Approximately 90% of the observed changes involve the scaphoid and occur in three distinct patterns:
    - The scapholunate advanced collapse (SLAC) pattern of wrist arthritis
    - Triscaphc (scapho-trapezium-trapezoid) arthritis
    - Combination of the latter two conditions

47. **Which pattern of degenerative arthritis is the most common?**
    SLAC.

48. **Describe the pathologic changes with the SLAC pattern of wrist arthritis.**
    There are several patterns of change. The most common is rotary subluxation of the scaphoid that occurs after an acute rupture or chronic attenuation of the scapholunate ligament. Palmar flexion of the scaphoid results in a decrease in surface area but an increase in joint contact forces between the radius and scaphoid. The articular cartilage of the radioscaphoid articulation is destroyed. The alteration in carpal mechanics and alignment results in carpal collapse and destruction of the capitolunate and finally the entire carpal and radiocarpal articulations.

49. **Which joint is usually preserved in the SLAC pattern of wrist arthritis?**
    Radiolunate.

50. **What are the surgical options for the SLAC pattern of wrist arthritis?**
    Surgical treatment depends on disease stage and, in particular, the status of the proximal capitate and radiolunate articular surfaces. The two most common procedures are scaphoid

excision and "four-corner" (capitate-lunate-hamate-triquetrum) arthrodesis (SLAC reconstruction operation) or a proximal row carpectomy (PRC). If the lunate fossa of the distal radius and proximal capitate are well preserved, several studies have shown that PRC is preferred to a four-corner arthrodesis. If the proximal capitate is destroyed and the lunate fossa is preserved, a four-corner arthrodesis is recommended. Wrist arthrodesis should be performed if both articular surfaces are damaged.

51. **What are the surgical treatment options for triscaphe arthritis?**
    The most common procedure is an arthrodesis of the triscaphe articulation. If concomitant degenerative arthritis of the TMC joint is present (pantrapezial arthritis), however, resection of the entire trapezium and reconstruction of the volar beak ligament are necessary. It is recommended that these procedures be combined with an interpositional arthroplasty of the TMC articulation and an arthrodesis of the scaphotrapezoid articulation.

52. **What are the risk factors for developing degenerative arthritis after a fracture of the distal radius?**
    Patients with an intra-articular fracture involving the radiocarpal and/or distal radioulnar joints or with fractures that have healed with excessive dorsal tilt and radial shortening are more at risk of developing degenerative arthritis. Intra-articular fractures with >2 mm of articular incongruity or gap between fragments are at increased risk for the development of degenerative arthritis. However, intra-articular fractures are high-energy injuries, and some patients with nondisplaced or minimally (<2 mm) displaced fractures still develop degenerative arthritis. Malunited fractures result in an alteration of radiocarpal joint mechanics with abnormal joint loading and subsequent articular degeneration.

53. **A middle-aged laborer presents with progressive pain at the mid-dorsum of his dominant hand. Examination reveals tenderness to palpation at the base of the third metacarpal, and aspiration of the area fails to yield any fluid. What is the preferred treatment?**
    The best treatment for a carpal boss is excision with preservation of the adjacent carpometacarpal joint cartilage.

54. **What is the so-called natural history of the neglected scaphoid nonunion?**
    A scaphoid nonunion advanced collapse (SNAC) pattern can develop from the carpal collapse and altered kinematics. Loss of scaphoid stabilization of both carpal rows results in proximal capitate migration, malrotation of the scaphoid with radioscaphoid incongruity, and the development of posttraumatic osteoarthritis.

55. **Which surgical procedure to alleviate wrist pain may be performed in conjunction with a limited wrist arthrodesis?**
    Resection of the posterior interosseous nerve (PIN) at the wrist. The PIN lies on the floor of the fourth extensor compartment and has only sensory function.

56. **What procedure is essential for maintaining strength and preventing subsidence of the first metacarpal after trapezium excision for TMC joint osteoarthritis?**
    Reconstruction of the volar beak ligament.

57. **Arthrodesis of the TMC joint for isolated osteoarthritis has been advocated for the young laborer in question 53. What are the drawbacks of this procedure?**
    - Risk of nonunion
    - Technical difficulties
    - Decreased thumb adduction

- Inability to place the hand flatly on surfaces
- Development of scaphotrapezial and trapeziotrapezoid arthritis

58. **In long-term follow-up care for a patient, how does a ligament reconstruction and tendon interposition (LRTI) procedure compare to an arthrodesis of the TMC joint?**
The results of both procedures are comparable at long-term follow-up visits, even for young patients.

59. **What are the treatment options for hyperextension instability of the thumb MCP joint associated with TMC joint arthritis?**
When performing an LRTI procedure for TMC joint arthritis, the surgeon must address the secondary deformity at the MCP joint. A total of 30 degrees or less of hyperextension is treated with temporary pinning of the MCP joint in slight flexion. More than 30 degrees of hyperextension requires volar plate advancement if the joint is preserved or an arthrodesis if degenerative changes are present.

60. **A 65-year-old nurse has pain at the base of her dominant thumb, and radiographs demonstrate changes consistent with early OA of the TMC joint. What are the nonoperative treatment modalities?**
   - NSAIDs
   - Oral analgesic medication
   - Splinting
   - Corticosteroid injection of the joint
   - Resistive thenar muscle strengthening

61. **After failing nonoperative treatment for thumb CMC osteoarthritis, a patient undergoes an LRTI procedure using half of the flexor carpi radialis (FCR) tendon. What long-term improvement in grip and key pinch may be predicted?**
Long-term follow-up studies have demonstrated a 92.5 % improvement in grip strength and a 50% improvement in key and chuck pinch strength, with maximal improvement obtained at approximately 2 years postoperatively.

62. **Which diagnostic modality for assessing the severity of arthritis in the TMC joint is gaining in popularity?**
TMC joint arthroscopy.

BIBLIOGRAPHY

1. American Academy of Orthopaedic Surgeons: Orthopaedic Knowledge Update 7. Rosemont, IL, AAOS, 2002.
2. Arthritis Foundation: Primer on the Rheumatic Diseases, 10th ed. Atlanta, Arthritis Foundation, 1993.
3. Cobby M, Cushnaghan J, Creamer P, et al: Erosive arthritis: Is it a separate disease entity? Clin Radiol 42:258–263, 1990.
4. Farino GC, Goitz RJ: Thumb trapeziometacarpal arthritis: Is there a role for arthroscopy? Curr Opin Orthop 13(4):256–259, 2002.
5. Green DP, Hotchkiss R, Pederson WC (eds): Green's Operative Hand Surgery, 4th ed. New York, Churchill Livingstone, 1999.
6. Hartigan BJ, Stern PJ, Kiefhaber RR: Thumb carpometacarpal arthritis: Arthrodesis compared with ligament reconstruction and tendon interposition. J Bone Joint Surg 83(A):1470–1478, 2001.
7. Krakauer JD, Bishop AT, Cooney WP: Surgical treatment of scapholunate advanced collapse. J Hand Surg 19A:751–759, 1994.
8. Leibovic SJ, Strickland JW: Arthrodesis of the proximal interphalangeal joint of the finger: Comparison of the use of the Herbert screw with other fixation methods. J Hand Surgery 19A:181–188, 1994.

9. Moritomo H, Tada K, Yoshida T, Masatomi TJ: The relationship between the site of nonunion of the scaphoid and scaphoid nonunion advanced collapse. Bone Joint Surg 81B:871–876, 1999.
10. Pellegrini VD Jr: Osteoarthritis at the base of the thumb. Orthop Clin North Am 23:83–90, 1992.
11. Spector TD, Cicuttini F, Baker J, et al: Genetic influences on osteoarthritis in women: A twin study. BMJ 312:940–943, 1996.
12. Tomaino MM, Miller RJ, Cole I, Burton RI: Scapholunate advanced collapse wrist: Proximal row carpectomy or limited wrist arthrodesis with scaphoid excision? J Hand Surg 19A:134–142, 1994.
13. Tomaino MM, Pellegrini VD Jr, Burton RI: Arthroplasty of the basal joint of the thumb. Long-term follow-up after ligament reconstruction with tendon interposition. J Bone Joint Surg 77A:346–355, 1995.
14. Viegas SF: Limited arthrodesis for scaphoid nonunion. J Hand Surg 19A:127–133, 1994.
15. Zhang Y, Nui J, Kelly-Hayes M, et al: Prevalence of symptomatic hand osteoarthritis and its impact on functional status among the elderly: The Framingham study. Am J Epidemiol 156:1021–1027, 2002.

RHEUMATOID ARTHRITIS
Mark N. Halikis, MD, William M. Weiser, MD, and Julio Taleisnik, MD

1. **By what criteria is the diagnosis of rheumatoid arthritis (RA) made?**
   Four of the following seven criteria must be met:
   1. Morning stiffness (lasting 1 hour and located around joints) present for at least 6 weeks
   2. Arthritis of three or more joint areas for at least 6 weeks
   3. Arthritis of hand joints present for at least 6 weeks
   4. Symmetric arthritis present for at least 6 weeks
   5. Rheumatoid nodules
   6. Seropositive rheumatoid factor
   7. Radiographic changes (erosive changes or decalcification around joints)

2. **How does the pattern of joint involvement in the hand differ between RA and osteoarthritis (OA)?**
   - **RA:** Metacarpophalangeal (MCP) and proximal interphalangeal (PIP) joints are characteristically involved. Enlargement of the PIP joints is referred to as **Bouchard's nodes,** which are more characteristic of RA.
   - **OA:** Distal interphalangeal (DIP) joints are more characteristically involved. Enlargement of the DIP joints is referred to as **Heberden's nodes**, which are characteristic of OA.

3. **How do radiographic findings differ in patients with RA and patients with OA?**
   See Table 8-1.

TABLE 8-1. RADIOGRAPHIC FINDINGS IN RA VERSUS OA	
Rheumatoid Arthritis	Osteoarthritis
Marginal erosions	Marginal osteophytes
Periarticular osteopenia	Subchondral sclerosis
Joint malalignment, subluxation	Malalignment, subluxation less common
Joint ankylosis	Ankylosis less common

4. **What is the process that leads to deformity and destruction of joints in patients with RA?**
   1. Synovial hyperplasia and inflammatory cell infiltration form an expansive pannus.
   2. A lytic process mediated by the pannus destroys articular cartilage and causes periarticular osteopenia.
   3. Further expanding pannus stretches and disrupts joint capsules and ligaments.
   4. Joints become unstable and are deformed by surrounding musculotendinous forces.

5. **What are the indications for surgical intervention in RA of the hand?**
   - **Pain relief:** This is the most predictable outcome of surgery.
   - **Improve/prevent deformity:** Many patients have a deformity without debilitating pain or dysfunction; this is rarely considered an indication on its own except in cases of impending, irreparable deformity.

- **Improve function:** This is usually related to pain relief and improvement in deformity, both of which result in improved function.
- **Correction of (or prevention of impending) tendon rupture:** This procedure is usually done on an urgent basis. Tenosynovectomy is usually also performed (*see* questions 8 and 9).

6. What are rheumatoid nodules?
   They are subcutaneous masses of a collagenous capsule surrounding a fibrous center that itself is also surrounded by chronic inflammatory cells. As the rheumatoid nodules expand, avascularity may lead to central necrosis. The nodules may erode through overlying skin and form draining sinuses.

7. Where do rheumatoid nodules commonly occur?
   They are usually over bony prominences on the extensor surfaces. The most common site is over the olecranon and extensor surface of the forearm. They may appear on the extensor surface of the fingers, where they are unsightly and often tender, or occasionally on the palmar aspect of the fingers, where they may impinge on digital nerves or alter finger motion. Excision can provide symptomatic relief and improve function and appearance.

8. What is tenosynovitis; how does it present, and where does it commonly occur?
   - Tendons in synovial-lined sheaths are subject to the same pathologic process that synovial lined joints experience in patients with RA. **Synovial hyperplasia** and **inflammatory cell infiltration** result in swelling of tendon sheaths and synovial invasion of tendons.
   - **Presenting symptoms** include swelling along the course of tendon sheaths, loss of motion (active motion loss greater than passive), crepitance, triggering, tendon ruptures, or compressive neuropathy. Tenosynovitis may be the first sign of RA.
   - In the hand, the **most common sites** are the dorsal—involving the digital extensor tendons, the wrist flexors, and digital flexor tendons at the wrist and within the flexor tendon sheath.

9. How is tenosynovitis treated?
   - **Medical and drug treatment:** When caught early, medications may be prescribed by a rheumatologist to control or arrest progression of tenosynovitis.
   - **Corticosteroid injection:** Injection in tendon sheaths used judiciously can control flares of localized tenosynovitis.
   - **Rest and splinting:** Splints can be used to rest the areas affected as well as prevent deformity caused by the tenosynovial destruction of surrounding structures.
   - **Tenosynovectomy:** For cases in which the treatments just described are unsuccessful over a 4- to 6-month period, and if nerve compression or impending or actual tendon rupture exists, surgical excision of involved tenosynovium may be necessary.

10. What is caput ulnae syndrome?
    Described by Backdahl in 1963, it is end-stage destruction of the distal radioulnar joint. The **piano key sign** may be elicited. When the ulnar head is depressed manually, it will rebound dorsally as pressure is released. Caput ulnae syndrome can lead to attritional rupture of the EDQ or EDC tendons, a condition known as **Vaughn-Jackson syndrome.** Failure of the ligamentous structures of the ulnar wrist and distal radioulnar joint result in:
    - Dorsal subluxation of the ulnar head
    - Supination of the carpus on the radius
    - Volar subluxation of the extensor carpi ulnaris (ECU) tendon

KEY POINTS: CHARACTERISTICS OF RHEUMATOID ARTHRITIS OF THE HAND

- Tenosynovitis
- MCP joint ulnar deviation and synovitis
- Unstable, swollen, painful distal ulna
- Subcutaneous nodules on extensor surfaces
- Radiographic evidence of erosions, especially at the MCP joints, distal ulna, and periscaphoid region

11. **What are some common techniques used to treat distal radioulnar joint (DRUJ) instability is in the patient with RA?**
    Treatment is aimed at relieving pain, removing the prominence of the distal ulna, relocating the ECU tendon dorsally, increasing range of motion, and preventing tendon rupture.
    - **Darrach procedure**: Resection of the distal portion of the ulna and stabilization of the ulnar stump through capsular and retinacular reefing
    - **Sauve-Kapandji procedure**: Fusion of the distal ulna to the sigmoid notch of the radius and removal of a section of the ulna just proximal to the fusion to allow forearm rotation
    - **Resection hemiarthroplasty**: Removal of the articular portion of the ulnar head while preserving the attachments of the radius to the ulna and reefing the DRUJ capsule

12. **Why do tendon ruptures occur in the patient with RA?**
    - **Attrition** by abrasion over rough bony prominences
    - **Infiltration** of adjacent unchecked synovium
    - **Ischemia** due to external pressure by expansive synovium

13. **Which tendons commonly rupture in the patient with RA?**
    - **Ulnar finger extensors:** Rupture usually begins with the extensor disiti minimi or extensor disitorum communis to the small finger and progresses in a radial direction to the ring and long fingers due to attrition over the ulnar head. This is known as **Vaughn-Jackson syndrome** (Fig. 8-1).

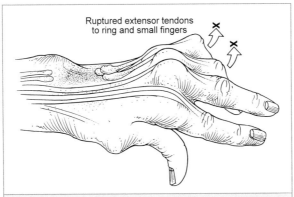

Figure 8-1. Extensor digitorum communis tendon ruptures to the ring and small fingers over the dislocated ulna. Also known as Vaughan-Jackson syndrome. (Courtesy of the Indiana Hand Center.)

- **Flexor pollicis longus (FPL):** This is the most common flexor tendon to rupture. The flexor digitorum profundus (FDP) to the index finger may also rupture with the FPL, both due to attrition because of a spur in the scaphoid. This is known as **Mannerfelt syndrome**.
- **Extensor pollicis longus (EPL):** Rupture results from attrition at Lister's tubercle.

14. How are tendon ruptures treated?
    - **Direct repair:** This treatment is possible in the first 4–6 weeks after rupture but is difficult and unreliable due to tendon substance loss and contracture.
    - **Tendon grafting:** Interpositional grafts placed at the site of rupture may be employed. They require healing at the proximal and distal junctions and are prone to adhesion formation.
    - **Tendon transfers:** This procedure is generally the preferred method. Good technique allows early motion for better results, and contracture of the tendon is not an issue.

15. What changes typically occur in the rheumatioid wrist?
    - Synovial pannus formation at the volar radiocarpal ligaments causes loss of ligament integrity. This destabilizes the scaphoid and causes scaphoid rotation and carpal collapse. The carpus subluxates ulnarly and volarly.
    - Destruction of the ulnar ligaments and DRUJ occur due to pannus formation. This leads to the caput ulnae syndrome (see question 10). The carpus also supinates relative to the radius.
    - The changes in the carpal position lead to radial deviation of the metacarpals (Fig. 8-2).

16. What surgical options are available for treatment of the rheumatoid wrist?
    - **Synovectomy:** Removal of synovial tissue from all crevices in the wrist is difficult, has limited utility, and is associated with uncertain long-term results.
    - **Limited arthrodesis:** Often performed at the radiocarpal level, when possible, this procedure provides pain relief, offers a stable foundation for hand function, and preserves some wrist motion.
    - **Total arthrodesis:** This is the most reliable method for pain relief and stabilization of the wrist but sacrifices wrist motion.
    - **Arthroplasty:** Joint replacement of the wrist with a silicone or metal and plastic combination has had a history of a relatively high complication rate. Research continues to improve this option.

17. What is the typical deformity of the MCP joints, and what factors contribute to it?
    The classic deformity pattern is ulnar deviation and palmar subluxation.
    - Intra-articular synovial pannus causes capsular expansion, loss of volar plate integrity, and collateral ligament incompetence.
    - Cartilage destruction leads to loss of geometric stability of the joint surfaces.
    - Radial deviation of the carpus and metacarpals leads to ulnar deviation of the fingers, known as the "Z deformity" or "Z collapse pattern" (see Fig. 8-2).

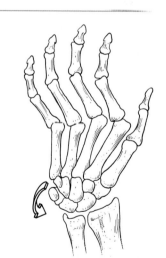

Figure 8-2. Characteristic Z deformity of the hand and wrist in rheumatoid arthritis. The wrist translates ulnarly, the metacarpals angle radially, and the fingers deviate ulnarly at the MCP joints. (Courtesy of the Indiana Hand Center.)

- Attenuation of the radial sagittal bands may lead to ulnar subluxation of the extensor tendons (Fig. 8-3).
- Contracture of the ulnar intrinsic tendons maintains the fingers in ulnar deviation.

18. **What surgical options are available for the treatment of the MCP joints in the patient with RA?**
    - **Synovectomy and extensor tendon realignment:** Removal of the synovial pannus from the joint and recentralization of the ulnarly subluxated extensor tendon may be employed in the patient with deformity but minimal radiographic changes. Results are uncertain and largely temporary; therefore, indications for this procedure are limited.
    - **Arthroplasty:** Joint replacement can predictably relieve pain and often improves hand cosmesis. It can improve function in selected patients. This is the most common surgical treatment.

19. **Which two finger deformities commonly occur in the patient with RA?**
    - **Swan-neck deformity:** This is characterized by PIP joint hyperextension and DIP joint flexion (Fig. 8-4).
    - **Boutonnière deformity:** This is characterized by PIP joint flexion and DIP joint hyperextension. This deformity may also include MCP joint hyperextension (Fig. 8-5).

20. **What classification system is used for swan-neck deformities?**
    Classification of swan-neck deformities is based on PIP joint mobility and radiographic changes.
    - **Type 1:** PIP joint is flexible in all positions.
    - **Type 2:** PIP joint flexion is position dependent. The PIP joint is usually flexible when the MCP joint is flexed and ulnarly deviated. The PIP joint is inflexible when the MCP joint is extended and radially deviated due to secondary contractures of the intrinsic tendons that inhibit PIP flexion.
    - **Type 3:** PIP joint flexion is limited in all positions.
    - **Type 4:** There is radiographic destruction of the PIP joint with limitation of motion in all directions.

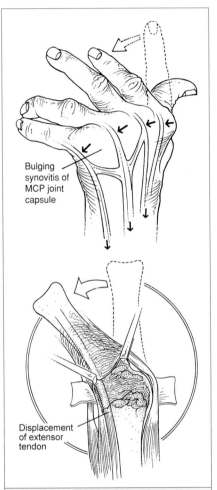

Figure 8-3. Ulnar subluxation of the extensor tendons. Synovial pannus destabilizes the MCP joint by eroding or stretching its ligamentous support. Ulnar subluxation of the extensor tendons contributes to the deforming forces. (Courtesy of the Indiana Hand Center.)

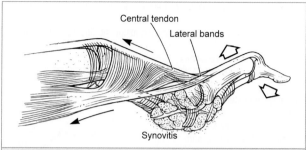

Figure 8-4. The swan-neck deformity. This deformity is characterized by PIP joint hyperextension and DIP joint flexion. (Courtesy of the Indiana Hand Center.)

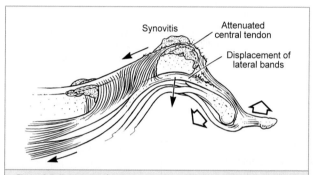

Figure 8-5. The boutonnière deformity. This deformity is characterized by PIP joint flexion, DIP joint hyperextension, and MCP joint hyperextension. The lateral bands sublux and become fixed volarly. (Courtesy of the Indiana Hand Center.)

21. **What classification system is used for a boutonnière deformity?**
Classification of boutonnière deformities is based upon flexibility of the PIP and DIP joints and radiographic changes at the PIP joint (Table 8-2).

TABLE 8-2. CLASSIFICATION OF BOUTONNIÈRE DEFORMITIES

	PIP Joint	DIP Joint
Type 1	Reducible, extension lag	Uninvolved
Type 2	Reducible, extension lag	Intrinsic-intrinsic plus positive*
Type 3	Irreducible, fixed flexion	Full flexion range
Type 4	Irreducible, fixed flexion	Fixed in hyperextension
Type 5	Irreducible, fixed flexion, destroyed articular surface	Fixed in hyperextension

*The intrinsic-intrinsic plus test is positive if the DIP joint cannot be flexed when the PIP joint is extended but can be flexed when the PIP joint is flexed.

22. **What is the modified Nalebuff classification of rheumatoid thumb deformities?**
    - **Type 1:** This type is known as the boutonnière deformity. The MCP joint is flexed and the interphalangeal (IP) joint is hyperextended. This is the *most common* thumb deformity (Fig. 8-6A).
    - **Type 2:** This condition is the same as type 1 with associated carpometacarpal (CMC) joint dislocation or subluxation (very uncommon).
    - **Type 3:** This type is known as the swan-neck deformity. The CMC joint is dislocated; the MCP joint is hyperextended; the IP joint is flexed, and the metacarpal is adduced. This is the second most common thumb deformity (Fig. 8-6B).
    - **Type 4:** This condition is known as gamekeeper's thumb. It is characterized by incompetence of the ulnar collateral ligament of the MCP joint (uncommon).
    - **Type 5:** This type is the same as type 3 but with some exceptions: the CMC joint is stable, and the metacarpal is not adducted (uncommon).

23. **What conditions result in loss of finger extension in the rheumatoid patient?**
    - **Tendon rupture:** This rupture can be caused by invasive synovitis and attrition over the dorsally dislocated ulnar head, usually starting with the small finger and progressing radially.
    - **Tendon subluxation:** The finger extensors at the level of the MCP joints dislocate between the metacarpal heads. The patients are unable to actively extend the MCP joints.
    - **Posterior interosseous nerve (PIN) palsy:** The PIN innervates the finger extensors and can be compressed by significant radiocapitellar synovitis, resulting in loss of finger extension.

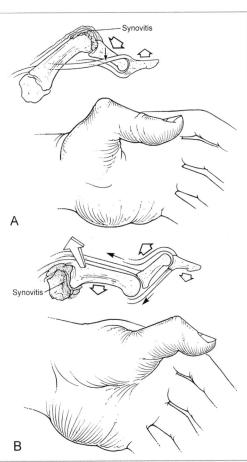

Figure 8-6. *A,* The boutonnière deformity of the thumb consists of flexion at the MCP joint and hyperextension of the interphalangeal joint. *B,* The swan-neck deformity of the thumb consisting of metacarpal adduction contracture, MCP joint hyperextension, and interphalangeal joint flexion. (Courtesy of the Indiana Hand Center.)

24. **How are the causes of sudden loss of finger extension differentiated?**
    - **Rupture versus subluxation:** The MCP joints are passively extended, and the patient is asked to maintain extension. In the case of tendon rupture, the patient is unable to hold the MCP joints in extension. In subluxation, the patient will be able to maintain the MCP joints in extension; patients with subluxation are unable to attain but able to maintain MCP joint extension.
    - **Rupture versus palsy:** If the wrist is flexible, it is flexed maximally. In a PIN nerve palsy, the fingers will passively extend because of the **tenodesis effect**. In the case of tendon rupture, the fingers will not extend with passive wrist flexion because of loss of the tenodesis effect.

25. **Which tendon transfers are used to treat a loss of finger extension due to tendon and extensor digitorum communis (EDC) rupture?**
    1. **One finger:** Typically the small finger with rupture of the extensor digiti quinti (EDQ) and extensor digitorum communis (EDC)
        - Side-to-side transfer to ring EDC *or*
        - Extensor indicis proprius (EIP) transfer to small extensors (Fig. 8-7A)
    2. **Two fingers:** Rupture of EDC to ring and small extensors
        - Side-to-side transfer of EDC ring to long *and*
        - EIP transfer to EDC of small (Fig. 8-7B)
    3. **Three fingers:** Rupture of extensors to long, ring, and small fingers
        - Transfer of flexor digitorum sublimus (FDS) of ring to long, ring, and small extensors *or*
        - Transfer of FDS to ring and small extensors and side-to-side transfer of long to index extensors (Fig. 8-7C)
    4. **Four fingers:** Rupture of extensors to index, long, ring, and small fingers
        - Transfer of FDS to index and long extensors *and*
        - Transfer of FDS to ring and small extensors (Fig. 8-7D)

26. **How is rupture of the EPL treated?**
    - Interpositional tendon graft (less common)
    - EIP transfer to the EPL (more common)

27. **Along with tendon transfers, grafting, or repair, what must be addressed in the treatment of tendon ruptures in RA?**
    - **Tenosynovectomy:** For cases in which invasive synovitis has caused mechanical disruption of the tendon or ischemia leading to rupture, the offending synovial pannus must be removed.
    - **Removal of bony prominences:** Bony spicules as seen in the scaphoid, ulna, and radius must be addressed by removal. In the case of a prominent distal ulna, as in caput ulnae syndrome, resection of the distal ulna is necessary.

28. **How is rupture of the FPL treated?**
    - **Interpositional graft:** If the distal stump is long enough and mobile, a tendon graft can be performed.
    - **FDS tendon transfer:** Usually the ring FDS is transferred to the FPL stump.
    - **Full-length graft:** If the distal stump is too short, the distal portion of the FPL can be replaced by a tendon graft from the proximal stump to the distal phalanx of the thumb.

29. **What are the diagnostic criteria for juvenile rheumatoid arthritis (JRA)?**
    1. Age of onset <16 years
    2. Arthritis in one or more joints (swelling of effusion) *or* the presence of two or more of the following: limitation of range of motion, tenderness or pain on range of motion, and increased heat
    3. Duration of disease 6 weeks to 3 months
    4. Exclusion of other rheumatic diseases

    Unlike adult-onset RA, rheumatoid factor (IgM) is found in only 20% of patients with JRA and is seen almost exclusively in patients ≥8 years of age at the time of onset.

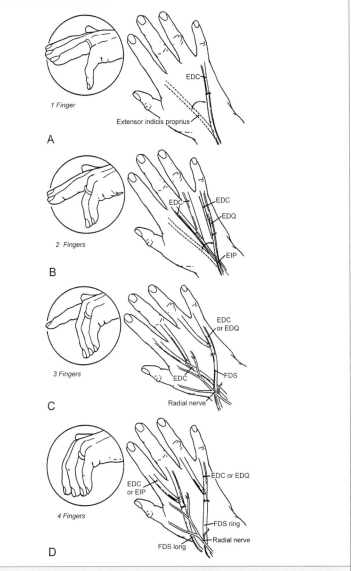

Figure 8-7. *A,* Loss of extension of the small finger. An alternative to side-to-side transfer of a small extensor to the ring EDC is transfer of the EIP to the EDC or the EDQ of the small finger. *B,* Loss of extension of the ring and small fingers. The ring finger EDC is transferred side-to-side to the long finger EDC and the EIP to the EDC of the small finger EDC or EDQ. *C,* Loss of extension of the long, ring, and small fingers. The long finger FDS is transferred to the ring and small finger extensors, and the EDC of the long finger is transferred side-to-side to the EDC of the index finger. *D,* Loss of extension of four fingers. The FDS of the long finger is transferred to the index and long finger extensors, and the ring finger FDS is transferred to the ring and small finger extensors. EDC = extensor digitorum communis, EIP = extensor indicis proprius, EDQ = extensor digiti quinti, FDS = flexor digitorum superficialis. (From Green DP, Hotchkiss R, Pederson WC [eds]: Operative Hand Surgery, 4th ed. New York, Churchill Livingstone, 1999, pp 1675, 1677, and 1679.)

30. **What are the three major types of JRA?**
    The three major types are based on presentation at onset and for the first 6 months of the disease process:
    - Systemic onset
    - Polyarticular onset
    - Pauciarticular onset

31. **What are the characteristics of systemic onset JRA?**
    Also known as Still's disease, systemic onset accounts for 20% of JRA cases. Characteristics include:
    - Intermittent high fevers
    - Transient arthritis associated with fevers
    - Hepatosplenomegaly, lymphadenitis, uveitis
    - Leukocytosis and anemia
    - Rheumatoid factor (RF) not present
    - Severe, chronic arthritis in only 25% of patients

32. **What are the characteristics of polyarticular onset JRA?**
    Polyarticular onset JRA loosely resembles adult-onset RA and accounts for 40% of all JRA cases. Characteristics include:
    - Five or more joints involved in a symmetric pattern
    - Small and large joints involved
    - No systemic manifestations present
    - RF usually not present (when present in 25% of cases, heralds chronic, severe disease)
    - Most likely to continue into adulthood

33. **What are the characteristics of pauciarticular onset JRA?**
    Pauciarticular onset accounts for 40% of all JRA cases. Characteristics include:
    - Involvement of fewer than five joints in an asymmetric pattern
    - Involvement of large joints

    Patients with pauciarticular onset JRA are subdivided into two groups based on gender. The predominantly female group has the following characteristics:
    - Age of onset at <6 years
    - Antinuclear antibody (ANA) positive, rheumatoid factor (RF) negative
    - Lower extremities more commonly involved
    - Predominance of iridocyclitis

    The predominantly male group has the following characteristics:
    - Median age of onset at 10 years
    - ANA- and RF-negative, association with HLA-B27
    - Lower extremity involvement most common
    - Risk for sacroiliitis

34. **Which findings in the hand distinguish JRA from adult RA?**
    Unlike adult RA, JRA has the following characteristics:
    - The wrist ulnarly deviates, and MCP joints radially deviate.
    - Hand size is small.
    - Tubular bones are often narrow and short (especially ring and small finger metacarpals and ulna).

BIBLIOGRAPHY

1. Arnett FC, Edworthy SM, Bloch DA, et al: The American Rheumatism Association 1987 Revised Criteria for the Classification of Rheumatoid Arthritis. Arthritis Rheum 31:315–324, 1988.
2. Bachdahl M: The caput ulnae syndrome in rheumatoid arthritis. Acta Rheum Scand Suppl 5:1–75, 1963.
3. Blank JE, Cassidy C: The distal radioulnar joint in rheumatoid arthritis. Hand Clin 12(3):499–513, 1996.
4. Heywood AWB, Learmonth ID, Thomas M: Cervical spine instability in rheumatoid arthritis. J Bone Joint Surg 70-B:702–707, 1988.
5. Kaye JJ: Radiographic methods of assessment (scoring) of rheumatoid disease. Rheum Dis Clin 17(3):457–470, 1991.
6. Kelley WN, Harris ED Jr, Ruddy S, et al (eds): Textbook of Rheumatology. Philadelphia, W.B. Saunders Company, 1993.
7. King JA, Tomaino MM: Surgical treatment of the rheumatoid thumb. Hand Clin 17:275–289, 2001.
8. Massoarotti EM: Medical aspects of rheumatoid arthritis—Diagnosis and treatment. Hand Clin 12(3):463–475, 1996.
9. Rizio L, Belsky MR: Finger deformities in rheumatoid arthritis. Hand Clin 12(3):531–540, 1996.
10. Shmerling RH: Synovial fluid analysis. Rheum Dis Clin 20(2):503, 1994.
11. Stirrat CR: Metacarpophalangeal joints in rheumatoid arthritis of the hand. Hand Clin 12(3):515–529, 1996.
12. Taleisnik J: Rheumatoid arthritis of the wrist. Hand Clin 5:257–278, 1989.

OTHER ARTHRITIDES

Michael I. Vender, MD, and Prasant Atluri, MD

1. **Volar subluxation and ulnar drift of the metacarpophalangeal (MCP) joints with normal appearing radiographs are characteristic of what disease process?**
   Systemic lupus erythematosus (SLE) may resemble the deformities seen in rheumatoid arthritis (Fig. 9-1). Radiographs typically reveal joint subluxations or dislocations without erosive changes or joint space narrowing.

2. **True or false: Typical deformities in SLE are passively correctable.**
   True. The supporting ligaments of the hand and wrist become lax and contribute to joint deformities. Volar plate laxity and tendon subluxation also allow subluxation and dislocation. The deformities often can be passively corrected and maintained by splints until later in the disease when they may become fixed.

3. **Describe the typical patient with SLE.**
   In contrast to patients with rheumatoid arthritis, patients with SLE are predominantly young women (female-to-male ratio is 9:1). Average age of onset is between 15 and 25 years.

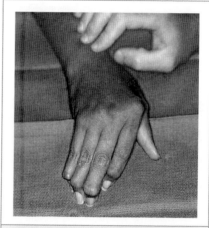

Figure 9-1. The right hand of a 30-year-old woman with a 5-year history of systemic lupus erythematosus.

4. **True or false: Extensor tendon rupture associated with SLE is typically due to invasive tenosynovitis.**
   False. Dorsal wrist tenosynovitis is uncommon in patients with SLE. When tendon ruptures occur, they generally are attritional ruptures secondary to distal ulna instability.

5. **True or false: Soft tissue procedures are effective in correcting the digital deformities in SLE.**
   False. Soft tissue realignment has a history of failure in long-term outcome studies. Either implant arthroplasty or joint fusion is usually performed despite normal-appearing joint cartilage (Fig. 9-2).

6. **What is the most common bacterial agent in septic arthritis of the hand and wrist?**
   Staphylococcus aureus is the most common organism identified in septic arthritis of the hand or wrist.

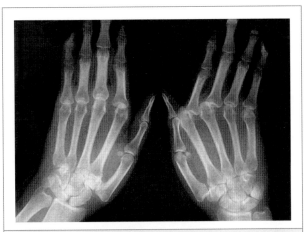

Figure 9-2. Radiograph of hand deformities in a patient with systemic lupus erythematosus.

KEY POINTS: SYSTEMIC LUPUS ERYTHEMATOSUS

1. Volar subluxation and ulnar drift of MCP joints are characteristic of the condition.
2. Deformities typically are passively correctable.
3. Significant pain may be present despite normal-appearing radiographs.

7. **Which bacteria are characteristic of septic arthritis of the MCP joints due to a human bite wound?**
   Eikenella corrodens is associated with 7–29% of human bite wounds. It is susceptible to Penicillin G, variably susceptible to cephalosporins, and resistant to penicillinase penicillins.

8. **True or false: Patients with psoriatic arthritis test positive for rheumatoid factor.**
   False. Psoriatic arthritis is a seronegative spondyloarthropathy. These entities have a negative rheumatoid factor on serologic testing.

9. **True or false: Most patients with psoriasis eventually develop arthritis at some time during the course of their disease.**
   False. Most patients with psoriasis do not develop signs and symptoms consistent with psoriatic arthritis. Although study results vary, approximately 20% of all patients with psoriasis eventually develop the clinical and radiographic features of psoriatic arthritis.

10. **What time of year is more appropriate for performing surgery on the patient with psoriatic arthritis?**
    The skin condition of psoriasis improves in the summer months, probably due to increased exposure to ultraviolet light, which is an important part of treatment. *S. aureus* has been cultured from psoriatic skin lesions. Thus, elective surgical procedures should be scheduled in the warmer months.

11. **What are the common nail findings of psoriatic arthritis?**
    Pitting is a classic finding and represents involvement of the proximal matrix. Other nail findings include leukonychia and crumbling (Fig. 9-3).

12. **What are the characteristic radiographic findings of psoriatic arthritis?**
    Osteolysis is common with bone destruction and widening of the joint spaces in longstanding cases. Bone proliferates along the margins of the distal side of the joint. The proximal bone tapers and produces a "pencil-in-cup" appearance when associated with distal changes.

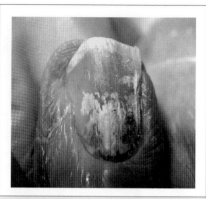

Figure 9-3. Psoriatic nail findings. (From Concannon MJ, Hurov J: Hand Pearls. Philadelphia, Hanley & Belfus, 2002.)

13. **Which disease has a higher incidence of tenosynovitis and tendon ruptures—psoriatic or rheumatoid arthritis?**
    Rheumatoid arthritis. Although diseased synovium is common to both conditions, fibrosis appears to be the dominant feature in psoriatic patients.

14. **True or false: Psoriatic arthritis causes asymmetric hand involvement.**
    True. Asymmetric involvement may also be present in the hand with adjacent digits apparently unaffected. Involved fingers shorten or demonstrate significant deformity.

15. **Which procedures are most commonly performed for the hand deformities of psoriatic arthritis?**
    Prophylactic procedures such as tendon repairs, joint releases, synovectomies, and tendon transfers are rarely performed. Osteotomies, arthroplasties, and fusions are often necessary, even early in the disease process.

KEY POINTS: PSORIATIC ARTHRITIS

1. Sausage digits and nail pitting are common.
2. About 20% of patients with psoriasis develop psoriatic arthritis.
3. The "pencil-in-cup" deformity is often seen on radiographs.
4. The rheumatoid factor (RF) is negative.

16. **A tophus is seen in which inflammatory condition of the hands?**
    Gout. Tophi are deposits of sodium urate in joints, periarticular tissues, and other parts of the body. Whereas acute gout results from joint crystals, chronic gout leads to extra-articular deposits that develop into small masses.

17. **Calcification of the triangular fibrocartilage is consistent with which disease process?**
    Simple calcification through the triangular fibrocartilage and articular cartilage of the wrist is known as chondrocalcinosis. It is indicative of calcium pyrophosphate dihydrate deposition

disease (CPPD or pseudogout) and may be present without any inflammation or joint destruction.

18. **Hyperuricemia contributes to which disease process in the hand?**
Excessive levels of uric acid in the blood stream result in a tendency toward precipitation of sodium urate in articular cartilage, synovium, and other tissues and can lead to gout.

19. **Which disease affects the distal interphalangeal (DIP) joints of the hands?**
Osteoarthritis and psoriatic arthritis classically affect the DIP joints. Rheumatoid arthritis typically spares the DIP joints.

20. **Arthritis mutilans or "telescoping" of the digits is seen in which disease?**
Psoriatic arthritis. It also may be seen in an aggressive form of seropositive rheumatoid arthritis.

21. **Radiographic absorption of the tufts of the distal phalanges indicates which disease?**
Scleroderma is a systemic disease with various manifestations in the hand. Some of the classic findings include loss of the tuft of the distal phalanx on radiographs, accompanied by atrophy of the soft tissues in the pulp and subcutaneous calcinosis.

22. **What are the typical physical findings in the hands of a patient with scleroderma?**
The patient with scleroderma often presents with taut and waxy appearing skin, flexion contractures of the proximal interphalangeal joints, and extension contractures of the MCP joints. Features consistent with Raynaud's phenomenon, such as cold fingers with characteristic color changes and calcinosis cutis, may also be noted. Calcinosis cutis is the deposition of calcium deposits within the digits. This condition can be painful and may require surgical excision.

23. **True or false: A septic joint in the hand or wrist is a surgical emergency.**
True. Although the DIP joint contains little joint fluid, bacterial infection results in cartilage destruction just as in the larger joints of the hand and wrist. The amount of cartilage destruction depends on the duration of infection. Therefore, immediate surgical drainage combined with appropriate intravenous antibiotics is the treatment of choice. The antibiotic regimen is best determined by wound cultures obtained at the time of joint debridement.

24. **A 38-year-old man presents to the emergency department with a 24-hour history of pain, swelling, and progressive loss of wrist motion. He believes that he sustained a puncture wound to the dorsum of the wrist while cleaning out his garage the day before. What is the recommended management?**
Careful physical examination should attempt to determine whether the infection represents a cellulitis or septic arthritis. The distinction can be quite difficult. The patient's temperature should be obtained. A complete blood count (CBC) with differential should be ordered. Wrist radiographs are also essential. The differential diagnosis includes cellulitis, septic arthritis, and possibly gout or pseudogout. Radiographic findings consistent with gout include periarticular calcifications. Calcification of the triangular fibrocartilage complex (TFCC) is consistent with chondrocalcinosis. A foreign body may be detected on radiographs. Although an elevated temperature and elevated white blood cell count are not specific for infection, they suggest the presence of an inflammatory process. A definitive diagnosis of septic arthritis is based on aspiration of the joint and analysis of the aspirate. Gram stain, culture, and crystal analysis should be obtained routinely. If overlying cellulitis is present, aspiration of the joint should not be performed through the site of infection because of possible secondary contamination of the joint itself and subsequent development of septic arthritis.

25. **What is the most effective surgical procedure for preventing further bone loss in the hands of patients with arthritis mutilans?**
   The most effective procedure is joint arthrodesis. Arthritis mutilans involves progressive severe joint destruction and occurs in patients with rheumatoid or psoriatic arthritis. Local joint disease usually can be arrested after resection of the joint, and skeletal stability may be restored via arthrodesis with supplemental bone graft.

26. **Arthritis associated with inflammatory bowel disease is clinically similar to what other type of arthritis?**
   The features of Crohn's disease and other inflammatory bowel conditions may resemble rheumatoid arthritis. Recommended treatment (synovectomy, joint arthrodesis, or implant arthroplasty) is similar for both types of arthritis.

BIBLIOGRAPHY

1. Nalebuff EA: Surgery of psoriatic arthritis of the hand. Hand Clin 12:603–614, 1996.
2. Nalebuff EA: Surgery of systemic lupus erythematosus of the hand. Hand Clin 12:591–602, 1996.

TENDINITIS AND TENOSYNOVITIS
Peter J.L. Jebson, MD

1. **What is de Quervain's disease?**
   Stenosing tenosynovitis of the first dorsal compartment at the wrist, involving the abductor pollicis longus (APL) and extensor pollicis brevis (EPB) tendons.

2. **What is the typical clinical presentation of de Quervain's disease?**
   It presents most commonly in women between 30 and 50 years of age. Symptoms include pain and tenderness localized to the dorsoradial aspect of the wrist. Pain is aggravated by thumb movement with radiation over the dorsum of the thumb and radial forearm.

3. **What is Finkelstein's test?**
   A provocative test used to diagnose de Quervain's disease. The patient flexes the thumb into the palm of the hand and then makes a fist over the thumb. When the wrist is passively deviated in an ulnar direction by the examiner, the APL and EPB are placed under maximal tension, which causes pain localized to the first dorsal compartment (Fig. 10-1).

4. **List the differential diagnoses for patients with presumed de Quervain's disease.**
   - Arthritis of the first carpometacarpal (CMC) joint
   - Scaphoid fracture
   - Arthrosis involving the radiocarpal or intercarpal joints
   - Intersection syndrome
   - Neuritis of the superficial radial nerve (Wartenberg's disease)

5. **How is de Quervain's disease treated?**
   - **Nonoperative:** Resting and avoiding aggravating activities, immobilizing the thumb and wrist, and receiving local steroid injection and systemic nonsteroidal anti-inflammatory drugs (NSAIDs)
   - **Operative:** Surgical release of the first dorsal compartment

6. **Why does nonoperative management frequently fail in de Quervain's disease?**
   The presence of significant anatomic variation of the contents of the first

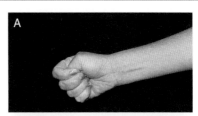
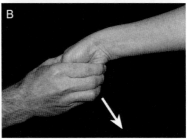

Figure 10-1. Finkelstein's test for de Quervain's disease. *A,* The patient clasps the thumb within the palm. *B,* The examiner then ulnarly deviates the hand *(arrow).* If pain is reproduced, the test is positive and suggestive of de Quervain's disease. (From Concannon MJ: Common Hand Problems in Primary Care. Philadelphia, Hanley & Belfus, 1999.)

dorsal compartment has been implicated as the reason for the frequent failure of nonoperative treatment. Anomalous tendons as well as multiple tendon slips of the APL, separate tendon compartments, and multiple subcompartments have all been described.

7. **What are the most common complications associated with release of the first dorsal compartment?**
   - Laceration of the superficial radial nerve and its terminal divisions with the formation of a painful neuroma
   - Tendon subluxation with painful snapping
   - Traction neuritis of the superficial radial nerve (Fig. 10-2)
   - Incisional tenderness
   - Tendon adhesions
   - Failure to release all subcompartments with persistent or recurrent symptoms
   - Reflex sympathetic dystrophy

8. **Define intersection syndrome.**
   Tenosynovitis of the second dorsal compartment.

9. **How is intersection syndrome differentiated from de Quervain's disease?**
   Pain and swelling are located more proximal in intersection syndrome than in de Quervain's disease. The area is usually 4 cm proximal to the wrist joint where the muscle bellies of the EPB and APL cross the extensor carpi radialis longus and brevis tendons (Fig. 10-3).

10. **Describe the treatment of intersection syndrome.**
    - **Nonoperative:** Resting and avoiding aggravating activities, modifying workload and type, splinting of the wrist in slight dorsiflexion, and receiving local corticosteroid injection
    - **Operative:** Release of the second dorsal compartment combined with tenosynovectomy (rarely required)

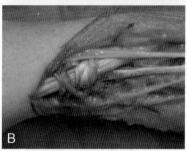

Figure 10-2. Tendon subluxation. *A,* Volar subluxation of the first dorsal compartment tendons after excision of the extensor retinaculum. *B,* Reconstruction of the retinaculum using a distally based slip of the brachioradialis tendon.

11. **What is a trigger finger?**
    Stenosing tenovaginitis of the thumb or finger flexor tendons that results in hand pain, catching of the finger/thumb during motion, or locking of the finger/thumb in flexion.

12. **What causes a trigger finger?**
    Triggering is due to the disproportion between the flexor digitorum superficialis (FDS) or flexor pollicis longus (FPL) tendons and the surrounding sheath. Pain is often reported at the proximal interphalangeal (PIP) joint of the finger or the interphalangeal (IP) joint of the thumb, but the

pathology is usually localized to the region of the first annular (A1) pulley of the flexor tendon sheath.

13. **What histologic changes occur in the A1 pulley of the patient with a trigger finger?**
Fibrocartilaginous metaplasia of the inner gliding layer.

14. **Which finger is most commonly involved?**
The ring finger followed by the thumb and long finger.

15. **How is a trigger finger classified?**
The following list is based on physical findings.
    - **Grade 0:** Mild crepitus, no triggering
    - **Grade 1:** Abnormal or uneven movement
    - **Grade 2:** Clicking or triggering but no locking
    - **Grade 3:** Locked trigger finger, passively correctable
    - **Grade 4:** Locked digit, possible PIP flexion contracture

Trigger finger is also classified as **nodular** (presence of a discrete nodule near the A1 pulley) or **diffuse** (diffuse fullness along the flexor sheath extending distal to the A1 pulley).

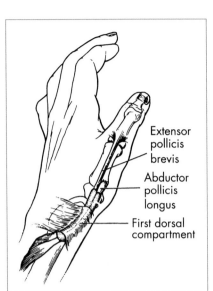

Figure 10-3. Intersection syndrome. Swelling is seen in the first dorsal extensor compartment, where the abductor pollicis longus (APL) and extensor pollicis brevis (EPB) cross the common radial wrist extensors. (From Brotzman SB: Clinical Orthopaedic Rehabilitation. St. Louis, Mosby, 1996.)

16. **How is trigger finger treated?**
    - **Nonoperative:** Modifying activities, receiving local steroid injection into the flexor tendon sheath, splinting, and taking NSAIDs are possible treatments.
    - **Operative:** This option involves open or percutaneous release of the A1 pulley. If a bulbous enlargement of the flexor tendon distal to the A1 pulley is noted to cause persistent triggering, reduction flexor tenoplasty (excision of a central portion of the tendon) may be necessary.

KEY POINTS: TENDINITIS AND TENOSYNOVITIS

1. Fibrocartilaginous metaplasia of the inner gliding layer of the fibroosseous flexor tendon sheath is noted in patients with trigger finger.
2. Nonoperative treatment is thought to be unsuccessful in de Quervain's disease because the first dorsal compartment often has subcompartments and multiple tendon slips.
3. Excision of the first dorsal compartment retinaculum should not be performed to treat de Quervain's disease because of postoperative symptomatic volar tendon subluxation.
4. Urgent operative treatment is recommended for extensor pollicis longus (EPL) tenosynovitis because of possible tendon rupture.
5. The ring finger is most commonly involved in trigger finger.

17. **What is the most common complication of an open trigger finger release?**
    Digital nerve injury.

18. **What structure is most at risk during a percutaneous trigger thumb release?**
    The radial digital nerve.

19. **How does the treatment of a trigger finger differ in the patient with rheumatoid arthritis?**
    The A1 pulley should be preserved to prevent flexor tendon bowstringing and ulnar deviation of the digits. A flexor tenosynovectomy should be performed; if a nodule is present within the flexor tendon, reduction flexor tenoplasty may be necessary. Alternatively, one of the two FDS slips may be excised to decompress the sheath and create room for the flexor tendons to pass freely beneath the pulleys.

20. **For which age group and gender is idiopathic trigger finger most commonly encountered?**
    Middle-aged women.

21. **Describe the clinical presentation of flexor carpi radialis (FCR) tunnel syndrome.**
    - Volar radial forearm and wrist pain
    - Tenderness along the FCR sheath
    - Swelling at the volar-radial wrist crease and along the FCR sheath
    - Increased pain with resisted wrist flexion and radial deviation

22. **Describe the treatment options for FCR tunnel syndrome.**
    - **Nonoperative:** Resting, avoiding aggravating activities, splinting, taking NSAIDs, and receiving local corticosteroid injection
    - **Operative:** Decompression of the tunnel by incising the roof and completely freeing the FCR tendon

23. **What are the typical findings in the patient with flexor carpi ulnaris (FCU) tenosynovitis?**
    Volar ulnar wrist and forearm pain accentuated with resisted wrist flexion and ulnar deviation.

24. **Describe the symptoms of a patient with extensor pollicis longus (EPL) tenosynovitis.**
    - Pain with thumb motion
    - Swelling along the EPL tendon sheath
    - Tenderness
    - Crepitus on the dorsum of the distal radius and forearm just distal to Lister's tubercle

25. **Why is urgent operative treatment recommended for EPL tenosynovitis?**
    To prevent tendon rupture.

26. **Define Linburg's syndrome.**
    Linburg's syndrome, or symptomatic restrictive thumb-index flexor tenosynovitis, involves hypertrophic tenosynovium between the flexor pollicis longus (FPL) and digital flexor tendons in the distal volar forearm; the syndrome can also involve tenosynovitis of the FPL and index flexor digitorum profundus (FDP), with an associated anomalous tendinous interconnection between the two tendons (Linburg's anomaly). Patients complain of distal volar forearm pain that is aggravated when distal finger joint flexion (most commonly the index finger DIP joint) is blocked as the thumb is actively flexed into the palm (Linburg's test).

27. **What is the pathognomonic sign for Linburg's anomaly?**
    Simultaneous index finger DIP joint flexion when the thumb is actively flexed across the palm.

28. **Which condition frequently coexists with Linburg's syndrome?**
    Carpal tunnel syndrome.

29. **A 32-year-old woman presents with painful snapping at the radial aspect of the wrist 6 weeks after undergoing excision of the first dorsal compartment retinaculum for de Quervain's tenosynovitis. What is the diagnosis? How is it treated?**
    Painful subluxation of the first extensor compartment tendons. Treatment involves reconstruction of the retinaculum using a distally based brachioradialis tendon flap.

30. **Which patient population has a high incidence of secondary trigger finger?**
    Patients with diabetes mellitus. Patients who have diabetes and trigger finger have the following clinical features:
    - Multiple digits are commonly involved.
    - Nonoperative treatment is often unsuccessful.
    - Associated flexion contracture of the PIP joint is common.
    - Recurrence of triggering despite release of the A1 pulley is well recognized.

31. **A 54-year-old man of northern European descent presents with recurrent triggering of the left ring finger 3 weeks after undergoing open release. After splinting and injections fail, a decision is made to reoperate. During surgery, the surgeon notes that the A1 pulley has been completely released, there is no triggering at the A2 or A3 pulleys, and no intratendinous nodule is noted. What is the most likely cause of recurrent triggering?**
    Failure to release the palmar aponeurosis pulley. Palmar pulley triggering is often seen in patients with early Dupuytren's contracture. Treatment involves surgical release.

32. **A 19-year-old college student who is part of the rowing team presents with dorsoradial forearm pain 2 days after competing in a regatta. Swelling and tenderness are noted 6 cm proximal to Lister's tubercle. Palpable crepitus is present in the same area during simultaneous thumb and wrist motion. What are your recommendations?**
    The patient has **intersection syndrome** that can result from repetitive activity such as rowing (oarsman's wrist). Treatment involves resting, avoiding aggravating activities, immobilizing the wrist with a splint or cast, taking anti-inflammatory medication, and receiving a corticosteroid injection. Surgical treatment is indicated for those patients who do not respond to a minimum of 3 months of nonoperative treatment. The technique involves release of the second dorsal compartment and excision of the inflamed synovium or bursa between the muscle bellies of the APL and EPB and wrist extensor tendons.

33. **What complications are associated with the use of corticosteroid injections to treat tendinitis/tenosynovitis in the hand?**
    - Fat necrosis, subcutaneous atrophy, depigmentation of skin
    - Infection
    - Tendon rupture
    - Nerve laceration
    - Transient rise in blood and urine glucose levels in diabetic patients

34. **A healthy 48-year-old man presents to the emergency room 24 hours after constructing a storage shed in his backyard. His complaints suggest acute onset of severe volar ulnar wrist pain. He denies any penetrating trauma. He is afebrile, and the erythrocyte sedimentation rate, white blood cell count, and

radiographs of the wrist are normal. The distal volar forearm feels warmer than the other arm, and erythema and tenderness are present over the FCU tendon insertion. What is the diagnosis?

Acute calcific tendinitis. The FCU is most commonly involved. Misdiagnosis is quite common. Treatment involves resting, splinting, and taking anti-inflammatory medication.

35. **What is the success rate of a single corticosteroid injection for an idiopathic trigger finger?**
Between 40% and 90%.

36. **In the patient with trigger finger, which factors are associated with an unfavorable response to injection?**
    - Presence of multiple trigger fingers
    - Symptoms >6 months
    - Associated diabetes mellitus

37. **What are the most common anatomic variations noted in the first dorsal extensor compartment?**
    - Separate ulnar and radial fibro-osseous tunnels
    - One EPB tendon, two APL tendon slips

38. **What are the success rates for the various treatment modalities used in de Quervain's disease?**
    - **Splint immobilization:** 30%
    - **Single corticosteroid injection:** 62%
    - **Two corticosteroid injections:** 80%
    - **Surgical release:** 90%

39. **During a surgery for de Quervain's disease, the superficial radial nerve is inadvertently lacerated. What are the treatment options?**
Nerve repair or resection with translocation of the nerve end away from the surgical site.

BIBLIOGRAPHY

1. Bain GI, Turnbull J, Charles MN, et al: Percutaneous A1 pulley release: A cadaveric study. J Hand Surg 20A:781–784, 1995.
2. Gabel G, Bishop AT, Wood MB: Flexor carpi radialis tendinitis. Part II: Results of operative treatment. J Bone Joint Surg 76A:1015–1018, 1994.
3. Grundberg AB, Reagan DS: Pathologic anatomy of the forearm: Intersection syndrome. J Hand Surg 10A:299–302, 1985.
4. McMahon M, Craig SM, Posner MA: Tendon subluxation after de Quervain's release: Treatment by brachioradialis flap. J Hand Surg 16:30–32, 1991.
5. Patel MD, Bassini LB: Trigger fingers and thumb: When to splint, inject, or operate. J Hand Surg 17A:110–113, 1992.
6. Pope DF, Wolfe SW: Safety and efficacy of percutaneous trigger finger release. J Hand Surg 20A:280–283, 1995.
7. Sampson SP, Badalamente MA, Hurst LC, et al: Pathobiology of the human A-1 pulley in trigger finger. J Hand Surg 16:714–721, 1991.
8. Sampson SP, Wisch D, Badalamente MA: Complications of conservative and surgical treatment of de Quervain's disease and trigger fingers. Hand Clin 10:73–82, 1994.
9. Sherman P, Lane L: The palmar aponeurosis as a cause of trigger finger. J Bone Joint Surg 11:1753–1754, 1996.
10. Witt J, Pess G, Gelberman RA: Treatment of de Quervain's tenosynovitis: A prospective study of the results of injection of steroids and immobilization in a splint. J Bone Joint Surg 73A:219–222, 1991.

INFECTIONS

H. Stephen Maguire, MD

1. **What are the clinical signs that suggest suppurative flexor tenosynovitis?**
   Allen Buchner Kanavel (1874–1938), an American surgeon, described the four classic findings of pyogenic flexor tenosynovitis:
   1. Fusiform swelling of the digit
   2. Semiflexed posture of the digit
   3. Pain with passive extension of the digit
   4. Tenderness along the flexor tendon sheath

2. **What is the treatment for an acute suppurative flexor tenosynovitis?**
   Early pyogenic flexor tenosynovitis (within 24–48 hours) can occasionally be treated with antibiotics, immobilization, elevation of the hand, and frequent reevaluation. If the symptoms fail to resolve or progress, operative irrigation and possibly debridement is required. Often, two incisions are made, one proximally and one distally over the flexor tendon sheath, followed by sheath irrigation.

3. **What is the most common infecting organism in suppurative flexor tenosynovitis?**
   Staphylococcus aureus is the most common organism identified in an acute skin and soft tissue infection of the hand as well as in suppurative flexor tenosynovitis. Antibiotic therapy, however, may need to be modified, depending on the results of the wound culture.

4. **What is a paronychia?**
   An infection of the tissue surrounding the eponychial fold. It is the most common infection of the hand overall. *S. aureus* is the most common infecting organism. Of all of the fingertip areas, the hyponychium is the most resistant to infection.

5. **How is a chronic paronychia characterized?**
   A chronic paronychia is characterized by a chronically indurated, retracted, and rounded cuticle with recurring episodes of pain, swelling, and inflammation. It is not necessarily associated with purulent drainage. *Candida albicans* is the organism responsible in 70–90% of cases. There is often a bacterial superinfection, usually caused by *Pseudomonas aeruginosa*.

6. **How is a chronic paronychia treated?**
   The area should be kept dry, and topical antifungals are applied. Treatment may require eponychial marsupialization, a surgical procedure in which the fibrotic and chronically infected tissue of the eponychium is excised and the resultant wound left open. The tissue is sent for histopathology to rule out a malignancy that may mimic a chronic infection.

7. **Which organism is usually found in a chronic wound infection following an abrasion or laceration sustained while working with a fish tank?**
   Mycobacterium marinum. *M. marinum* is an atypical mycobacterial species that thrives in fresh water and seawater environments. Treatment consists of antituberculous therapy and surgical debridement. Tissue samples should be sent for culture using Lowenstein-Jensen medium at 32°C.

INFECTIONS

8. **What organism is frequently associated with an infected animal bite?**
   Pasteurella multocida is often seen in conjunction with an animal bite, particularly a cat bite. The infection is treated with amoxicillin/clavulanate or penicillin.

9. **What organism is frequently associated with a human bite wound?**
   Eikenella corrodens. The infection is treated with a penicillinase-resistant penicillin or ampicillin.

10. **A dental hygienist presents with a swollen fingertip and a hemorrhagic coalescence of resolving vesicular lesions of 10- to 12-day duration. She notes intense throbbing and pain. Which infecting organism should be considered first?**
    The patient has a herpetic whitlow or infection with the herpes simplex virus type 1. Serious infections are treated with acyclovir. Avoid incision and drainage to prevent possible superinfection with bacteria.

11. **What organism is most frequently isolated in the patient with necrotizing fasciitis?**
    Group A *Streptococcus pyogenes*.

12. **List the general principles of wound management to minimize the likelihood of a postoperative wound infection.**
    - Meticulous debridement of devitalized tissue to decrease the bacterial count to $<10^5$ bacteria per mm^3
    - Elimination of dead space
    - Copious wound irrigation

13. **What is a felon?**
    A felon is a closed-space infection of the distal volar pulp. It is treated by incision and drainage over the area of fluctuance along with antibiotics. The location of the incision is dictated by the location of any fluctuation, the avoidance of digital vessels and nerves, the avoidance of the distal flexor tendon sheath, and the potential for a disabling scar.

14. **What organism should be covered with prophylactic antibiotics while using leeches to decrease venous congestion in a replanted digit?**
    Aeromonas hydrophilia is a gram-negative rod that lives symbiotically in *Hirudo medicinalis* (the leech). The organism helps break down hemoglobin in the leech intestine. It can cause a rapidly progressive infection with gas in the soft tissues that resembles clostridial myonecrosis. *Aeromonas* is also associated with a "dead fish" smell. Treatment consists of trimethoprim-sulfamethoxazole or a third-generation cephalosporin application.

15. **An immunosuppressed patient with prior episodes of gouty arthritis presents with a swollen, red, painful wrist; an erythrocyte sedimentation rate of 45; a white blood cell count of 12,500; a uric acid level of 8; and a temperature of 38.1°C. Plain films are negative. What diagnostic imaging study may help differentiate between a recurrent gouty flare and septic arthritis?**
    An indium or technetium-labeled leukocyte bone scan.

16. **What group of organisms are considered "great masqueraders"?**
    Mycobacterial infections. A mycobacterial infection should be included in the differential diagnosis of all indolent masses or inflammatory processes that are "sterile" yet fail to respond to the usual medical and surgical therapy.

17. **Yellow sulfur granules draining from a sinus are highly suggestive of what organism?**
    Actinomyces israelii. The granules are actually dense colonies of the microorganism.

18. **Hansen's disease (leprosy) is first seen in which structure?**
Mycobacterium leprae is first isolated from the Schwann cells of the peripheral nerves. The ulnar nerve is the first nerve to be involved in the upper extremity.

19. **What are the different types of deep subfascial space infections of the hand?**
    - Subaponeurotic
    - Midpalmar
    - Thenar
    - Interdigital (web)
    - Parona's space

20. **What are the characteristics of a thenar space infection?**
A thenar space infection is manifested by marked swelling of the thenar eminence with the thumb held in abduction. The thenar space is located radial to the vertical septum, between the third metacarpal and the long finger profundus tendon, and it extends to the lateral edge of the adductor pollicis. The infection should be drained through separate dorsal and volar incisions.

KEY POINTS: INFECTIONS OF THE HAND

1. Paronychia, an infection of the tissue surrounding the eponychial fold, is the most common infection of the hand overall. *S. aureus* is the most common infecting organism.
2. Of all the areas of the nail bed, the hyponychium is the most resistant to infection.
3. The different types of deep subfascial space infections of the hand are (1) subaponeurotic, (2) midpalmar, (3) thenar, (4) interdigital (web space), and (5) Parona's space.
4. *Mycobacterium marinum* should be suspected in any infection following exposure to freshwater and saltwater environments.
5. *Mycobacterium marinum* must be cultured on Lowenstein-Jensen medium at 32°C.
6. The clinical findings of a suppurative flexor tenosynovitis are fusiform swelling, tenderness along the sheath, pain on passive extension, and semiflexed posture of the finger.
7. Necrotizing fasciitis is a severe limb and life-threatening soft tissue infection that is typically seen in patients with a history of IV drug abuse or alcoholism.

21. **Where is Parona's space?**
Parona's space is a potential space bounded by the pronator quadratus dorsally, the digital flexors, flexor pollicis longus, and flexor carpi ulnaris.

22. **What is a collar button abscess?**
An infection of the interdigital subfascial web space. The fingers adjacent to the involved web space are usually abducted and painful. The term "collar button" is derived from the hourglass shape of the abscess with both a dorsal and volar component separated by the superficial transverse metacarpal ligament. This infection requires separate palmar and dorsal incisions to be adequately drained.

23. **What is the most common cause of septic arthritis in the wrist?**
Penetrating trauma.

24. **A 19-year-old woman presents with septic arthritis of the wrist. She has experienced migratory polyarthralgias and is sexually active. You suspect what kind of infection?**
Neisseria gonorrhoeae should be considered in any sexually active young patient with septic arthritis and migratory polyarthralgias.

25. **How common is osteomyelitis in the hand?**
    The hand is involved in 10% of all cases of osteomyelitis. Osteomyelitis accounts for 1–6% of all hand infections. The most common infecting organism overall is *Staphylococcus aureus*.

26. **How is osteomyelitis of the hand treated?**
    Adequate debridement of all infected, necrotic, and devitalized tissue is essential. The residual defect requires skeletal reconstruction after an appropriate course of systemic antibiotics. Amputation is reserved for chronic cases that have failed appropriate treatment and when extensive tissue damage precludes the satisfactory return of function despite surgical reconstruction.

27. **What type of hand infection appears to be unique to the diabetes patient?**
    Subepidermal abscess.

28. **A 46-year-old man with a history of intravenous drug use presents with a 24-hour history of painful forearm swelling and erythema. He is febrile and states that the swelling and erythema have progressed rapidly. What is the most likely diagnosis?**
    Necrotizing fasciitis.

29. **Describe the treatment for necrotizing fasciitis of the upper extremity.**
    Treatment includes early recognition; aggressive surgical debridement of all necrotic skin, subcutaneous tissue, and muscle; and the administration of broad spectrum antibiotics. After the initial debridement, the wound is left open, and repeat debridements are performed every 24 to 48 hours as needed. Amputation may be necessary to control a severe infection. Hyperbaric oxygen (HBO) has been recommended for recalcitrant, severe infections, but its efficacy remains controversial.

30. **What is gas gangrene?**
    Gas gangrene, also known as clostridial myonecrosis, is a rare anaerobic infection that develops in devitalized, contaminated wounds. It is most commonly caused by *Clostridium perfringens*, which is a gram-positive rod. Other toxin-producing *Clostridium* species may also be responsible.

BIBLIOGRAPHY

1. Barbieri RA, Freeland AE: Osteomyelitis of the hand. Hand Clin 14:589–603, 1998.
2. Gilbert DN, Moellering RC, Sande MA: The Sanford Guide to Antimicrobial Therapy 2002. Hyde Park, VT, Antimicrobial Therapy, 2002.
3. Gonzales MH: Necrotizing fasciitis and gangrene of the upper extremity. Hand Clin 14:635–645, 1998.
4. Gonzales MH, Kay T, Weinzeig N, et al: Necrotizing fasciitis of the upper extremity. J Hand Surg 21A:689–692, 1992.
5. Jebson PJL, Louis DS (eds): Hand Infections. Hand Clin 14:557–566, 1998.
6. Kanavel AB: Infections of the Hand. Philadelphia, Lea & Febiger, 1912.
7. Keyser JJ, Lettler JW, Eaton RG: Surgical treatment of infections and lesions of the perionychium. Hand Clin 6:137–153, 1990.
8. Neviaser RJ: Acute infections. In Green D, Hotchkiss R, Pederson W (eds): Green's Operative Surgery, 4th ed. New York, Churchill-Livingstone, 1999, pp 1033–1047.
9. Patel MR: Chronic infections. In Green D, Hotchkiss R, Pederson W (eds): Green's Operative Surgery, 4th ed. New York, Churchill-Livingstone, 1999, pp 1048–1093.

EXTRAVASATION, FOREIGN BODIES, AND HIGH-PRESSURE INJECTION INJURIES

Cassandra Robertson, MD, and Douglas P. Hanel, MD

EXTRAVASATION INJURY

1. **List the three categories of agents responsible for an extravasation injury.**
   - **Vesicant:** An agent that results in local tissue necrosis
   - **Irritant:** An agent that produces pain and inflammation without necrosis
   - **Nontoxic:** An agent without vesicant or irritant effects

2. **Describe five mechanisms of cellular destruction in extravasation injuries.**
   - **Osmolality:** Hyperosmolality or hypoosmolality
   - **Local tissue ischemia:** Intense vasoconstriction due to vasoactive agents
   - **Direct cellular toxicity:** Inhibition of vital cellular physiochemical mechanisms, such as DNA replication
   - **Mechanical compression:** Circulation affected, perhaps resulting from a mechanical infusion pump or injection of fluid into a closed space during a crisis, such as for cardiac resuscitation
   - **Bacterial colonization:** Abnormal proliferation of normal skin organisms

3. **How common is extravasation of vesicant chemotherapy agents?**
   The incidence ranges from 0.2% to 6% of all infusions administered.

4. **Which commonly used chemotherapy agents are classified as vesicants?**
   Doxorubicin, vincristine, vinblastine, adriamycin, cisplatin, mitomycin, and mithramycin.

5. **Give examples of other solutions and agents that produce severe tissue necrosis after extravasation.**
   Phenytoin, norepinephrine, dopamine, albumin, lipid agents, hyperalimentation substances, and solutions containing hypertonic dextrose and/or calcium salts, such as calcium chloride or gluceptate.

6. **What proportion of vesicant extravasation injuries result in ulceration?**
   About one third.

7. **Which factors determine the degree and extent of tissue damage from an extravasation injury?**
   - Chemical structure of the agent
   - Site of extravasation
   - Concentration, volume, and duration of extravasation
   - Host response (cellular and humeral immunity)
   - Delay in recognition and initiation of treatment
   - Type of treatment

8. **Describe the presentation and clinical course of an extravasation injury in the hand or forearm.**
   See Table 12-1. In the case of vasoconstrictor agents, the extremity may appear cool and cyanotic with poor capillary refill.

TABLE 12-1. CLINICAL FEATURES OF AN EXTRAVASATION INJURY

Severity of Injury	Extravasation	Presentation	Clinical Course
Mild	Small volume, possibly a vesicant or irritant	Minimal soft tissue swelling and pain	Asymptomatic within 1 or 2 days with simple conservative treatment
Moderate	Approximately 1 to 5 cc of a vesicant or irritant	Erythema typically less than 10 cm diameter, moderate local tenderness, blistering infrequent	Improves rapidly with conservative treatment, although may be symptomatic for 1 month or more
Severe	Larger volume and/or high-pressure vesicant	Extremely painful, marked edema, blistering/erythema	Skin sloughing in 5–7 days, painful full-thickness skin ulcerations that may become infected and may involve deep tissue necrosis

KEY POINTS: COMPLICATIONS OF A SEVERE EXTRAVASATION INJURY

1. Significant soft tissue defect
2. Sepsis
3. Compartment syndrome
4. Nerve compression syndromes
5. Permanent nerve deficits
6. Sympathetic dystrophy
7. Permanent joint stiffness
8. Amputation

9. **Describe the conservative treatment of an extravasation injury.**
    - Immediately stop the infusion. Leave the catheter in place if an antidote is to be infused.
    - Attempt to remove extravasated solution.
    - Elevate extremity and cover with a light dressing.
    - Apply cold compresses for vasoconstriction to reduce doxorubicin toxicity; apply warm compresses, decreasing the concentration, to promote absorption (indicated for vinca alkaloids).
    - Splint with the wrist in 30–40 degrees of extension, the metacarpophalangeal (MCP) joints in 90 degrees of flexion, and the interphalangeal (IP) joints in full extension.
    - Perform frequent wound inspections and evaluations.
    - Assist patient with range of motion exercises.

10. **Name the specific antidote for those agents that have a recognized specific antidote.**
    See Table 12-2. No conclusive evidence supports the use of corticosteroids or sodium bicarbonate for extravasation injury.

    TABLE 12-2. SPECIFIC ANTIDOTES

Agent	Antidote
10% dextrose, calcium, potassium, aminophylline, nafcillin, radiocontrast media, parenteral nutrition solutions, vinca alkaloids	Hyaluronidase (Wydase): Enzyme that allows rapid diffusion of fluid into the interstitial space, 150–1500 U subcutaneous injection
Vasopressors	Phentolamine (Regitine): Enzyme that allows rapid diffusion of fluid into the interstitial space
Nitrogen mustard	Subcutaneous sodium thiosulfate (1:6,000,000)
Mitomycin C	Topical dimethyl sulfoxide (DMSO) in 90% alpha tocopherol succinate (every 8 hours for 7 days)
Anthracyclines (e.g., doxorubicin)	DMSO (see previous antidote), dexrazoxane IV (1000 mg/m^2 within 5 hours, then 1000 mg/m^2 on day 2, then 2,500 mg/m^2 on day 3)

11. **What is the necrosis interval?**
    The time elapsed from the time of extravasation to the point of irreversible tissue damage. Surgery performed during this time is likely to improve the clinical outcome. The necrosis interval is 72 hours for many vesicants, 6 hours for radiological contrast agents, and 4–6 hours for vasopressor agents.

12. **When is early surgery indicated for an extravasation injury?**
    Early aggressive surgical debridement is indicated for massive extravasation with persistent severe pain, arterial compromise, compartment syndrome, or rapidly spreading skin necrosis. For doxorubicin, excision must be radical, including margins of normal skin and subcutaneous tissue to reach healthy tissue and possibly sacrificing involved tendons and fascia. Ultraviolet light helps detect doxorubicin intraoperatively, and fluorescein dye can help delineate viable tissues.

13. **When early surgery is not indicated for an extravasation injury, what role does surgery play?**
    Following demarcation, wide excision of all involved tissues to viable margins is recommended. Skin grafting is performed after a granulation tissue bed has developed, or coverage is achieved with local or free soft tissue flaps. Reconstructive procedures or amputation may be necessary in some cases. Early rehabilitation consisting of active range-of-motion exercises and edema control should be initiated to prevent residual stiffness.

14. **What percentage of severe injuries will need later reconstructive surgery?**
    Approximately 50%.

15. **Which host characteristics are recognized as risk factors for extravasation injury?**
    - Generalized vascular disease with diminished local blood flow
    - Inability to communicate the pain produced by extravasation (e.g., infant, young child, comatose patient, patients under general anesthesia)

- Impaired venous circulation or lymphatic drainage (e.g., patients with vena cava syndrome or mastectomy and axillary dissection)

16. **What are some of the precautions taken to avoid an extravasation injury?**
    - Chemotherapy should be administered by experienced personnel with maximal dilution of the drug.
    - A subcutaneous device is often indicated for administration of chemotherapeutic agents.
    - The use of large, central veins for resuscitative efforts is best if possible.
    - The use of long, smooth-tipped polyethylene catheters instead of sharp, metal-tipped needles is recommended.
    - The catheter should be placed in the forearm whenever possible. Areas over joint surfaces, tendons, nerves, or vessels (such as the dorsum of the hand or radial wrist) should be avoided.
    - Multiple perforations or repeated use of the same vein, administration of an infusion under pressure, and use of lower extremity infusion sites, especially in the elderly and patients with vascular diseases, are not recommended.
    - Use of an extremity that is insensate or has impaired circulation should be avoided.
    - All tight clothing and jewelry that may cause a tourniquet effect should be removed.
    - Personnel should monitor for signs and symptoms of extravasation.

17. **A 35-year-old man sustains a fluoric acid burn to his right hand when he is exposed to an industrial rust remover at work. What is the appropriate treatment?**
    - Remove all exposed clothing (wear gloves, eye protection, and other safety equipment as needed).
    - Wash thoroughly under running water for 15–30 minutes.
    - Massage calcium gluconate gel into the skin.
    - Note that an intravenous Bier's block, using calcium gluconate (and 5000 U heparin), may be indicated.
    - If pain persists, consider nail removal and intra-arterial infusion of calcium gluconate (in an intensive care unit setting).

18. **What complications may occur following an intra-arterial calcium gluconate infusion?**
    - Arrhythmias (especially in digitalized patients)
    - Bradycardia
    - Depression of neuromuscular function
    - Flushing
    - Tingling sensations
    - Cardiac arrest

FOREIGN BODIES

19. **What are the three most commonly retained foreign bodies in the hand?**
    Wood, glass, and metal account for approximately 95% of retained foreign bodies in the hand (Fig. 12-1).

20. **What diagnostic imaging methods are used to detect and localize a foreign body?**
    Plain radiography, xeroradiography, computed tomography (CT), magnetic resonance imaging (MRI), fluoroscopy, and high-resolution ultrasound.

21. **What types of foreign bodies are detected with diagnostic imaging methods?**
    Glass is almost always visible on plain radiographs and can be identified easily on xeroradiography, ultrasound, CT, and MRI. Plastic and wood are difficult to identify on plain radiographs

and xeroradiography. Ultrasound, CT, or MRI provides optimal visualization of wood. Ultrasound or MRI is recommended for the detection of plastic. The ability to visualize gravel on plain radiography and xeroradiography is variable, but all gravel types are easily detected by CT. Metallic foreign bodies are easily visible on plain radiographs.

22. **What are the preferred imaging methods for nonradiopaque foreign bodies?**
MRI and ultrasound. Real-time high-resolution ultrasound is both highly sensitive (94%) and specific (99%) for evaluating the presence of foreign bodies in the hand. Because it is accurate, reliable, and cheaper than MRI, a combination of plain radiographs and high-resolution ultrasound has become the preferred strategy to diagnose and localize virtually all suspected foreign bodies. Real-time high-resolution ultrasound should be performed by an experienced ultrasonographer.

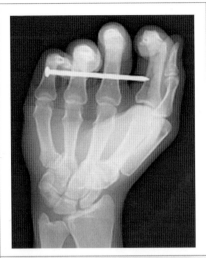

Figure 12-1. The left hand of a 30-year-old construction worker who drove a nail through all four fingers. Since the nail was not barbed, it was easily withdrawn in a retrograde fashion. There was no significant injury to surrounding structures.

23. **Which imaging method should not be used for localizing retained gravel? Why?**
MRI. Ferromagnetic streak artifacts produced by gravel obscure accurate visualization and localization.

24. **What complications result from a retained foreign body?**
Allergic and inflammatory reactions, infection, and migration with neurovascular and/or tendinous injury.

25. **Describe the management of a suspected foreign body in the hand.**
Perform a careful history and focused physical exam, determining the type of foreign body and ruling out associated infection or injury to surrounding structures. Assess tetanus status and treat appropriately. Antibiotics may be considered. Multiplanar radiographs are obtained for all patients. If the foreign body is identified within the superficial tissues, removal can be performed in the emergency department. If the foreign body is deep, blind exploration is discouraged; the object should be removed surgically. A deep, nonradiopaque foreign body can be localized by a real-time high-resolution ultrasound, and needles can be placed to guide the surgeon to the embedded object(s) if necessary. Failure to identify a suspected retained foreign body is unlikely with this approach, but CT or MRI may be indicated in patients with persistent signs and symptoms.

26. **Describe the general principles for removing a fishhook.**
Fishhooks are almost always superficial. Removal techniques are aimed at minimizing additional tissue damage. The choice of how to remove a fishhook depends on the type of hook, its depth of penetration, and the local anatomy. More than one technique may need to be applied sequentially. Tetanus status should be assessed and treated appropriately, but prophylactic

antibiotics are generally not needed for uncomplicated cases. Retrograde removal is simplest and is usually attempted first. Eye protection should be worn.

27. **Describe the technique for retrograde removal of a fishhook.**
    Cleanse the area, attain adequate anesthesia, and remove all material attached to the hook. Uninvolved points should be cut off or taped. Retrograde removal requires applying downward pressure on the shank of the hook, advancing the hook slightly to disengage the barb(s), and finally backing the hook out along its path of entry. Any resistance should alert the physician to stop and consider other methods of removal. Additional techniques that may be applied involve inserting an 18-gauge needle along the shank of the hook, beveling toward the concave side to cover the barb during removal, or inserting a #11 blade along the same line, creating a path for the barb. Alternatively, a suture (or fishing line) can be tied around the center of the bend of the hook and pulled firmly in line with the shank to remove the hook while the eye (or distal end of the hook) is depressed simultaneously to free the barb(s) from the patient's tissue.

28. **When can anterograde removal be useful?**
    When a barbed tip is close to the skin, anterograde removal can be considered. The tip can be advanced through the skin and the tip and barb cut off before the rest of the hook is removed in a retrograde fashion. Alternatively, for hooks with multiple barbs, the eye can be cut off once the tip is exposed and the hook advanced and removed in an anterograde fashion.

29. **How should a nail (from a nail gun) be removed?**
    Many nails have barbs. Their presence may necessitate cutting off the head of the nail (with large pin cutters) and removing the nail in the same direction it entered. The wounds are left open. Appropriate tetanus prophylaxis and a short course of antibiotics are administered.

HIGH-PRESSURE INJECTION INJURY

30. **What is a high-pressure injection injury of the hand?**
    The accidental inoculation of foreign material or a substance into the hand and/or digits via a pressurized industrial device, such as a paint or spray gun.

31. **What are the three most common devices responsible for high-pressure injection injury?**
    Grease gun, spray gun, and diesel fuel injector.

32. **Name the most common site of a high-pressure injection injury.**
    The tip of the nondominant index finger.

33. **What is the most common injected material?**
    Grease.

34. **Summarize the pathophysiology of a high-pressure injection injury.**
    - **Direct pressure effect:** At normal temperatures, a pressure on the order of 7 atmospheres is necessary to penetrate the skin. Commercial paint spray guns operate at a pressure of 30 times this amount, and diesel engine fuel injectors may exceed this pressure by 100-fold.
    - **Chemical effect:** Solvents and oil-based paints produce an intense inflammatory response, leading to further vessel thrombosis and tissue necrosis. Water-based latex paints are less irritating. Grease produces a more chronic inflammatory response with characteristic "oleogranulomas" and fistulas. Cement produces both a chemical and thermal burn (exothermic production of calcium hydroxide).

35. **Describe the clinical presentation of a high-pressure injection injury of the hand.**
    The inciting injury can be quite painless, and the hand or finger may appear relatively innocuous. A small entrance wound exuding the injected material may be noted. Within hours, however, the digit or hand may become increasingly painful and swollen. Tissue distention and vessel compression manifest as pallor and loss of sensation, potentially resulting in ischemic necrosis and gangrene. The patient may become febrile as a result of systemic absorption of the material with leukocytosis, lymphangitis, and/or lymphadenitis (Fig. 12-2).

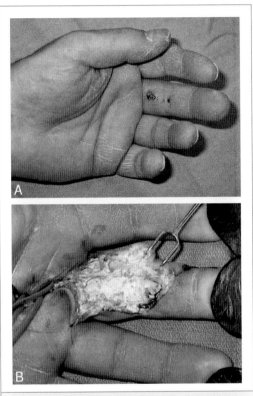

Figure 12-2. *A*, Small puncture wound over the proximal phalanx of the middle finger of a 52-year-old painter. This puncture wound could be mistaken for a minor injury. *B*, Subsequent surgical exploration revealing widespread infiltration of white paint. Note that the surgical debridement required was even more extensive than shown in this figure.

36. **What is the role of radiographs?**
    Radiographs detect fractures and determine the location of radiopaque material (e.g., lead-based paint, grease containing lead as a thickener), guiding the surgical approach.

37. **Which injection injuries require only conservative management?**
    Water, air, and low-volume vaccines (e.g., chicken vaccine) generally cause no serious damage. Medical treatment consisting of the administration of wide-spectrum antibiotics and tetanus prophylaxis is usually sufficient.

38. **How should a high-pressure injection of a toxic substance into the hand be managed?**
A high-pressure injection of a toxic substance into the hand is a surgical emergency that requires expedient recognition and treatment. Broad-spectrum antibiotics (including gram-positive and gram-negative coverage) and tetanus prophylaxis should be given. The surgical technique includes wide exposure, debridement of all devitalized tissue and injected material, preservation of neurovascular structures, irrigation, and placement of a drain or open packing. Serial debridements are performed as necessary. Early postoperative rehabilitation improves functional outcome. Delayed surgical treatment can increase tissue damage and may be contraindicated. Amputation may be necessary in late-presenting cases or after treatment failure.

39. **Is there a role for systemic corticosteroids in the treatment of high-pressure injection injuries?**
No. Although in theory corticosteroids may diminish the intense inflammatory response associated with high-pressure injection injuries, this effect has not been consistently demonstrated in clinical experience. It is believed that the risks of bacterial superinfection and compromised wound healing outweigh the proposed benefits.

KEY POINTS: FACTORS ASSOCIATED WITH POOR PROGNOSIS AND INCREASED RISK OF AMPUTATION

1. Higher injection pressure and volume of injected material
2. Injection into the finger region
3. Injection into the flexor tendon sheath
4. Toxicity of substance (solvents and thinners > oil based paints > oil > grease)
5. Spread of substance (related to injection site, pressure and volume, and viscosity)
6. Delay in treatment (more than 6 hours from time of injury leads to a higher amputation rate)

40. **How are high-pressure injection injuries best prevented?**
Most injuries occur from workers mishandling equipment toward the end of their shift. Education, the use of gloves resistant to high-pressure jets, and early treatment can reduce the incidence of and morbidity associated with a high-pressure injection injury.

41. **A 62-year-old farmer presents to the emergency department with a swollen, discolored, and painful long finger 9 hours after "doing some tractor maintenance." Physical examination demonstrates that the finger is bluish, cold, and insensate to the level of the base of the proximal phalanx. What is the recommended treatment?**
The history and findings are consistent with a high-pressure injection injury resulting in ischemia and digital necrosis. The recommended treatment is amputation. The level of amputation is determined by the extent of tissue damage and contamination. Ray resection with delayed primary closure in 48–72 hours may be preferable, but the level of amputation and timing of wound closure are best determined by the status of the wound.

Acknowledgment

This chapter is based extensively on the previous work by Peter J.L. Jebson, MD, and Dean S. Louis, MD. The authors gratefully acknowledge their contribution.

BIBLIOGRAPHY

1. Bray PW, Mahoney JL, Campbell JP: Sensitivity and specificity of ultrasound in the diagnosis of foreign bodies in the hand. J Hand Surg 20A:661–666, 1995.
2. Caddick JF, Rickard RF: A molten metal, high-pressure injection injury of the hand. J Hand Surg 29B:87–89, 2004.
3. Gammons M, Jackson E: Fishhook removal. Am Fam Physician 63(1):2231–2234, 2001.
4. Hatzifotis M, Williams A, Muller M, Pegg S. Hydrofluoric acid burns. Burns 30(2):156–159, 2004.
5. Mirzayan R, Schnall SB, Chon JH, et al: Culture results and amputation rates in high-pressure paint gun injuries of the hand. Orthopedics 24(6):587–589, 2001.
6. Schrijvers DL: Extravasation: A dreaded complication of chemotherapy. Ann Oncol 14(Suppl 3):iii26–iii30, 2003.
7. Tredget EE, Jarman A, Gabriel V: Management of thermal and extravasation injuries of the upper extremity. In Trumble TE (ed): Hand Surgery Update 3. Rosemont, American Society for Surgery of the Hand, 2003, pp 426–429.
8. Valentino M, Rapisarda V, Fenga C: Hand injuries due to high-pressure injection devices for painting in shipyards: Circumstances, management, and outcome in twelve patients. Am J Ind Med 43:539–542, 2003.
9. Vasilevski D, Noorbergen M, Depierreux M, Lafontaine M: High-pressure injection injuries to the hand. Am J Emerg Med 18:820–824, 2000.

FROSTBITE

Stephen D. Trigg, MD

1. **What is frostbite?**
   The thermal injury to living tissues that results from prolonged exposure to subfreezing temperatures.

2. **From experimental studies of frostbite, what appears to be the mechanism of injury?**
   Cellular death due to the direct formation of ice crystals within the extracellular fluid, which produces an osmotic gradient resulting in cellular dehydration and electrolyte imbalance. Blood flow in the affected tissues is also impaired by endothelial cell freezing, which produces vasoconstriction and blood shunting. Increased blood viscosity results in sludging and stasis.

3. **How is frostbite classified?**
   Frostbite is classified as either superficial or deep. In **superficial** frostbite, the injury is confined to the skin alone and is characterized by large clear blisters with minimal tissue loss. **Deep** frostbite results in subdermal tissue damage and is anesthetic after thawing. Hemorrhagic blisters after rewarming are associated with deep frostbite.

4. **What environmental and physiologic factors can influence frostbite injury?**
   - Increased wind velocity (wind chill) proportionally accelerates heat loss from exposed skin and is a major factor in the development of frostbite.
   - Moisture and skin contact with ice or metal are factors that encourage the onset of frostbite.
   - Physiologic factors include physical conditioning as well as associated injuries that may affect mobility.
   - Cold habituation also may influence the development of frostbite.
   - Alcohol abuse should be considered as a major associated physiologic factor. Intoxication impairs a person's judgment of the risks associated with being exposed to cold temperatures and his or her assessment of the need for protective clothing and avoidance of wind exposure.

5. **Which measures are included in first aid to a patient with frostbite?**
   - Assessing immediately for shock, dehydration, and the possibility of associated internal organ or major limb trauma
   - Resuscitation with warm intravenous fluids and insulation with blankets or a body warmer

6. **Describe the initial first aid management of the frost-bitten hand.**
   - Application of a nonrestrictive, bulky, conforming dressing
   - Strict avoidance of thawing and refreezing, which is paramount to minimize continued tissue injury
   - Avoidance of massaging the extremity to increase local blood flow

7. **After initial first aid, what is the accepted early management of a frost-bitten hand?**
   Rapid rewarming of the frozen part, with restoration of the normal core body temperature.

8. **Define *rapid rewarming*. How is it accomplished?**
   Immersion for 30 minutes in a water bath at 40–44°C until vasodilatation or flushing is noted about the frost-bitten areas.

9. **During the rapid rewarming process, what symptoms should be anticipated?**
   The thawing of frozen tissues is intensely painful. Therefore, administration of analgesic and sedative medication is often required.

10. **Which adjunct medications have been advocated in the initial treatment of frostbite?**
    - Pharmacologic agents that improve blood flow and/or minimize further tissue injury have been advocated. Examples include low-molecular-weight dextran, heparin, antifibrinolytic agents (urokinase and streptokinase), tissue plasminogen activator (TPA), arachidonic acid cascade inhibitors, antiprostaglandin inhibitors, and systemic corticosteroids.
    - TPA has shown promise in limited clinical studies to improve blood flow to frost-bitten digits through blood clot lysis within the injured vessels. Low-molecular-weight dextran and the anticoagulants streptokinase and urokinase have been used experimentally or in clinical studies that were too limited to prove their effectiveness.

11. **Describe the role of antibiotics in the acute management of frostbite.**
    Although their role is controversial, antibiotics should be given only if an observable infection develops.

12. **After rewarming of the frost-bitten hand, what physical changes should be anticipated around the injured tissues?**
    Surface blebs (blisters) usually appear within 6–24 hours following rewarming. White blisters should be debrided. Hemorrhagic blisters should be drained but left intact. Topical antibiotics (e.g., sulfadiazine [Silvadene]) should be applied if rupture of the blisters occurs.

13. **Which radiologic studies are useful in the early assessment and management of the frost-bitten hand?**
    A technetium-99 bone scan is an accurate and sensitive method to assess the return of digital blood flow.

KEY POINTS: FROSTBITE

1. The mechanism of injury in frostbite is direct cellular death from intracellular dehydration caused by the formation of ice crystals within extracellular fluid.
2. The severity of frostbite is worsened by exposed skin, skin contact with ice or metal, associated dehydration, exposure to increased wind velocity (wind chill), and alcohol use.
3. The frost-bitten hand should be treated with *rapid* rewarming in a water bath maintained at 40–44°C.
4. A technetium-99 bone scan is a useful diagnostic tool to assess the return of blood flow in the frost-bitten extremity.
5. Premature debridement or amputation of a frost-bitten part should be avoided until clear demarcation between viable and nonviable tissue has occurred.

14. **In addition to the protection and care of surface blebs, what other basic care should be implemented in the early management of the frost-bitten hand?**
    - Limb elevation to minimize edema
    - Hand therapy, including initiation of a gentle, active range-of-motion protocol if tolerated
    - Functional splinting to prevent digital and/or wrist contractures

15. **When is early surgical debridement of frost-bitten tissues indicated?**
    Premature debridement of frost-bitten tissue or amputation of an affected part before the demarcation of necrotic and viable tissues should be avoided because it may jeopardize the surgical outcome and appropriate amputation level. The process of demarcation may require several weeks to months. During this process, the patient may exhibit signs of emotional distress, embarrassment, and loathing of the appearance of the blackened, mummified digits. Therefore, it is important to educate the patient about the injury. In rare cases, a circumferential, constrictive eschar may form and impair blood flow to more distal tissues. When a constrictive eschar is identified, an escharotomy is indicated.

16. **Which surgical principles should be considered in the management of mummified, frost-bitten digits?**
    - Maximal length should be preserved to improve overall function.
    - Any asymmetric but viable skin for local flap coverage should be preserved. The use of a rotation, pedicle, or free flap should be considered for coverage of essential tissues (e.g., nerves, tendons).
    - A release and/or a partial resection of intrinsic muscles should be performed to correct contractures and improve motion.

17. **What is a common late sequela of frostbite in the hands of a skeletally immature patient?**
    Premature complete or asymmetric epiphyseal-plate closure, resulting in length and/or angular deformities.

18. **Which other symptoms are commonly associated with a late sequela of frostbitten hands?**
    - Cold sensitivity with or without vasospasm
    - Digit pain with cold exposure
    - Sympathetic nervous system hyperactivity, including vasoconstriction and hyperhidrosis

19. **A 15-year-old boy presents 4 months after a significant frostbite injury to his nondominant hand. He complains of distal interphalangeal (DIP) joint arthralgia and an associated loss of motion. Plain radiographs are normal. What is the cause of his complaints?**
    He is developing premature closure of the epiphyseal plates. Before the deformities are recognized, the patient often complains of joint pain and decreased motion. Radiographic evidence of an epiphyseal injury may not become obvious for as long as 6–12 months after the frostbite injury. The distal epiphyses are involved more frequently. It is difficult to predict which skeletal abnormalities may result based on the degree of soft tissue injury from frostbite. Significant epiphyseal injury has been reported after minor cases of frostbite.

20. **What other late radiographic findings may be noted in the patient with a frostbite injury?**
    Degenerative arthritis.

21. **What is the most common deformity encountered in the skeletally immature frostbitten hand?**
    Radial deviation of the small finger DIP joint.

22. **What other treatment modalities are used to treat postfrostbite cold intolerance?**
    - Calcium channel blockers and nonselective beta blockers have been used with variable success.
    - Digital and/or regional sympathectomies have been performed in limited studies with inconclusive results.

BIBLIOGRAPHY

1. Cauchy E, Marsigny B, Allamel G, et al: The value of technetium-99 scintigraphy in the prognosis of amputation in severe frostbite injuries of the extremities: A retrospective study of 92 severe frostbite injuries. J Hand Surg. 25A(5):969–978, 2000.
2. House JH, Fidler MO: Frostbite of the hand. In Green DP, Hotchkiss RN, Pederson WC (eds): Green's Operative Hand Surgery, 4th ed. New York, Churchill Livingstone, 1999, pp 2061–2067.
3. Mehta RC, Wilson MA: Frostbite injury: Prediction of tissue viability with triple-phase bone scanning. Radiology 170:511–514, 1989.
4. Page RE, Roberson GA: Management of the frostbitten hand. Hand 15:185–191, 1983.
5. Salimi Z, Vas W, Tang-Barton P, et al: Assessment of tissue viability in frostbite by 99mTc pertechnetate scintigraphy. Am J Roentgenol 142:415–419, 1984.
6. Skolnick AA: Early data suggest clot-dissolving drug may help save frostbitten limbs from amputation. JAMA 267:2008–2010, 1992.
7. Su CW, Lohman R, Gottlieb LJ: Frostbite of the upper extremity. Hand Clin 16:235–247, 2000.
8. Twomey JA, Heim-Duthoy K: Tissue plasminogen activator (TPA) in the treatment of severe frostbite. Proceedings of the American Burn Association, Albuquerque, NM, 1995.
9. Urschel JD: Frostbite: Predisposing factors and predictors of poor outcome. J Trauma 30:340–342, 1990.
10. Valnicek SM, Chasmar LR, Clapson JB: Frostbite in the prairies: A 12-year review. Plast Reconstr Surg 92:633–641, 1993.
11. Vogel JE, Dellon AL: Frostbite injuries of the hand. Clin Plast Surg 16:565–576, 1989.

CHAPTER 14

BURNS

Vera C. van Aalst-Barker, MD

1. **Define the four degrees of burn injury.**
   - A **first-degree burn** (partial thickness) is characterized by pain and erythema.
   - A **second-degree burn** (partial thickness) is characterized by redness, pain, and blistering and can be superficial or deep. Part of the dermis is viable.
   - A **third-degree burn** (full thickness) is "leather-like" with pale skin without capillary refill and can be insensate.
   - A **fourth-degree burn** (full thickness) involves not only the entire skin but also the underlying structures, such as subcutaneous fat, muscle, bone, and/or tendon.

2. **How does a (superficial) partial-thickness burn wound heal?**
   Healing occurs from dermal appendages (hair follicles, sweat glands), generally within 3 weeks, and may result in scarring and pigmentation changes.

3. **How does a full-thickness burn wound heal?**
   Because of the lack of dermal appendages, the wound must heal by contracture and epithelialization from the wound edges, resulting in unstable scars and contracture of joints. Excision of burn eschar and skin grafting is often the preferred treatment.

4. **What are the three zones of burn injury? What do they represent?**
   - **Zone of coagulation:** This is the central area that is composed of nonviable tissue.
   - **Zone of stasis:** This zone surrounds the zone of coagulation. In this area, the initial injury was not severe enough to result in permanent damage. Over the first 72 hours, microvascular sludging, thrombosis of vessels, and tissue edema can result in progressive tissue necrosis. In this zone, appropriate early intervention can have the most profound effect on minimizing the extent of the burn injury.
   - **Zone of hyperemia:** This is the outermost zone, and it is entirely viable.

5. **Name the frequently used topical agents used in burn care along with their advantages and disadvantages.**
   - **Silver sulfadiazine cream (Silvadene)** is bacteriostatic, has poor burn eschar penetration, and is not painful on application. The main side effect is neutropenia.
   - **Mafenide acetate cream (Sulfamylon)** is bacteriostatic, penetrates burn eschar, and has a broad spectrum of activity against gram-negative organisms. Disadvantages include pain on application and inhibition of carbonic anhydrase, leading to metabolic acidosis.
   - **Silver nitrate solution (0.5%)** has a broad spectrum of activity and is painless. Disadvantages include poor eschar penetration and leaching of sodium, potassium, chloride, and calcium across the eschar. It also stains the surrounding skin and environment (black).

6. **List the criteria for patient transfer to a burn center.**
   - Second- and/or third-degree burns covering > 10% of total body surface area (TBSA) in patients younger than 10 or older than 50 years
   - Second- and/or third-degree burns covering > 20% of TBSA in patients between 10 and 50 years old
   - Significant burns of the face, hands, perineum, genitalia, feet, or major joints

- Full-thickness burns covering > 5% TBSA in any patient
- Significant electrical or chemical injury
- Any inhalation injury, concomitant trauma, or significant preexisting medical disorder

7. **What are escharotomies? How are they performed when vascular compromise is detected in the burned hand?**
   In circumferential hand burns, vascular compromise develops due to tissue edema and swelling in an inelastic compartment (burned skin). Escharotomy is the release of burned skin by incising it with a scalpel or electrocautery. Escharotomies are best performed on the midlateral sides of the fingers (ulnar side for the index, middle, and ring fingers and radial side for the small finger and thumb) and via two longitudinal incisions over the dorsum of the hand.

8. **What is tangential excision? For which type of hand burn is it indicated?**
   Tangential excision is serial excision of burn eschar, parallel to the skin surface, down to viable tissue as indicated by bleeding. This technique is used for deep second-degree burns.

9. **How do you decide whether a hand burn needs skin grafting?**
   The decision whether to perform surgery is not always easy. Most burns display a mixture of depth of injury. It might take several days to determine the thickness of the burn, especially if the burn was caused by chemicals or grease. If the burned area is not expected to heal in 2 weeks, the wound should be debrided, and skin grafting should be performed as soon as possible.

10. **What types of skin grafts are used to cover the hand?**
    Unmeshed split-thickness skin grafts are used most often. Unmeshed skin grafts ("sheet grafts") provide superior appearance and less scar tissue than meshed grafts. If donor sites are limited, cadaveric dermis (Allograft) or dermal substitutes may be considered. Fibrin glue can be used to improve adherence of the skin graft and help with hemostasis.

11. **In what position should the hand be immobilized after skin grafting? How is this position maintained?**
    The "anticlaw" position (or "position of protection"). In this position, the wrist extension is between 10–30 degrees, the metacarpophalangeal (MCP) joints are flexed at 90 degrees, and the proximal interphalangeal (PIP) joints and distal interphalangeal (DIP) joints are extended. The thumb should be abducted and slightly extended. The "fist" position has been advocated to allow the greatest amount of skin to be grafted to the dorsum of the hand. The position is commonly maintained with splints. In deeply burned hands, Kirschner wires can be helpful to keep the joints appropriately positioned.

12. **What is the optimal management of a palmar burn?**
    Palmar burns are less common and occur mostly in children. The palmar skin is thick and rarely sustains a full-thickness burn. If palmar burns have not healed within 3–4 weeks, consideration is given to cover the wound with a full-thickness skin graft.

KEY POINTS: TREATMENT OF HAND BURN IN THE FIRST 24–48 HOURS

1. Clean the hand.
2. Check for pulses (with Doppler) every hour, especially in circumferentially burned hands.
3. Give tetanus prophylaxis as indicated.
4. Give adequate pain control.
5. Cover the hand with a topical agent and dressing, and change at least twice a day.
6. Elevate the burned extremity (above heart level).
7. Prescribe active and/or passive finger range of motion of the burned hand.

13. **What are the most common hand deformities following a burn injury?**
    - Web space syndactyly
    - Digital flexion contracture
    - Boutonnière deformity
    - Dorsal skin deficiency
    - Nail deformities

14. **How does a boutonnière deformity commonly develop after burn injury?**
    Boutonnière deformity (PIP joint flexion, DIP joint hyperextension) develops after injury to the central slip (or paratenon covering the central slip). The flexed position of the PIP joint allows the lateral bands to move volar to the PIP joint axis. The DIP joint undergoes compensatory hyperextension as the tension of the intrinsic tendons is transferred to the DIP.

15. **How do you prevent or reduce hand contractures after a burn injury?**
    - Mobilization of the hand before skin grafting and soon (after 5–7 days) postoperatively
    - Early grafting for deep burns
    - Splinting in the safe "position of protection"

16. **How do you prevent or reduce hypertrophic scars after burn injury?**
    - External compression (25 mmHg) with compression garments/gloves, splints, silicone sheets, or bandages (like Coban) for up to two years after the burn injury
    - Intralesion injection of triamcinolone

17. **In evaluating a patient with an electrical burn, what needs to be examined other than the point of contact?**
    - The patient must be examined for an exit point. Unless the exit point is broad enough to dissipate the energy, a second burn often results.
    - All extremities must be examined for possible compartment syndrome secondary to deep occult tissue damage.
    - Systemic effects should be kept in mind and considered in the examination. These include, but are not limited to, deeper than expected necrosis, myocardial damage, and renal failure.

18. **In electrical burns, which radiographic test can be used to determine tissue viability (including bone)?**
    A technetium-99 bone scan.

19. **What is the initial treatment for most chemical burns?**
    Unless a specific neutralizing agent is known, most chemical burns should be irrigated copiously with water as soon as possible.

20. **Which acid burn classically continues to damage tissue long after exposure? How should it be treated?**
    Hydrofluoric acid continues to burn until the acid is completely neutralized. All blisters should be removed. Liberal application of calcium gluconate gel (2.5%) should be instituted as soon as possible. If indicated, calcium gluconate can also be given subcutaneously (10%) or intra-arterially (1%, in D5W) *with caution*.

BIBLIOGRAPHY

1. Achauer BM: The burned band. In Green DP, Hotchkiss RN, Pederson W (eds): Green's Operative Hand Surgery, 4th ed. New York, Churchill Livingstone, 1999.
2. Barret JP, Desai MH, Herdon DN: The isolated palm burn in children: Epidemiology and long-term sequelae. Plast Reconstr Surg 105:581, 2000.

3. Burm JS, Oh SJ: Fist position for skin grafting on the dorsal hand II: Clinical use in deep burns and burn scar contractures. Plast Reconstr Surg 105:581, 2000.
4. Kasdan ML, Amadio PC, Bowers WH: Technical Tips for Hand Surgery, 1st ed. Philadelphia, Hanley Belfus, 1994.
5. Logsetty S, Heimbach DM: Modern techniques for wound coverage of the thermally injured upper extremity. Hand Clin 16:205–214, 2000.
6. Minanov OP, Peterson HD: Burn injury. In Georgiade GS, Riefkohl R, Levin LS (eds): Plastic, Maxillofacial, and Reconstructive surgery, 3rd ed. Baltimore, Williams & Wilkins, 1997.
7. Reilly DA, Gardner WL: Management of chemical injuries to the upper extremity. Hand Clin 16:215–224, 2000.
8. Robson MC, Barnett RA, Leith IO, Hayward PG: Prevention and treatment of postburn scars and contracture. World J Surg 16:87–96, 1992.
9. Salisbury RE: Acute care of the burned hand. In McCarthy JG, May JW Jr, Littler JW (eds): Plastic Surgery. Philadelphia, W.B. Saunders, 1990.
10. Tredget EE: Management of the acutely burned upper extremity. Hand Clin 16:187–203, 2000.

CHAPTER 15

COMPARTMENT SYNDROME
Joseph J. Thoder, MD

1. **Define a compartment syndrome.**
   A pathologic process in which elevated tissue pressure within a confined anatomic space compromises vascular flow at the capillary level, thus preventing adequate tissue oxygenation.

2. **What anatomic variable must exist for compartment syndrome to occur?**
   Compartment syndrome requires a closed anatomic space that most commonly involves the deep fascia. External compression, such as a tight postoperative dressing or cast, may also contribute to the development of compartment syndrome.

3. **Describe the physiology of compartment syndrome.**
   Although various circumstances and theories have been described, the arteriovenous gradient theory serves as the best model:

   $$LBF = P_a - P_v/R$$

   In this equation, LBF is local blood flow, Pa is local arterial pressure, Pv is local venous pressure, and R is local vascular resistance. As local tissue pressure rises with increased edema, the pressure on and within the local venous system also rises. The result is a decrease in the local arteriovenous gradient. Rising pressure within the closed space eventually inhibits small-vessel flow, resulting in irreversible damage to the contents of the compartment.

4. **What variable other than pressure is important in determining the effect of compartment syndrome?**
   Time. The effect of elevated pressure becomes progressively more severe with increasing duration of applied pressure.

5. **How are compartment syndromes classified?**
   Compartment syndromes can be subdivided into **acute** and **chronic** forms:
   - **Acute** compartment syndrome is a progressive process that commonly occurs following an inciting traumatic event (e.g., crush injury or vascular injury) and can lead to irreversible tissue changes.
   - **Chronic** exertional or recurrent compartment syndrome occurs in a compartment following exertion.

6. **What are the most common causes of an *acute* compartment syndrome?**
   Compartment syndrome can be precipitated by an increase in either extrinsic or intrinsic pressure:
   - **Extrinsic pressure** causes include crush injury, tight dressing/cast, tight surgical wound closure, inappropriate or extreme prolonged limb positioning, and thermal or electrical burn.
   - **Intrinsic pressure** causes include snakebite, reperfusion after limb ischemia > 4–6 hours, edema or hemorrhage, and spontaneous hematoma (coagulopathy).

7. **What are the long-term sequelae of untreated compartment syndrome?**
   - Myonecrosis
   - Muscle fibrosis
   - Muscle and soft tissue contractures

8. **What is Volkmann's ischemic contracture?**
   Volkmann described the sequelae of a compartment syndrome following a supracondylar fracture of the humerus in a child with the subsequent development of a severely contracted and functionless forearm. The forearm is typically fixed in pronation, the wrist is flexed, and the hand is postured is in the "claw" position with the metacarpophalangeal (MCP) joints hyperextended and the proximal interphalangeal (PIP) and distal interphalangeal (DIP) joints flexed. In addition to the fixed contractures, the hand is also insensate.

9. **What is the most common *and* most important symptom of an impending compartment syndrome?**
   Pain out of proportion to the injury.

10. **How is the diagnosis of compartment syndrome established?**
    Compartment syndrome is a **clinical diagnosis**. The diagnosis can be confirmed by actual measurement of compartment pressures. Clinical findings include a swollen, tense, tender compartment; sensory deficits; and eventual motor weakness. The most reliable finding is severe pain on passive stretch of involved muscles. A high index of suspicion is the best safeguard. Pain is classically persistent, progressive, and unrelieved by immobilization.

11. **What are "the five Ps of compartment syndrome"?**
    - **P**ain
    - **P**allor
    - **P**aresthesias
    - **P**aralysis
    - **P**ulselessness (which is usually not noted until late)

12. **How do you examine the various muscle compartments in the hand and forearm for a suspected compartment syndrome?**
    - **Flexor compartment of the forearm:** Pain on passive extension of the fingers
    - **Extensor compartment of forearm:** Pain on passive flexion of the fingers
    - **Intrinsic hand compartments:** Pain on passive abduction and adduction of the fingers or pain with interphalangeal joint flexion with the MCP joints held in extension
    - **Adductor compartment:** Pain on passive palmar abduction of the thumb
    - **Thenar muscles:** Pain on passive radial abduction of the thumb
    - **Hypothenar muscles:** Pain on passive extension and adduction of the small finger

KEY POINTS: COMPARTMENT SYNDROME

1. Compartment syndrome of the upper extremity occurs most commonly following a severe crush injury or limb revascularization.
2. Pain out of proportion to the injury is the most common finding in the patient with an actual or impending compartment syndrome.
3. Pain on passive stretch of the muscles in the involved compartment is the most sensitive clinical sign of a compartment syndrome.
4. The most common cause of compartment syndrome in the upper extremity of children is a supracondylar fracture of the distal humerus.

13. **What diagnostic test may be used to confirm a compartment syndrome?**
    Compartment pressure measurement with a commercial device.

14. **How is compartment pressure clinically useful?**
    A compartment pressure is particularly useful in patients with equivocal clinical findings and in patients who are unresponsive or unable to cooperate with the clinical examination (e.g., patients who are obtunded or have a closed head or spinal cord injury).

15. **Which pressure level is consistent with the diagnosis of compartment syndrome?**
    An absolute value of > 30–40 mmHg or a value within 20 mmHg of the diastolic blood pressure. The latter value is particularly useful in the hypotensive patient.

16. **What is the treatment of an acute compartment syndrome?**
    Emergent fasciotomy of all affected compartments.

17. **How is the wound managed after a fasciotomy?**
    Wound closure is delayed until all necrotic devitalized tissue has been debrided and the compartment pressures have normalized. Skin grafting may be necessary to achieve wound closure.

18. **When is a "prophylactic" fasciotomy indicated?**
    A prophylactic fasciotomy is indicated (i) after an arterial repair if ischemia time has exceeded 4–6 hours; (ii) in patients with equivocal clinical findings of compartment syndrome at the time of surgical stabilization of an associated fracture; and (iii) in patients with equivocal clinical findings who are unresponsive or unreliable or are unable to communicate or comply with a clinical examination.

19. **What is the role of a carpal tunnel release in the patient with a compartment syndrome in the upper extremity?**
    A carpal tunnel release should be performed routinely in every patient who undergoes a fasciotomy of the volar forearm and/or hand compartments.

20. **Name the compartments of the forearm.**
    Volar, dorsal, and mobile wad.

21. **How many separate hand compartments are there?**
    There are 10 hand compartments:
    - Four dorsal interosseous compartments
    - Three volar interosseous compartments
    - Adductor compartment
    - Thenar compartment
    - Hypothenar compartment

22. **Describe the surgical approach for compartment syndrome of the forearm.**
    Fasciotomy of the forearm is first performed by decompressing the volar compartment because the dorsal and mobile wad compartment pressures will often improve after a volar decompression. Although various skin incisions have been described, a curvilinear incision along the volar–ulnar border or the forearm is a simple approach with an acceptable cosmetic result. Curved incisions are necessary at the flexion creases of the wrist and elbow. The incision

begins medially about 2 cm proximal to the medial epicondyle and crosses the antecubital fossa obliquely to reach the volar aspect of the mobile wad. The incision is curved ulnarly to the midforearm and can be continued distally to incorporate a standard carpal tunnel release. The entire volar compartment must be released completely. The compartment pressure can be measured to confirm a satisfactory decompression. It is also feasible to release the dorsal compartment through the volar incision. Alternatively, a separate straight longitudinal dorsal incision can be used to decompress the dorsal compartment and mobile wad (Figs. 15-1 and 15-2.).

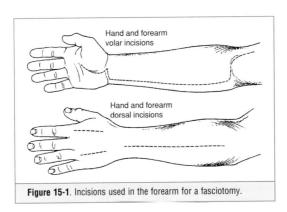

Figure 15-1. Incisions used in the forearm for a fasciotomy.

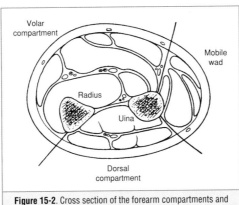

Figure 15-2. Cross section of the forearm compartments and the dissection planes.

23. **How should the various compartments of the hand be released?**
    The dorsal and volar interosssseous and adductor compartments are released through two dorsal longitudinal incisions over the second and fourth metacarpals; the thenar and hypothenar compartments are approached through longitudinal incisions over the radial border of the first metacarpal and the ulnar border of the fifth metacarpal, respectively; the carpal tunnel is released through a standard open approach utilizing a volar longitudinal incision between the thenar and hypothenar creases (Figs. 15-3 and 15-4).

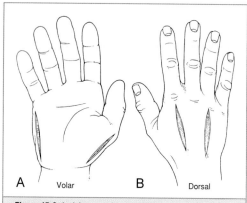

Figure 15-3. Incisions used in the hand for a fasciotomy.
A, Incisions for hypothenar and thenar compartments.
B, Incisions for interosseous compartments.

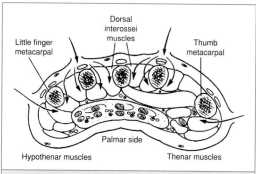

Figure 15-4. Cross section of the hand at the level of the metacarpals demonstrating the plane of dissection to decompress the interosseous, thenar, and hypothenar compartments.

24. **Describe the surgical approach used for a digital fasciotomy.**
    Midaxial incision on the ulnar border of the index, long, and ring fingers and on the radial border of the thumb and little finger.

BIBLIOGRAPHY

1. Amendola A, Twaddle B: Compartment syndromes. In Browner (ed): Skeletal Trauma. Philadelphia, W.B. Saunders, 1998, pp 365–389.
2. Chan PS, Steinberg DR, Pepe MD, Beredjiklian PK: The significance of the three volar spaces in forearm compartment syndrome: A clinical cadaveric correlation. J Hand Surg 23A:1072–1081, 1998.
3. Hastings H, Misamore G: Compartment syndrome resulting from IV regional anesthesia. J Hand Surg 12A:559–562, 1987.
4. Matsen FA III: Compartment syndrome: A unified concept. Clin Orthop 113:8–14, 1975.

5. Naidu SH, Heppenstall RB: Compartment syndrome of the forearm and hand. Hand Clin 10:13–27, 1994.
6. Pedowitz RA, Toutoughi FM: Chronic exertional compartment syndrome of the forearm flexor muscles. J Hand Surg 13A:694–696, 1988.
7. Rorabeck CH, Clarke KM: The pathophysiology of the anterior tibial compartment syndrome: An experimental investigation. J Trauma 3:299–305, 1978.
8. Rowland S: Fasciotomy. In Green DP, Hotchkiss R, Pederson W (eds): Green's Operative Hand Surgery, 4th ed. New York, Churchill Livingstone, 1999, pp 689–710.
9. Spinner M, Aiache A, Siver L, et al: Impending ischemic contracture of the hand. Plast Reconstr Surg 50:341–349, 1992.
10. Szabo RM, Gelberman RH: Peripheral nerve compression—Etiology, critical pressure threshold, and clinical assessment. Orthopedics 7:1461, 1984.
11. Volkmann R: Die ischaemischen muskellahmungen und Kontrakturen. Zentralbl Chir 8:801–803, 1881.

CHAPTER 16

DUPUYTREN'S DISEASE

Zvi Margaliot, MD, MS

1. **Who is credited with describing Dupuytren's disease?**
   In 1831, Baron Guillaume Dupuytren described the disease, its histopathology, and treatment. Sir Henry Cline, however, described the condition earlier in 1777, the year of Dupuytren's birth.

2. **Define Dupuytren's disease.**
   It is a benign fibroproliferative disorder of the superficial palmar and digital fasciae that results in the formation of subcutaneous nodules, cords, and digital flexion contractures. Dupuytren's disease belongs to the general class of diseases known as fibromatoses.

3. **Describe the classic clinical findings of Dupuytren's disease in the hand.**
   - Subcutaneous palmar nodules
   - Palmar and digital cords
   - Flexion contractures (Fig. 16-1)
   - Palmar pitting (dermal involvement)
   - Web space contractures
   - Adduction contracture of the thumb

4. **Which components of the palmar and digital fasciae are affected by Dupuytren's disease?**
   - The contracture affects the superficial (subcutaneous) fascia along the lines of natural tension (which are primarily longitudinal).
   - Diseased fasciae may form discrete nodules fixed to overlying skin or continuous cords.

5. **Outline the natural history of Dupuytren's disease.**
   In the majority of patients, the first presentation involves a palmar nodule, followed by the formation of cords and the gradual development of a flexion contracture that initially involves the MCP joint. The PIP joint is then affected, and the DIP joint is rarely affected. Disease progression is typically insidious over years or decades.

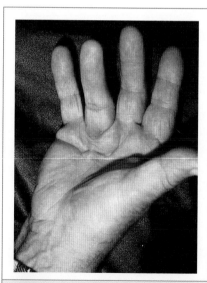

Figure 16-1. Clinical features of Dupuytren's disease. Note the pit in the palm of the hand, the web space contractures, and the cord formation with flexion contracture of the ring finger and, to a lesser degree, the small finger.

6. **Describe the epidemiology of Dupuytren's disease.**
   - It is primarily a disease of Caucasian males of northern European descent.
   - The disease is much less frequent but has been reported in Hispanics, Native Americans, African Americans, and Asians (in decreasing order).
   - Men develop the disease more frequently than women at a ratio of 10:1.
   - Incidence increases with age, but prevalence peaks in the sixth and seventh decades.

7. **What percentage of patients with Dupuytren's disease have a positive family history?**
   Between 10% and 44%.

8. **What is the anatomic distribution of Dupuytren's disease?**
   - The ring finger is involved in 70% of cases, followed by (in decreasing frequency) the small and long fingers, thumb, and index finger.
   - The right hand is slightly more frequently affected than the left, but bilateral involvement is seen in over half of patients.

9. **What is Dupuytren's diathesis?**
   A term used by Hueston to describe patients with:
   - Disease onset before 40 years of age
   - Disease presence in first- and second-degree relatives
   - Disease involvement in an ectopic site, such as the feet or genitalia
   - Bilateral disease at presentation
   - Involvement of the radial side of the hand

10. **What is the significance of Dupuytren's diathesis?**
    The presence of a diathesis is a marker for:
    - Aggressive disease with early progression to severe joint contractures
    - Higher risk of recurrence after surgery

11. **List the locations of ectopic disease outside the palmar fascia.**
    - Plantar (sole of foot) fibromatosis (Ledderhose's disease)
    - Dorsum of PIP joints (Garrod's knuckle pads)
    - Penile fibromatosis (Peyronie's disease)

12. **What is the etiology of Dupuytren's disease?**
    Although a genetic cause is suspected, the etiology of Dupuytren's disease is not known at this time.

13. **What is the proposed mode of inheritance of Dupuytren's disease?**
    Family history studies suggest an autosomal dominant inheritance with variable penetrance, but sporadic (noninherited) cases are more common. The responsible gene has not been isolated.

14. **What clinical conditions have a recognized association with Dupuytren's disease?**
    - Diabetes mellitus (up to 60% of diabetics develop a mild form of contracture)
    - Smoking and related conditions
    - Chronic pulmonary disease
    - Human immunodeficiency virus infection
    - Alcohol consumption (independent of the presence of liver disease)
    - Seizure disorders and the use of anticonvulsant medication

15. **What is the relationship between occupation and the development of Dupuytren's disease?**
    An association between occupation and Dupuytren's disease has not been definitively identified.

16. **List the differential diagnosis of Dupuytren's contracture.**
    - **Palmar nodule:** Epidermoid inclusion cyst, ganglion cyst, skin callus/hyperkeratosis, soft tissue giant cell tumor, soft tissue sarcoma
    - **Flexion contracture:** Locked trigger finger, intrinsic joint contracture, camptodactyly

17. **Describe the predominant cellular components of Dupuytren's fascia.**
    - Palmar nodules consist primarily of proliferating myofibroblasts (fibroblasts with contractile properties).
    - Cords are predominantly acellular bands of collagen.

18. **What is the role of myofibroblasts in the formation of Dupuytren's contractures?**
    There is evidence that myofibroblasts have a defect in the normal cellular response to applied tension, causing an exaggerated myofibril contraction. Cellular contraction, accompanied by deposition of extracellular collagen, leads to progressive contracture along the lines of applied tension.

19. **What is the primary biochemical abnormality present in diseased fasciae?**
    - An increase in the ratio of type III to type I collagen is present compared to normal fasciae.
    - Abnormalities in fibroblast growth factor (FGF), platelet-derived growth factor (PDGF), and transforming growth factor (TGF-beta) have been implicated.

20. **List the stages of Dupuytren's disease.**
    - **Proliferative:** Proliferation of myofibroblasts and formation of nodules
    - **Involutional:** Alignment of myofibroblasts along lines of tension and deposition of collagen
    - **Residual:** Acellular, thickened cords and contractures

KEY POINTS: DUPUYTREN'S DISEASE

1. Dupuytren's disease is a fibroproliferative disorder (fibromatosis) of the superficial palmar and digital fasciae.
2. The classic findings of Dupuytren's disease are palmar nodules, skin pitting, cords, and flexion contractures of the digits.
3. Dupuytren's disease most commonly involves the small and ring fingers.
4. Dupuytren's disease occurs most commonly in Caucasian males of northern European and Scandinavian ancestry.
5. The etiology of Dupuytren's disease is unknown, but a genetic role with an autosomal dominant, variable penetrance pattern of inheritance is recognized.
6. Dupuytren's disease is associated with diabetes mellitus, alcoholism, epilepsy, smoking, chronic pulmonary disease, and HIV infection.

21. **List the components of normal palmar and digital fasciae and their diseased counterparts.**
    The normal fasciae form bands that become cords in Dupuytren's disease (Table 16-1).

TABLE 16-1.	NORMAL PALMAR AND DIGITAL FASCIAE AND THEIR DISEASED COUNTERPARTS	
	Normal Fasciae	Diseased Fasciae
Palm	Pretendinous band	Pretendinous cord
Web	Natatory band	Natatory cord
Digit		Central cord
	Spiral band	Spiral cord
	Lateral digital sheet	Lateral cord, spiral cord
	Retrovascular band	Retrovascular cord

22. **Which four fascial structures contribute to the spiral cord?**
    - Pretendinous band
    - Spiral band
    - Lateral digital sheet
    - Grayson's ligament

23. **What is the clinical significance of the spiral cord?**
    With a progressive flexion contracture of the PIP joint, the spiral cord displaces the neurovascular bundle to the midline centrally, proximally, and superficially, placing it at risk for injury during surgery.

24. **List the possible clinical findings in the patient with Dupuytren's disease and the corresponding fascial structures.**
    See Table 16-2.

TABLE 16-2. CLINICAL FINDINGS IN DUPUYTREN'S DISEASE AND CORRESPONDING FASCIAL STRUCTURES	
Deformity	Responsible Fascial Structure
MCP flexion contracture	Pretendinous cord
MCP flexion contracture of small finger	Tendon of abductor digiti minimi
PIP flexion contracture	Central cord
	Spiral cord
	Lateral cord
DIP flexion contracture	Lateral cord
	Retrovascular cord
Web adduction contracture	Natatory cord
Thumb adduction contracture	Natatory cord
	Termination of transverse band of palmar aponeurosis

25. **What are the indications for surgical treatment of Dupuytren's disease?**
    The decision to proceed with surgical treatment should be based on the patient's functional limitations and the severity of joint contracture. MCP joint contracture >30 degrees or *any* PIP contracture is considered an indication for surgery. According to Hueston, the patient's inability to place the hand flat on a table is also considered an indication for surgical treatment (Hueston's tabletop test).

26. **What nonoperative treatment modalities are used to treat Dupuytren's disease?**
    - Enzymatic fasciotomy via the percutaneous injection of clostridial collagenase
    - Corticosteroid injection of nodules
    - Splinting

27. **Classify the surgical approaches used in treating Dupuytren's disease.**
    - **Fasciotomy:** Percutaneous or open
    - **Fasciectomy:** Limited (partial excision of diseased fasciae), regional (excision of diseased fasciae within an anatomic region), radical (removal of all diseased and normal fasciae), or dermofasciectomy (removal of diseased fasciae and overlying dermis and epidermis)

28. **What surgical approaches can be used in patients with established joint contracture?**
    - Capsulotomy
    - Implant arthroplasty
    - Arthrodesis
    - Amputation

29. **What structures may prevent the full restoration of digital extension after fasciectomy?**
    - Residual fascial cord
    - Skin deficiency
    - Shortened digital neurovascular bundles
    - Established periarticular capsular and/or ligamentous contracture
    - Flexor tendon sheath contracture

30. **List the techniques available for managing a skin deficiency following a fasciectomy.**
    - Primary closure with a local flap (e.g., multiple Z-plasties, V-Y advancement, cross finger flap)
    - Full-thickness skin grafting
    - Healing by secondary intention (McCash open palm and Burkhalter open finger techniques)

31. **What are the proposed advantages of the open palm technique?**
    Healing by secondary intention prevents hematoma formation with subsequent pain and edema, which are considered precursors to the development of reflex sympathetic dystrophy (RSD).

32. **What are the advantages of skin grafting following a palmar and/or digital fasciectomy?**
    The procedure of dermal excision and full-thickness skin grafting appears to prevent disease recurrence and is indicated for revision surgery and primary surgery in patients with Dupuytren's diathesis.

33. **Describe the postoperative management of a patient after palmar and digital fasciectomy.**
    1. A well-padded postoperative forearm-based splint with the fingers immobilized as well as the PIP joints in extension and MCP joints slightly flexed as tolerated

2. Early active motion, edema control, and wound management usually initiated within 5–10 days postoperatively
 3. Progressive static or dynamic splinting to maximize digital extension (as indicated):
    - The MCP joint contracture is not placed in a splint or is treated with short-term splinting only.
    - The PIP contracture requires 6 weeks of splint treatment followed by 3 months of nighttime splinting.
    - The open palm technique requires splinting until the wound has fully healed.
 4. Scar management (massage, desensitization) initiated after suture removal and continued for several months until wound maturation

34. **Which three components of the palmar fascia are involved in Dupuytren's contracture of the thumb?**
    - Pretendinous band
    - Natatory ligament
    - Transverse fibers of the palmar aponeurosis

35. **What are the common complications following a fasciectomy?**
    Intraoperative and early postoperative complications:
    - Arterial or nerve injury
    - Vascular compromise of the digit secondary to spasm or intimal injury following stretch and elongation
    - Hematoma formation
    - Skin loss
    - Delayed healing
    - Infection

    Late complications:
    - PIP joint stiffness
    - Dupuytren's flare
    - Recurrence of contracture
    - Complex regional pain syndrome

BIBLIOGRAPHY

1. Badalamente MA, Hurst LC: Enzyme injection as nonsurgical treatment of Dupuytren's disease. J Hand Surg 25A:629–636, 2000.
2. Bisson MA, Mudera V, McGrouther DA, Grobbelaar AO: The contractile properties and responses to tensional loading of Dupuytren's disease-derived fibroblasts are altered: A cause of the contracture? Plast Reconstr Surg 113:611, 2004.
3. Boyer MI, Gelberman RH: Complications of the operative treatment of Dupuytren's disease. Hand Clin 15:161–166, 1999.
4. Hurst LC, Badalamente MA: Nonoperative treatment of Dupuytren's disease. Hand Clin 15:97–107, 1999.
5. McCash CR: The open palm technique in Dupuytren's contracture. Br J Plast Surg 17:271–280, 1964.
6. McFarlane RM: On the origin and spread of Dupuytren's disease. J Hand Surg 27A(3):385–390, 2002.
7. Roush TF, Stern PJ: Results following surgery for recurrent Dupuytren's disease. J Hand Surg 25A:291–296, 2000.
8. Saar JD, Grothaus PC: Dupuytren's disease: An overview. Plast Reconstr Surg 106:125, 2000.
9. Saboeiro AP, Pokorny JJ, et al: Racial distribution of Dupuytren's disease in Department of Veterans Affairs patients. Plast Reconstr Surg 106:71, 2000.
10. Starkweather KD, Lattuga S, Hurst LC, et al: Collagenase in the treatment of Dupuytren's disease: An *in vitro* study. J Hand Surg 21A:490–495, 1996.
11. Strickland JW, Leibovic S: Anatomy and pathogenesis of the digital cords and nodules. Hand Clinics 7:645–657, 1991.

CHAPTER 17
CEREBRAL PALSY, STROKE, TRAUMATIC BRAIN INJURY, AND TETRAPLEGIA
Radford J. Hayden, PA-C

1. **What is cerebral palsy (CP)?**
   A nonprogressive injury to the developing central nervous system (CNS) that results in motor dysfunction, movement disorders, weakness, and functional impairment.

2. **How is CP classified?**
   According to its anatomic area of distribution. The different types are listed below.
   - **Diplegic:** Involvement of both lower extremities
   - **Hemiplegic:** Involvement of upper and lower extremities on same side
   - **Triplegic:** Involvement of one upper and both lower extremities
   - **Quadriplegic:** Involvement of all four limbs

3. **What are the treatment options for the CP patient with upper extremity involvement?**
   - Tone reduction modalities such as medications, injections, and/or neurosurgical procedures
   - Physical therapy
   - Splint/cast immobilization
   - Surgical reconstruction

4. **What are the most common pharmacologic agents used to treat spasticity in children with CP?**
   - **Baclofen:** Stimulates the GABA-b receptor and suppresses excitatory neurotransmitter release
   - **Diazepam (Valium):** Binds to GABA-a receptors and increases presynaptic inhibition
   - **Tizanidine:** Decreases presynaptic activity in excitatory neurons as an alpha–2 agonist
   - **Dantrolene:** Inhibits calcium release from the sarcoplasmic reticulum

5. **What are the most common pharmacologic agents used to treat dystonia (involuntary movement disorder) in children with CP?**
   - **Levodopa:** Acts as a dopaminergic
   - **Benzhexol:** Acts as an anticholinergic
   - **Baclofen:** Stimulates the GABA-b receptor and suppresses excitory neurotransmitter release
   - **Diazepam (Valium):** Binds to the GABA-a receptors and increases presynaptic inhibition

6. **What are the most common upper limb manifestations in the patient with spastic hemiplegia due to CP?**
   Internal rotation of the shoulder, elbow flexion, forearm pronation, wrist flexion, ulnar deviation, clenched fist or swan-neck deformities of the fingers, and a thumb-in-palm deformity.

7. **What is the preferred procedure for treating an elbow flexion contracture in the patient with CP?**
   Brachialis and biceps lengthening.

8. What are the surgical treatment options for the forearm pronation deformity in the patient with CP?
   - Pronator teres release
   - Pronator teres rerouting
   - Pronator quadratus release
   - Incision of the contracted lacertus fibrosis

9. Which upper extremity deformity is the greatest deterrent to good hand function in the patient with CP?
   Thumb-in-palm deformity.

10. What are the indications for surgical treatment in the CP patient with intrinsic spasticity of the hand?
    - Interference with hand function
    - Difficulties with personal hygiene

11. What are the surgical treatment options for the CP patient with intrinsic spasticity of the hand?
    - Intrinsic muscle origin release
    - Ulnar nerve motor neurectomy
    - Lateral band rerouting to correct hyperextension of the proximal interphalangeal (IP) joint (if the deformity is flexible) or tenodesis of the flexor digitorum superficialis (FDS)

12. What are the key clinical features that should be assessed in the CP patient with a thumb-in-palm deformity?
    - Spasticity of the flexors and adductors
    - Weakness of the extensor and abductor muscles
    - Hypermobility of the thumb metacarpophalangeal (MCP) joint
    - First web space contracture

13. What is the preferred surgical treatment for the thumb-in-palm deformity?
    - Release of the adductor pollicis and flexor pollicis brevis (FPB) origin or insertion can be performed. Release of the origin is recommended if the patient demonstrates voluntary motor control of the muscles. The insertion is released if the patient lacks voluntary control.
    - First web space release via a four-flap Z-plasty is also feasible.
    - Flexor pollicis longus (FPL) lengthening is possible.
    - Extensor pollicis longus (EPL) rerouting can be performed if good motor function is present.
    - If satisfactory function is not present, imbrication of the abductor pollicis longus and extensor pollicis brevis is performed to enhance extension and abduction of the thumb. However, if hypermobility of the thumb MCP joint is present, the hypermobility will be worsened with tendon imbrication, and an MCP joint arthrodesis or capsulodesis is usually necessary.

14. What operative procedures are recommended for the CP patient with digital flexor tendon tightness?
    - Flexor-pronator muscle slide
    - Superficialis-to-profundus (STP) tendon transfer
    - Fractional or Z-plasty tendon lengthening

15. When evaluating the CP patient with wrist and digital flexor tendon tightness, how does the surgeon decide which of the procedures described in question 14 to use?
    - The flexor-pronator muscle slide is used when significant finger function is not expected.
    - Fractional tendon lengthening is indicated in the functional hand. However, it should not be performed if the fingers cannot be fully extended with the wrist flexed. Z-plasty lengthening or

STP transfer should be considered in such patients. Z-plasty lengthening is used when fractional lengthening will not be sufficient or when severe contractures prohibit full digital extension with the wrist flexed.
- Z-plasty lengthening is typically used for correcting the thumb IP joint contracture.
- STP tendon transfer results in the greatest amount of lengthening. The procedure is performed when personal hygiene is the primary concern and function of the hand is not expected.

16. **A 9-year-old boy with CP and spastic hemiplegia presents with a palmar-flexed wrist and curling of the fingers into the palm. How is the surgeon able to determine whether the wrist flexion deformity is caused by spasticity or by contracture of the wrist or digital flexor tendons?**
The deformity may be caused by spasticity or contracture of the flexor carpi radialis, flexor carpi ulnaris, palmaris longus, and/or digital flexor tendons. To assess for the presence of digital flexor tendon tightness, passive wrist extension is compared with the fingers in full extension and flexion. If digital flexor tendon tightness is present, passive wrist extension is less when the fingers are extended than when they are held in flexion.

17. **What functional problems are encountered in the CP patient with weakness or absence of wrist extension?**
    - Lack of normal wrist extension may result from weakness of the wrist extensors, tightness or spasticity of the wrist flexor tendons, and/or volar capsular contracture.
    - The flexed wrist posture results in two functional problems. First, grip is weakened because of a loss of the mechanical advantage of the digital flexor tendons (i.e., loss of the tenodesis effect). Second, visualization of the fingers is obstructed by the flexed posture of the hand, resulting in compromised visual feedback.

18. **What are the major indications for performing surgery in the upper extremity of the CP patient?**
    - Functional impairment
    - Personal hygiene difficulties
    - Cosmesis

19. **How would you differentiate intrinsic spasticity from a fixed contracture in the CP patient with flexed MCP and extended IP joints?**
Perform a diagnostic nerve block of the motor branch of the ulnar nerve at the wrist.

20. **What are the key elements of the upper extremity physical examination in the patient with CP?**
    1. Determine whether voluntary motor activity is present.
    2. Differentiate muscle spasticity from muscle and/or joint contracture.
    3. Assess active and passive range of motion.
    4. Assess muscle strength.
    5. Examine the axilla, antecubital fossa, volar wrist flexion crease, palm, and interdigital spaces for hygiene problems.

21. **How long should a physician wait before recommending surgical reconstruction in a patient who has had a stroke?**
Spontaneous neurologic recovery occurs primarily within the first 6 months following a stroke. Generally, the patient is neurologically stable after 6 months, and thus decisions regarding surgical treatment should be deferred until then.

22. **How long should the physician wait before recommending surgical reconstruction in the tetraplegic patient?**
    Approximately 1 year. This allows spinal cord healing and physical and psychological stabilization.

23. **What are the most common musculoskeletal manifestations of spasticity following a stroke or traumatic brain injury (TBI)?**
    - Joint subluxation or dislocation
    - Contractures
    - Decubitus ulcers
    - Fracture malunion
    - Heterotopic ossification
    - Peripheral neuropathy

24. **How common is heterotopic ossification (HO) following TBI?**
    The incidence of periarticular HO following a TBI is approximately 11%. The hips are involved most commonly, followed by the shoulders and elbows. The incidence is dramatically increased (85%) in patients who have concomitant musculoskeletal injuries.

25. **What are the clinical findings in the patient with HO following TBI?**
    The characteristic features are warmth and erythema over the area, severe pain, and a rapidly decreasing range of joint motion.

26. **How is spasticity managed during the period of physiologic recovery?**
    - A combination of peripheral nerve blocks and splinting or casting can manage spasticity during this time period. Casting maintains muscle fiber length and diminishes muscle tone by decreasing sensory input. Serial casting is most effective when used to treat a contracture that has been present for less than 6 months.
    - Loss of joint motion secondary to spasticity may be effectively treated with a phenol or botulinum toxin peripheral nerve block. The block typically works for approximately 3–5 months.

27. **Describe the mechanism of action of a botulinum toxin block.**
    Botulinum toxin A is produced by *Clostridium botulinum*. The toxin inhibits calcium-mediated release of acetylcholine (Ach) at the neuromuscular junction by attaching to the presynaptic nerve terminal, thereby preventing the release of Ach from storage vesicles.

28. **How does a phenol block reduce spasticity?**
    Phenol denatures the protein membrane of a peripheral nerve when injected in or near a nerve bundle. However, it has to be in an aqueous concentration of 5% or more.

29. **A 28-year-old man presents with a traumatic brain injury and a thumb-in-palm deformity 3 months after a motorcycle accident. The deformity has persisted despite splinting and range-of-motion exercises. The palm of the hand is chronically macerated. What is the recommended treatment?**
    1. Nonoperative treatment in the form of a peripheral nerve block, which may be performed percutaneously or using the open (incisional) technique.
        - If excessive flexion of the IP joint is present, a selective motor block of the FPL is performed.
        - If the thumb is severely adducted, a block of the motor branches of the ulnar nerve at the level of the wrist is performed.
        - If the adduction deformity is also secondary to the thenar muscles, a block of the recurrent branch of the median nerve also should be performed.
    2. Casting or splinting is performed following a motor block(s) with intensive range of motion exercises.

30. What are the common clinical patterns of motor dysfunction in the upper extremity of the patient with a TBI?
    - Thumb-in-palm deformity
    - Clenched fist
    - Palmar-flexed wrist
    - Flexed elbow
    - Adducted/internally rotated shoulder

KEY POINTS: CP, STROKE, AND TBI

1. The major indications for performing surgery in the patient with CP are functional impairment, difficulties with personal hygiene, and cosmesis.
2. In the patient with CP, the thumb-in-palm deformity is the most limiting with respect to hand function.
3. Surgical treatment should not be performed earlier than 6 months after injury in the patient who has had a stroke.
4. Surgical reconstruction of the hand and upper extremity should not be performed until 1 year after injury in the tetraplegic patient.
5. Periarticular heterotopic ossification (HO) occurs in approximately 11% of all patients with a TBI.

31. Which specific muscle is usually the most spastic in the patient with a TBI and elbow flexion spasticity?
    The brachioradialis.

32. What is the pattern of muscle activity responsible for spastic flexion of the elbow in the patient with a TBI?
    Dynamic electromyographic (EMG) analysis has demonstrated a consistent pattern of muscle activity, including a normal phasic pattern of all three heads of the triceps, continuous spastic activity of the brachioradialis, and spasticity in one or both heads of the biceps.

33. In tetraplegic patients with a spinal cord injury (SCI) at the same level, what accounts for varying degrees of neurologic deficit?
    The size of the injured metamere (IM) is determined by both the direct physical neuronal damage and the amount of secondary swelling from:
    - Edema
    - Venous stasis
    - Spinal venous infarction
    - Compromise of arterial blood supply
    - Release of noxious substances, such as lipases, free radicals, and excitotoxins

34. Why is ulnar neuropathy at the elbow (cubital tunnel syndrome) common in stroke and brain-injured patients?
    Peripheral neuropathy involving the ulnar nerve at the elbow occurs in approximately 2–3% of all patients with TBI.
    The contributing factors are:
    - Chronic flexion of the elbow, which reduces the volume of the cubital tunnel
    - Leaning on the elbows to support the upper extremity, resulting in direct pressure on the ulnar nerve and cubital tunnel
    - Formation of heterotopic ossification

35. A 37-year-old woman presents with a wrist flexion deformity 9 months after a TBI. The wrist can be passively extended to 35 degrees short of neutral. Maceration is noted at the region of the wrist flexion crease. The patient has no volitional control of the limb. Radiographs reveal diffuse osteoporosis but no other abnormalities. What is the preferred treatment?
   - When the wrist flexion deformity is severe and hand function is absent or minimal, release of the wrist flexors combined with a wrist arthrodesis is recommended.
   - A carpal tunnel release is also usually performed.
   - Occasionally, a concomitant proximal row carpectomy may be necessary to help correct a severe flexion deformity.

36. In tetraplegic patients, when functional surgery of the upper limb is being considered, which classification should be referred to in the decision-making process?
   The International Classification for Surgery of the Hand in Tetraplegia (Table 17-1).

TABLE 17-1. INTERNATIONAL CLASSIFICATION FOR SURGERY OF THE HAND IN TETRAPLEGIA*

Sensibility O or Cu Group[†]	Motor Characteristics[‡]	Description Function
0	No muscle below elbow suitable for transfer	Flexion and supination of the elbow
1	BR	Flexion of the elbow
2	ECRL	Extension of the wrist (weak or strong)
3[§]	ECRB	Extension of the wrist
4	PT	Extension and pronation of the wrist
5	FCR	Flexion of the wrist
6	Finger extensors	Extrinsic extension of the fingers (partial or complete)
7	Thumb extensors	Extrinsic extension of the thumb
8	Partial digital flexors	Extrinsic flexion of the fingers (weak)
9	Lacks only intrinsics	Extrinsic flexion of the fingers
X	Exceptions	

O = oculo, Cu = cutaneous, BR = brachioradialis, ECRL = extensor carpi radialis longus, ECRB = extensor carpi radialis brevis, PT = pronator teres, FCR = flexor carpi radialis.
* The need for shoulder and triceps reconstruction is stated separately. This procedure may be required to make brachioradialis transfers function properly (see text).
[†] There is a sensory component to the classification. Afferent input is recorded, using the method described by Moberg, and precedes the motor classification. Both ocular and cutaneous input should be documented. When vision is the only afferent available, the designation is "oculo" (abbreviated O). Assuming there is a 10 mm or greater two-point discrimination in the thumb, the correct classification would be Cu; the Cu stands for the fact that the patient has useful cutaneous sensibility. If the two-point discrimination was greater than 10 mm (meaning inadequate cutaneous sensibility), the designation O would precede the motor group (example O 2).
[‡] Motor grouping assumes that all listed muscles are grade 4 or better, and a new muscle is added for each group; for example, a group 3 patient will have BR, ECRL, and ECRB rated at least grade 4.
[§] It is not possible to determine strength of ECRB without surgical exposure.

37. **What criteria are used to determine whether a functional surgical procedure should be performed in the patient who has had a stroke or TBI?**
   A **functional procedure** (tendon transfers, lengthening) should be used if the patient is able to:
   - Understand simple commands
   - Cooperate with postoperative rehabilitation
   - Remember what is taught during therapy sessions
   - Assimilate new activities into daily life
   - Move the extremity or selective muscles volitionally; has intact pain, light touch, and temperature sensation; and demonstrates spontaneous use of the extremity

38. **What criteria are used to determine whether a nonfunctional surgical procedure should be performed in the patient who has had a stroke or TBI?**
   A **nonfunctional procedure** (tenotomy, myotomy, neurectomy) should be used if the patient:
   - Does not obey commands
   - Cannot cooperate with rehabilitation
   - Has absent sensation and no spontaneous or volitional use of the extremity

39. **What is the preferred surgical approach in the patient with TBI and a "functional" spastic clenched fist?**
   The spastic clenched fist is quite common in patients with a TBI or stroke. The deformity results from the unmasking of the primitive grasp reflex.
   - If the patient has a "functional" deformity, fractional lengthening of the extrinsic digital flexor tendons (FPL, FDS, and flexor digitorum profundus [FDP]) is performed.
   - If the patient lacks volitional control and the hand is macerated and malodorous, more significant lengthening is required. An STP tendon transfer is performed in this clinical setting.

40. **A 19-year-old man with a TBI and a spastic clenched fist deformity undergoes STP tendon transfer. At the time of surgery, the finger deformity is corrected, but an intrinsic contracture is noted. How should the contracture be managed?**
   There is no dependable way to determine preoperatively the presence of significant intrinsic contracture in patients with a spastic clenched fist. Therefore, after the STP transfer, an ulnar motor neurectomy combined with a lateral band release is most effective in preventing recurrence. Because releasing the lateral band alone does not eliminate the volar interossei, a recurrent intrinsic-plus contracture can occur unless the interossei are denervated by the neurectomy.

41. **A 42-year-old man presents with a spastic thumb-in-palm deformity 10 months following a TBI. Examination demonstrates that the patient can follow commands and has volitional control of the limb with intact sensation. Dynamic EMG and diagnostic lidocaine injections demonstrate that he has volitional control of the thumb extensors and abductors. What is the recommended surgical approach?**
   - FPL lengthening to improve thumb extension, usually in conjunction with wrist or digital flexor tendon lengthening.
   - Thumb IP joint stabilization to improve stability during pinch.
   - Thenar muscle slide can be performed in the patient with a fixed adduction contracture. The origins of the flexor pollicis brevis, abductor pollicis brevis, and opponens pollicis are detached with preservation of the recurrent motor branch of the median nerve.
   - Adductor pollicis release from the third metacarpal.
   - First web space release.

BIBLIOGRAPHY

1. Botte MJ, Abrams RA, Bodine-Fowler SC: Treatment of acquired muscle spasticity using phenol peripheral nerve blocks. Orthopedics 18:151–159, 1995.
2. Botte MJ, Keenan MA, Gellman H, et al: Surgical management of spastic thumb-in-palm deformity in adults with brain injury. J Hand Surg 9A:174–182, 1989.
3. Coulet B, Allieu Y, Chammas M: Injured metemere and functional surgery of the tetraplegic upper limb. Hand Clinics 18:399–412, 2002.
4. Elliasson AC, Ekholm C, Carlstedt T: Hand function in children with cerebral palsy after upper limb tendon transfer and muscle release. Dev Med Child Neurol 40(9):612–621, 1998.
5. Garland DE: Surgical approaches for resection of heterotopic ossification in traumatic brain injured adults. Clin Orthop 263:59–70, 1991.
6. Hoffer MM, Lehman M, Mitani M: Surgical indications in children with cerebral palsy. Hand Clin 5:69–74, 1989.
7. Keenan MA, Ahearn R, Lazarus M, Perry J: Selective release of spastic elbow flexors in the patient with brain injury. J Head Trauma Rehabil 11(4):57–68, 1996.
8. Keenan MA, Haider TT, Stone LR: Dynamic electromyography to assess elbow spasticity. J Hand Surg 15A:607–614, 1990.
9. Kozin SH, Keenan MAE: Using dynamic electromyography to guide surgical treatment of the spastic upper extremity in the brain-injured patient. Clin Orthop 288:109–117, 1993.
10. Manske PR: Redirection of extensor pollicus longus in the treatment of spastic thumb-in-palm deformity. J Hand Surg 10A:553–560, 1985.
11. Manske PR, Langewisch KR, Strecker WB, Albrecht MM: Anterior elbow release of spastic elbow deformity in children with cerebral palsy. J Pediatr Orthop 21(6):772–777, 2001.
12. McDowell CL, Moberg EA, House JH: The second international conference on surgical rehabilitation of the upper limb in tetraplegia (quadraplegia). J Hand Surg 11A(4):604–608, 1986.
13. O'Flaherty S, Waugh MC: Pharmacologic management of the spastic and dystonic limb in children with cerebral palsy. Hand Clin 19:585–589, 2003.
14. Pomerance JF, Keenan MAE: Comprehensive correction of severe spastic flexion contractures in the nonfunctional hand. J Hand Surg 21B:828–833, 1996.
15. Revol M, Cormerais A, Laffont I, et al: Tendon transfers as applied to tetraplegia. Hand Clin 18:423–439, 2002.
16. Roth JH, O'Grady SE, Richards RS, Porte AM: Functional outcome of upper limb tendon transfers performed in children with spastic hemiplegia. J Hand Surg 18B:299–303, 1993.
17. Sakellarides HT, Mital MA, Matza RA, Dimakopoulos P: Classification and surgical treatment of the thumb-in-palm deformity in cerebral palsy and spastic paralysis. J Hand Surg 20A:428–431, 1995.
18. Van Heest AE, House J, Putnam M: Sensibility deficiencies in the hands of children with spastic hemiplegia. J Hand Surg 18A:278–281, 1993.
19. Van Heest AE, House J, Cariello C: Upper extremity surgical treatment of cerebral palsy. J Hand Surg 24(A):323–330, 1999.
20. Van Heest AE: Applications of Botulinum toxin in orthopaedics and upper extremity surgery. Tech Hand Upper Ext Surg 1:27–34, 1997.
21. Waters RL, Sie IH, Gellman H, Tognella M: Functional hand surgery following tetraplegia. Arch Phys Med Rehab 77:86–94, 1996.

CHAPTER 18
COMPLEX REGIONAL PAIN SYNDROMES
Dean S. Louis, MD, and Morton L. Kasdan, MD

1. **What was complex regional pain syndrome (CRPS) previously called?**
   CRPS is a clinical condition that is poorly understood. It was previously described as reflex sympathetic dystrophy (RSD), a term that has been discarded because it does not represent a proven etiology or pathophysiology. RSD will no doubt continue to be used as a diagnostic term because it has historical precedence and has been ingrained in the minds of health care professionals for the past four decades.

2. **What are the signs and symptoms of CRPS?**
   - Severe, burning, and constant pain
   - Swelling
   - Stiffness
   - Discoloration that varies from erythema to cyanosis and pallor due to differences in vasoconstriction and vasodilation
   - Vasomotor instability characterized by vasoconstriction and increased cold intolerance (Raynaud's-like phenomenon)
   - Pseudomotor disorder such as hyperhidrosis, usually in the first month but diminished in later stages
   - Hand temperature that increases with vasodilation and reduces with vasoconstriction
   - Osteoporosis indicated by "punch-out" lesions in the wrist and hand, also referred to as Sudek's atrophy
   - Trophic changes such as atrophy of the skin and subcutaneous tissue, causing the skin to have a shiny appearance with loss of skin creases
   - Thickening of and nodules along the palmar fascia

3. **What is the clinical presentation of a patient with CRPS?**
   A patient with CRPS usually presents with a history of trauma. The trauma can be trivial with no break in the skin or associated fractures. The presentation more commonly accompanies a major fracture, such as that of the distal radius. The previously noted clinical signs and symptoms are the usual progression.

4. **What are the three etiologic factors that must be simultaneously present to confirm a diagnosis of CRPS?**
   - Persistent painful lesions due to either trauma or disease
   - Diathesis typified by an emotionally fragile personality, causing the individual to exhibit fear, be suspicious of others, and become a chronic complainer who is insecure and unstable
   - Prolonged abnormal sympathetic reflex that can cause vasoconstriction, pallor, and possible localized tissue ischemia

5. **What major clinical observations and tests tend to confuse the diagnosis of CRPS?**
   Pain as a major presenting symptom that is characteristically out of proportion to the trauma involved (most common presentation). This is the major and distinguishing factor of this

condition. The other findings (listed in question 4) are important and support the diagnosis but are secondary to the initial complaint of pain and often occur later.

6. **What diagnostic tests are used to find objective evidence of CRPS?**
   A three-phase bone scan is used to highlight areas of increased radioisotope uptake. This nuclear medicine test consists of three phases:
   - **Phase 1** (early blood flow phase) shows a diffuse increase in perfusion throughout the hand and wrist.
   - **Phase 2** (blood pool or tissue phase) should demonstrate increased juxta-articular activity in all joints.
   - **Phase 3** (delayed metabolic phase) monitors uptake of radionuclide 2–4 hours after injection. This phase is the most important diagnostically with respect to the detection of CRPS (a sensitivity of 96% and a specificity of 98%). Such an accurate test to aid in the diagnosis of CRPS facilitates early treatment, allows monitoring of response to therapy, and helps to evaluate new treatment regimens.

7. **What other tests may be useful in diagnosing CRPS?**
   - Bilateral comparison radiographs that are used to look for osteoporosis (can be confused with disuse demineralization)
   - Isolated cold stress test, which is highly sensitive to vasomotor disturbances that occur in 80–90% of patients with CRPS

8. **How is CRPS treated?**
   CRPS should initially be treated with hand therapy to relieve symptoms or, at a minimum, to gauge the patient's motivation. For example, splinting may be used to decrease motions that stimulate pain. Splinting, however, should be kept to a minimum. Further treatments include elimination of the sympathetic reflex with sympatholytic drugs, such as phentolamine, local anesthetic blocks, and stellate ganglion blocks. A sympathectomy and possible referral to a psychologist/psychiatrist can also be forms of treatment.

9. **What are the most important predictors of functional recovery and pain relief in CRPS?**
   - Early recognition and treatment of CRPS
   - A compliant and highly motivated patient

10. **Describe the stress-loading approach for managing CRPS**
    The patient performs stressful exercises with increasing frequency and duration. Although pain and swelling initially may increase, they generally subside within a few days. The program consists of a scrubbing exercise, in which the patient scrubs the floor with a scrub brush using the affected hand, and a carrying exercise, in which the patient is told to carry a briefcase or purse in the affected hand with the arm extended. Physical therapy is reported to be cost effective compared to invasive procedures.

KEY POINTS: COMPLEX REGIONAL PAIN SYNDROME (CRPS)

1. The usual symptoms of CRPS are pain, swelling, stiffness, and discoloration.
2. The diagnosis must be confirmed before labeling the patient with CRPS.
3. A misdiagnosis (the wrong diagnostic label) can produce the nocebo effect.
4. Treatment should include removal of the inciting pathology, such as a fracture compressing a peripheral nerve.

11. **List the management techniques that are important in caring for patients with CRPS and chronic pain.**
   First and foremost, be sure of the diagnosis. A factitious disorder with a dysfunctional posture can be confused with CRPS. Diagnostic imaging is very helpful. The physician should show respect for the patient and his or her pain and encourage verbal rather than somatic expressions of feelings. The physician should give physiologic explanations for the patient's symptoms and also explain how these symptoms can be caused by emotions. It is important to show confidence in the treatment and to get the patient and the patient's significant others actively involved in the treatment. Finally, the physician should see the patient regularly to monitor treatment and progress. *It is important to set definite and time-limited goals.*

12. **Are pharmacologic agents, such as phentolamine (an alpha-adrenergic antagonist), useful by themselves to diagnose a sympathetic cause of CRPS?**
   Sympathetic blockage of alpha-adrenergic receptors using phentolamine or similar pharmacologic agents may not be useful in diagnosing a painful sympathetic origin because of the placebo effect. Verdugo and Ochoa found no difference in the diminished pain response in patients infused with phentolamine compared with patients infused with saline.

13. **Give examples of common misdiagnoses of CRPS.**
   - **Diabetic neuropathy:** Patients with diabetic complications may present with chronic painful polyneuropathies that have symptoms similar to CRPS, including cutaneous hyperalgesia, hyperthermia, palmar fascia nodules, and stiff hands.
   - **Spinal cord disease:** Various spinal cord lesions or central nervous system (CNS) degenerative diseases and benign tumors may present as a persistent burning pain in the upper extremities that are hypersensitive to touch. Other symptoms similar to CRPS may include hyperthermia and discoloration.
   - **Psychogenic problems:** Patients with psychological problems may be the most common victims of misdiagnosed CRPS. Factitious disorders often present with a dysfunctional posture.
   - **Others:** Patients with hypertrophic arthritis, psoriatic arthritis, carpal tunnel syndrome, alcoholism, epilepsy, or peripheral vascular disease may experience increased fibrosis and stiffness after injury or surgery.

14. **Why does cigarette smoking put patients at risk for developing CRPS and aggravate existing CRPS?**
   Smoking can stimulate the sympathetic nervous system centrally, causing increased plasma levels of epinephrine and norepinephrine. Peripherally, smoking decreases blood flow to the extremities by causing vasoconstriction, which leads to ischemia.

15. **Why do psychological conditions make patients more susceptible to developing CRPS?**
   Patients with CRPS are often fearful, suspicious, emotionally labile, dependent, passive-aggressive, insecure, and unstable. In addition, they often display more somatization, depression, interpersonal sensitivity, and anxiety than healthy people. The psychological profile of patients with CRPS can cause problems with respect to accurate diagnosis and prompt, appropriate treatment. Malingerers can easily feign CRPS. Psychological manifestations of depression include constant, unrelenting, and unbearable pain similar to that in patients with CRPS. Psychological conditions such as anxiety and depression also can impede the treatment of CRPS. These conditions exaggerate and prolong vasoconstriction by cortical and subcortical stimuli. Furthermore, excessive anxiety can produce somatic changes, largely through the sympathetic branch of the autonomic nervous system.

16. **Define *causalgia*.**
    Dorland's Medical Dictionary defines *causalgia* as a "burning pain, often accompanied by trophic skin changes, due to injury of a peripheral nerve, particularly the median nerve."

17. **What is the difference between CRPS types 1 and 2?**
    The clinical characteristics of CRPS type 1, for the most part, are the same as seen in CRPS type 2. The most significant difference between type 1 and type 2 is that, by definition, type 2 occurs following a known peripheral nerve injury, whereas type 1 occurs in the absence of any known nerve injury.

BIBLIOGRAPHY

1. American Society for Surgery of the Hand: Reflex sympathetic dystrophy. In Regional Review Courses in Hand Surgery. American Society for Surgery of the Hand, Rosemont, IL, 1996, pp 1–12.
2. Atkins RM: Complex regional pain syndrome. J Bone Joint Surg 85B1:100–106, 2003.
3. Janig W, Blumberg H, Boas R, Campbell J: The reflex sympathetic dystrophy syndrome: Consensus statement and general recommendations for diagnosis and clinical research. In Proceedings of the 6th World Congress on Pain, Elsevier Science, New York, 1991.
4. Koman LA, Smith TL, Smith BP, et al: The painful hand. Hand Clin 12:757–764, 1996.
5. Ochoa J: Reflex sympathetic dystrophy (RSD): A tragic error in medical science. Hippocrates' Lantern 3:1–6, 1995.
6. Ochoa JL: Truths, errors, and lies around "reflex sympathetic dystrophy" and "complex regional pain syndrome." J Neurol 246:875–879, 1999.
7. Severens JL, Oerlemans HM, Weegels AJPG, et al: Cost-effectiveness analysis of adjuvant physical or occupational therapy for patients with reflex sympathetic dystrophy. Arch Phys Med Rehabil 80:1038–1043, 1999.
8. Stutts JT, Kasdan ML, Hickey SE, Bruner A: Reflex sympathetic dystrophy: Misdiagnosis in patients with dysfunctional postures of the upper extremity. J Hand Surg 25A:1152–1156, 2000.
9. Verdugo RJ, Ochoa JL: Sympathetically maintained pain. Phentolamine block questions the concept. Neurology 44:1003–1009, 1994.

CHAPTER 19

KIENBÖCK'S DISEASE

Asheesh Bedi, MD

1. **What is Kienböck's disease?**
   Kienböck's disease, or lunatomalacia, is a clinical condition characterized by wrist pain, weakness, and loss of motion secondary to osteonecrosis of the lunate (Fig. 19-1).

2. **What is the most recent theory regarding the cause of Kienböck's disease?**
   Most theories have focused on avascular necrosis of the lunate secondary to compromised arterial supply and mechanical trauma. Recent theories have focused on the role of venous congestion and elevated intraosseous pressure. The exact cause and pathogenesis remain obscure.

3. **What is the natural history of Kienböck's disease?**
   The natural history of Kienböck's disease is uncertain. A deterioration in wrist range of motion and radiographic progression of the disease process with preservation of wrist function have been documented. The correlation between symptoms and radiographic changes is poor.

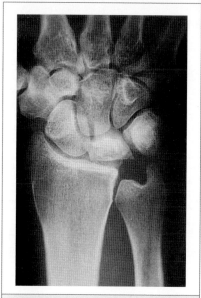

Figure 19-1. Fragmentation and collapse of the lunate consistent with Kienböck's disease.

4. **Describe the typical presentation of Kienböck's disease.**
   The typical patient is a young adult male with a painful, stiff, and weak wrist that may or may not occur following trauma. Dorsal wrist swelling may be present. Range of motion is typically limited. There is tenderness to palpation in the dorsum of the wrist directly over the lunate. Grip strength may be diminished.

5. **List the three basic patterns of *extra*osseous blood supply to the lunate.**
   - Single volar or dorsal vessel supplying the entire lunate
   - Several vessels, volar and dorsal, without a central anastomosis
   - Several vessels, volar and dorsal, with central anastomoses

6. **List the three basic patterns of *intra*osseous blood supply (Fig. 19-2).**
   - Y pattern (59% of patients)
   - I pattern (31%)
   - X pattern (10%)

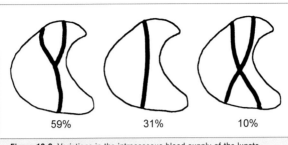

Figure 19-2. Variations in the intraosseous blood supply of the lunate.

7. **How does ulnar variance relate to Kienböck's disease?**
   Ulnar variance refers to the relationship between the distal articular surfaces of the radius and ulna. *Ulnar minus* is the term used when the articular surface of the distal radius is longer than that of the distal ulna. In 1928, Hulten noted that 74% of wrists with Kienböck's disease had an ulnar minus alignment, whereas only 23% of asymptomatic controls were ulnar minus. This variation in anatomy is thought to **increase load and shear stresses** on the lunate and predispose it to the development of Kienböck's disease. It is theorized that the increased load and shear stresses on the lunate, which result from a short ulna, predispose the lunate to injury during repetitive microtrauma.

8. **What radiographic view is required to determine the amount of ulnar variance?**
   Posteroanterior (PA) view of the wrist in neutral rotation.

9. **Define the *lunate at risk*.**
   The lunate at risk is associated with:
   - A short ulna (ulna minus)
   - A single *extra*osseous nutrient vessel
   - Poor *intra*osseous vascular anastomoses

   A combination of these features is believed to predispose the lunate to the development of Kienböck's disease.

10. **What are the diagnostic criteria for Kienböck's disease?**
    Kienböck's disease is a radiographic diagnosis based on the characteristic changes of increased density of the lunate (sclerosis) with cystic changes, fragmentation, collapse, and arthritic changes.

11. **What imaging studies are helpful when evaluating the patient with suspected Kienböck's disease?**
    - PA and lateral plain radiographs of the wrist are used to stage the disease and determine ulnar variance.
    - A technetium-99 bone scan, computed (CT) or conventional polytomography (Fig. 19-3), or magnetic resonance imaging (MRI)

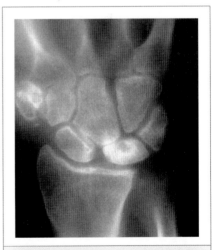

Figure 19-3. Conventional tomography of the wrist. This image demonstrates sclerosis and early collapse involving the radial half of the lunate.

may be used to identify early Kienböck's disease before the development of classic radiographic findings.

12. **What are the characteristic MRI findings noted in patients with Kienböck's disease?**
    Diffuse decreased signal throughout the lunate on T1- and T2-weighted images (Fig. 19-4).

13. **Following an MRI of the wrist, which clinical conditions must be differentiated from conditions indicating Kienböck's disease?**
    - Ulnocarpal abutment should not be mistaken for Kienböck's disease.
    - The cystic and avascular changes in the proximal ulnar aspect of the lunate as seen in ulnocarpal abutment should not be confused with the diffuse changes found in Kienböck's disease.

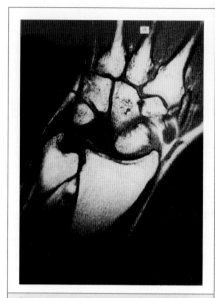

Figure 19-4. Early Kienböck's disease detected on an MRI. Note the loss of marrow signal on this T1-weighted image.

14. **Describe the radiographic features of Kienböck's disease.**
    The Lichtman classification of Kienböck's disease (Fig. 19-5) is as follows:
    - **Stage I:** Normal radiographs, signal intensity changes on MRI
    - **Stage II:** Lunate sclerosis, possible fracture lines
    - **Stage III:** Fragmentation and collapse of the lunate
        - □ **IIIA:** Normal carpal alignment and height
        - □ **IIIB:** Fixed scaphoid rotation, proximal capitate migration, loss of carpal height, radioscaphoid angle >60 degrees
    - **Stage IV:** Lunate collapse with radiocarpal or midcarpal arthrosis

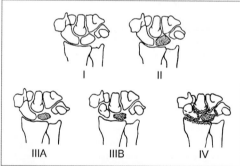

Figure 19-5. Lichtman classification of Kienböck's disease. (From Palmer AK, Benoit MY: Lunate fractures: Kienböck's disease. In Cooney WP, Linscheid RL, Dobyns JH [eds]: The Wrist: Diagnosis and Operative Treatment. St. Louis, Mosby, 1988.)

15. **How does the radiographic stage guide the treatment of Kienböck's disease?**
    - **Stage I:** Cast immobilization
        - □ **If ulna minus:** Radial shortening
        - □ **If ulna plus:** Vascularized bone graft, distal radius wedge or dome osteotomy, or capitate shortening

- **Stage II–IIIA**
  - ☐ **If ulna minus:** Radial shortening
  - ☐ **If ulna plus:** Vascularized bone graft, distal radius wedge or dome osteotomy, or capitate shortening
- **Stage IIIB:** Proximal row carpectomy or intercarpal arthrodesis
- **Stage IV:** Proximal row carpectomy or wrist arthrodesis

16. **What is the rationale for performing a joint-leveling procedure in a patient with Kienböck's disease?**
    To decrease the load across the lunate, providing pain relief, preventing further collapse and deterioration, and preventing development of arthrosis.

17. **Why is shortening of the radius preferred to lengthening of the ulna?**
    - **Shortening of the radius** is technically easier to perform and has a lower incidence of delayed union or nonunion (Fig. 19-6).
    - **Ulnar lengthening** requires the insertion of a bone graft with the associated morbidity and has a higher rate of delayed union or nonunion.

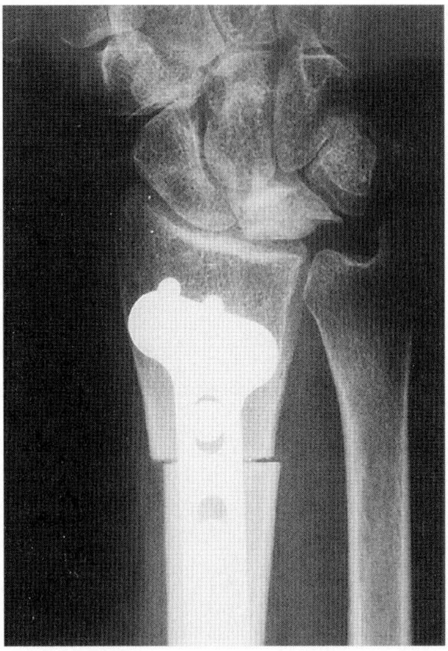

Figure 19-6. Radial shortening osteotomy for stage II Kienböck's disease.

18. **What is the rationale for performing a capitate shortening in a patient with Kienböck's disease?**
    Capitate shortening with or without a capitohamate arthrodesis is advocated in patients with ulna positive variance. Capitate shortening decreases the load on the lunate by 66%. However, the load at the scaphotrapezial joint is increased by approximately 150%. Additional concerns about a capitate shortening osteotomy include the development of nonunion and avascular necrosis.

19. **A 9-year-old boy presents with dorsal wrist pain. Radiographic findings are consistent with stage II Kienböck's disease. How should the boy be managed?**
    Kienböck's disease is *rare* in children. It is recommended that children under the age of 12 be treated with 6–12 weeks of cast immobilization. This approach is not as successful in older children; children aged 12 or older should be treated like adults.

KEY POINTS: KIENBÖCK'S DISEASE

1. Kienböck's disease, or avascular necrosis of the carpal lunate, is characterized by wrist pain, loss of motion, and weakness.
2. Kienböck's disease is associated with ulnar minus variance.
3. Plain radiographs may be normal in the early stages of Kienböck's disease.
4. The MRI findings of ulnocarpal abutment should be carefully differentiated from those of Kienböck's disease.
5. Kienböck's disease is rare in children.

20. **What is the role of silicone replacement arthroplasty in the treatment of Kienböck's disease?**

   There is no current indication for the use of a silicone implant because of implant collapse and fragmentation and the secondary development of particulate synovitis.

21. **Which vascularized bone graft is preferred in the treatment of Kienböck's disease?**

   The 4,5 intercompartmental supraretinacular artery (ICSRA) bone graft.

BIBLIOGRAPHY

1. Allan CH, Joshi A, Lichtman DM: Kienböck's disease: Diagnosis and treatment. J Am Acad Orthop Surg 9:128–136, 2001.
2. Delaere O, Dury M, Molderez A, Foucher G: Conservative versus operative treatment for Kienböck's disease. A retrospective study. J Hand Surg 23B:33–36, 1998.
3. Goldfarb CA, Hsu J, Gelberman RH, Boyer MI: The Lichtman classification for Kienböck's disease: An assessment of reliability. J Hand Surg 28A:74–80, 2003.
4. Iwasaki N, Minami A, Oizumi N, et al: Radial osteotomy for late-stage Kienböck's disease. Wedge osteotomy versus radial shortening. J Bone Joint Surg 84B:673–677, 2002.
5. Jafarnia K, Collins ED, Kohl HW III, et al: Reliability of the Lichtman classification of Kienböck's disease. J Hand Surg 25A:529–534, 2000.
6. Jensen CH: Intraosseous pressure in Kienböck's disease. J Hand Surg 18A:355–359, 1993.
7. Kaarela OI, Raatikainen TK, Torniainen PJ: Silicone replacement arthroplasty for Kienböck's disease. J Hand Surg 23B:735–740, 1998.
8. Kim TY, Culp RW, Osterman AL, Bednar JM: Kienböck's disease in children. Jefferson Orthop J 25:53–57, 1997.
9. Kristensen SS, Thomassen E, Christensen F: Kienböck's disease: Late results by nonsurgical treatment. A follow-up study. J Hand Surg 11B:422–425, 1986.
10. Schiltenwolf M, Martini AK, Mau HC, et al: Further investigations of the intraosseous pressure characteristics in necrotic lunates (Kienböck's disease). J Hand Surg 21A:754–758, 1996.
11. Sheetz KK, Bishop AT, Berger RA: The arterial blood supply of the distal radius and ulna and its potential use in vascularized pedicled bone grafts. J Hand Surg 20A:902–914, 1995.
12. Shin AY, Bishop AT: Treatment of Kienböck's disease with dorsal distal radius pedicled vascularized bone grafts. Atlas Hand Clin 4:91–118, 1999.
13. Tamai S, Yajima H, Ono H: Revascularization procedures in the treatment of Kienböck's disease. Hand Clin 9:455–466, 1993.
14. Zelouf DS, Ruby LK: External fixation and cancellous bone grafting for Kienböck's disease: A preliminary report. J Hand Surg 21A:746–753, 1996.

CHAPTER 20

WRIST ARTHROSCOPY
John V. Hogikyan, MD

1. **What are the most common indications for wrist arthroscopy?**
   - Persistent wrist pain
   - Intercarpal ligament tear
   - Tear or avulsion of the triangular fibrocartilage complex (TFCC)
   - Extrinsic radiocarpal wrist ligament injury
   - Osteochondral injury
   - Intra-articular loose body removal
   - Excision dorsal wrist ganglion

2. **How are the portals for radiocarpal wrist arthroscopy identified?**
   By the intervals between the six dorsal wrist extensor compartments through which they pass (e.g., portal 3–4 is between the tendons of the third and fourth dorsal compartments) (Fig. 20-1).

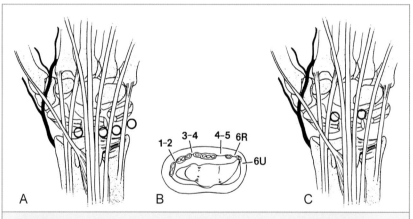

Figure 20-1. Portals used for wrist arthroscopy. *A*, Diagram of portals. *B*, Nomenclature used for identification of arthroscopic portals (entry sites). The nomenclature is based on the topographic anatomy of the extensor compartment; radiocarpal ports are named using the number of the adjacent extensor compartments and the letter R (for radial) or U (for ulnar). *C*, Midcarpal portals 0.75–1.25 cm distal to the radiocarpal joint. These portals are either radial or ulnar to the capitate. (With permission from Koman LA, Poehling GG, Toby EB, Kammire G: Chronic wrist pain: Indications for wrist arthroscopy. Arthroscopy 6(2):116–119, 1990.)

3. **What are the five portals used for radiocarpal arthroscopy?**
   The portals are 1–2, 3–4, 4–5, 6R, and 6U. *R* and *U* refer to the portals passing radial or ulnar, respectively, to the extensor carpi ulnaris (ECU) tendon in the sixth dorsal compartment.

4. **Which portals are most commonly used?**
   The 3–4 portal, which is used for visualization (i.e., placement of the arthroscope), and the 4–5 portal, which is used for instrumentation.

5. **Which structures are at risk when the 1–2 radiocarpal portal is used?**
   - Radial artery, as it passes through the anatomic snuffbox
   - Superficial radial and lateral antebrachial cutaneous nerves

6. **Which anatomic considerations should be kept in mind in establishing the 3–4 radiocarpal portal?**
   - The joint should be preinflated with saline or a local anesthetic to create an identifiable site for insertion of the trochar.
   - Injury to the extensor pollicis longus tendon should be avoided.
   - During introduction, the trochar should be angled about 10 degrees proximally in the sagittal plane to follow the natural angle of the distal radial articular surface and avoid inadvertent injury to the articular surfaces of the distal radius, proximal scaphoid, and lunate.

7. **Which ligamentous structure is immediately visualized following insertion of the arthroscope in the 3–4 radiocarpal portal?**
   The scapholunate ligament.

8. **Which structure is at risk when the 6U radiocarpal portal is used?**
   The dorsal sensory branch of the ulnar nerve.

9. **Through which portal is the lunotriquetral ligament best visualized?**
   The 6R radiocarpal portal.

10. **Which portals are used for arthroscopy of the midcarpal joint?**
    - The midcarpal radial (MCR) located 1 cm distal to the 3–4 radiocarpal portal
    - Midcarpal ulnar (MCU) located 1 cm distal to the 4–5 radiocarpal portal
    - Scaphotrapeziotrapezoid (STT)

11. **Which bones are visualized during midcarpal arthroscopy?**
    - **MCR portal:** The scaphocapitate, scapholunate, and occasionally the STT articulations
    - **MCU portal:** The lunocapitate, lunotriquetral, and triquetrohamate articulations
    - **STT portal:** The STT articulation

12. **Beginning radially and moving ulnarward, name the palmar extrinsic radiocarpal ligaments of the wrist as seen during wrist arthroscopy.**
    - Radioscaphocapitate
    - Long radiolunate
    - Radioscapholunate (ligament of Testut)
    - Short radiolunate
    - Ulnolunate
    - Ulnotriquetral

13. **Define the TFCC.**
    The TFCC has five components:
    - Articular disk that is of variable thickness and originates from the base of the ulnar styloid inserting on the sigmoid notch of the radius
    - Ulnocarpal ligaments
    - Ulnar collateral ligament
    - Dorsal and volar distal radioulnar joint ligaments
    - Floor of the ECU tendon sheath

14. **What are the major functions of the TFCC?**
    - Major stabilizer of the distal radioulnar joint (DRUJ)
    - Load-bearing structure between the carpus and distal ulna

15. **In the ulnar neutral wrist, what is the ratio of load transfer between the radiocarpal and ulnocarpal articulations?**
    The ratio is 80% radiocarpal and 20% ulnocarpal.

16. **What is the "trampoline test"?**
    A subjective assessment of the integrity of the TFCC determined during wrist arthroscopy. The center of the triangular fibrocartilage (TFC) is balloted by an arthroscopy probe. If the TFCC attachments are intact, the TFC should feel and appear firm and resilient like a trampoline. If probing the TFC fails to elicit the trampoline effect, a peripheral tear of the TFCC should be suspected.

17. **Describe the evaluation of the patient with a suspected traumatic TFCC tear.**
    - **History:** The patient typically complains of ulnar-sided wrist pain following a twisting injury or a fall on an outstretched wrist. Mechanical symptoms, such as popping, catching, or snapping, also may be present.
    - **Physical examination:** The area of tenderness should be localized. The stability of the DRUJ is assessed with the arm resting on a table, the elbow flexed at 90 degrees, and the forearm in neutral rotation. Both the injured and the uninjured wrists should be examined and the findings compared.
    - **Radiographs:** Lateral and neutral rotation posteroanterior views of the wrist should be taken. Assess ulnar variance, the presence of an ulnar styloid fracture, and the appearance of the proximal articular surface of the lunate.
    - **Diagnostic tests:** If a TFCC tear is suspected, a triple injection wrist arthrogram or MRI of the radiocarpal joint may be obtained. However, neither of these tests is as accurate as wrist arthroscopy, which may be used to both evaluate *and* treat a TFC tear.

18. **What determines whether a symptomatic tear of the TFC should be debrided or repaired?**
    - The location of the tear helps determine the course of action. The periphery of the TFC is vascular and therefore capable of healing after repair, whereas the central region is avascular and thus incapable of healing after repair.
    - In general, peripheral tears are repaired, and central tears are debrided.

19. **How are TFC abnormalities classified?**
    The TFC has been categorized into two types by Palmer.
    Class 1: Traumatic Tear
    - **1A:** Central perforation of the articular disk
    - **1B:** Avulsion from the ulnar attachment
    - **1C:** Avulsion or rupture of the ulnocarpal ligaments
    - **1D:** Avulsion from the sigmoid notch

    Class 2: Degenerative Tear
    - **2A:** TFCC wear
    - **2B:** TFCC wear, plus lunate and/or ulnar chondromalacia
    - **2C:** TFCC perforation, plus lunate and/or ulnar chondromalacia
    - **2D:** TFCC perforation, plus lunate and/or ulnar chondromalacia, and lunotriquetral ligament perforation
    - **2E:** TFCC perforation, plus lunate and/or ulnar chondromalacia, lunotriquetral ligament perforation, and ulnocarpal arthritis

20. **What is the most common type of traumatic TFC tear?**
    Type 1A (central perforation of the articular disk).

KEY POINTS: WRIST ARTHROSCOPY

1. Dorsal wrist arthroscopy portals are identified by the intervals between the extensor compartments through which they pass.
2. The radial artery and superficial radial and lateral antebrachial cutaneous nerves are at risk for injury when the 1–2 radiocarpal portal is used.
3. The scapholunate and lunotriquetral ligaments are best visualized via the 3–4 and 6R portals, respectively.
4. The most common traumatic TFC tear is the type 1A-central perforation.
5. Peripheral TFC tears are repaired, while central tears are debrided.

21. **How are traumatic TFC tears treated?**
    Treatment is based on the type (location) of the tear and length of the ulna relative to the radius (ulnar variance). If neutral or positive ulnar variance is present, it is recommended that a shortening osteotomy of the ulna or a wafer resection of the distal ulna be performed. The treatment is summarized in Table 20-1.

    TABLE 20-1. TREATMENT OF TRAUMATIC TFCC TEARS

Tear Type	Treatment
1A	Arthroscopic debridement
1B	Open or arthroscopic reattachment
1C	Arthroscopic debridement or open repair
1D	Open or arthroscopic reattachment

22. **What is ulnocarpal impaction syndrome?**
    - Ulnar-sided wrist pain with associated ulna-positive variance (the ulna is longer than the radius at the wrist) are symptoms.
    - The distal ulna impacts against the TFCC and ulnar carpal bones, particularly the proximal lunate and triquetrum.
    - Impaction results in erosion of articular cartilage, ligament attenuation, and inflammatory changes.

23. **What are the typical arthroscopic findings in the patient with ulnocarpal impaction syndrome?**
    - Synovitis in the ulnocarpal region
    - Thinning or an actual tear of the TFC
    - Erosive changes (chondromalacia) of the distal ulnar head and/or proximal lunate
    - Perforation of the lunotriquetral ligament

24. **How is ulnocarpal impaction syndrome treated?**
    - **Nonoperative:** Immobilization, activity modification, administration of nonsteroidal anti-inflammatory drugs (NSAIDs), corticosteroid injection
    - **Operative:** Arthroscopic debridement of TFCC and synovectomy, ulnar shortening osteotomy, removal of the distal 2–3 mm of the ulna via arthroscopic or open methods (wafer procedure), or resection of the distal ulna (Darrach procedure).

25. **During arthroscopic-assisted reduction and fixation (AARF) of a complete scapholunate ligament disruption, which portal is preferred for assessing the adequacy of reduction?**
The radial midcarpal (RMC) portal.

26. **What is the incidence of complications associated with wrist arthroscopy?**
Approximately 2%.

27. **Other than for disorders of the TFCC and interosseous ligaments, what other disorders or conditions are amenable to wrist arthroscopy?**
    - AARF of fractures of the scaphoid and distal radius
    - Arthroscopic proximal row carpectomy and radial styloidectomy
    - Resection of dorsal wrist ganglion
    - Synovectomy
    - Laser or thermal chondroplasty
    - Irrigation and debridement of a septic wrist

28. **What are the advantages of an arthroscopic dorsal wrist ganglion resection compared with the open excision technique?**
    - Avoidance of a dorsal scar
    - Less stiffness postoperatively

29. **What are the disadvantages of arthroscopic resection of a dorsal wrist ganglion?**
    - Inability to remove the actual ganglion sac, which may result in a higher recurrence rate
    - Residual fullness over the dorsum of the wrist

30. **Why is it recommended that AARF of distal radius fractures be performed 2–7 days after injury?**
To minimize bleeding from the fracture site, which obscures critical visualization of the articular surfaces and fracture fragments and makes the procedure more lengthy and difficult.

31. **Which steps are taken to minimize fluid extravasation into the soft tissues during AARF of a distal radius fracture?**
    - Application of a compressive dressing
    - Use of a lactated Ringer's solution (rapidly absorbed from the soft tissues)
    - Use of a gravity inflow portal instead of a mechanical pump
    - Maintenance of a separate outflow portal to decrease fluid extravasation

32. **What is the incidence of intracarpal soft tissue and osteochondral injuries in association with a distal radius fracture as detected during arthroscopy?**
Between 45% and 75% of all distal radius fractures are associated with an intracarpal soft tissue or osteochondral lesion. Tears of the TFCC are the most common and occur in approximately 50% of displaced intra-articular distal radius fractures.

33. **What is the major contraindication to arthroscopic synovectomy in the patient with rheumatoid arthritis? Why?**
Dorsal tenosynovitis. The weakened extensor tendons are at risk for injury during the establishment of arthroscopy portals.

34. **What is the most common technical difficulty encountered with arthroscopic resection of a dorsal wrist ganglion?**
   Difficulty visualizing the ganglion stalk.

35. **Which structures are at risk during arthroscopic resection of a dorsal wrist ganglion?**
   The extensor carpi radialis longus and brevis tendons are at risk for injury during the capsulectomy.

36. **What is the reported recurrence rate following arthroscopic resection of a dorsal wrist ganglion?**
   Between 0% and 7%.

37. **What are the reported advantages of performing arthroscopic-assisted percutaneous fixation of a scaphoid fracture?**
   - Less wrist stiffness postoperatively
   - Preserved scaphoid vascularity
   - Ability to assess for associated injuries during procedure
   - Improved visualization of fracture reduction

38. **How are intercarpal ligament injuries classified?**
   The Geissler grading system is most commonly used:
   - **Grade 1:** Attenuation/hemorrhage of interosseous ligament can be seen from the radiocarpal joint; no incongruency of carpal alignment in midcarpal space exists.
   - **Grade 2:** Attenuation/hemorrhage of interosseous ligament can be seen from the radiocarpal joint, and incongruency/step-off can be seen from the midcarpal space. A slight space between carpals may be noted.
   - **Grade 3:** Incongruency/step-off of carpal alignment can be noted in both the radiocarpal and midcarpal spaces. A probe may be passed through the gap between carpals.
   - **Grade 4:** Incongruency/step-off of carpal alignment can be noted in both the radiocarpal and midcarpal spaces. Gross instability with manipulation is noted. A 2.7-mm arthroscope may be passed through the gap.

39. **Which volar portals have been described in wrist arthroscopy?**
   - **Volar radial (VR):** This portal is established in the floor of the flexor carpi radialis tendon sheath. The trochar enters the radiocarpal joint between the radioscaphocapitate and long radiolunate ligaments.
   - **Volar ulnar (VU):** This portal is established along the ulnar edge of the finger flexor tendons at the proximal wrist crease.

40. **What are the purported advantages of the volar wrist arthroscopy portals?**
   Improved visualization of the dorsal wrist capsule, dorsal radiocarpal ligament, and palmar subregions of the scapholunate and lunotriquetral interosseous ligaments.

BIBLIOGRAPHY

1. Atik TL, Baratz ME: The role of arthroscopy in wrist arthritis. Hand Clin 15:489–494, 1999.
2. Bednar MS, Arnoczky SP, Weiland AJ: The microvasculature of the triangular fibrocartilage complex: Its clinical significance. J Hand Surg 16A:1101–1105, 1991.
3. Chung KC, Zimmerman NB, Travis MT: Wrist arthroscopy versus arthrography: A comparative study of 150 cases. J Hand Surg 21A:591–594, 1996.

4. Culp RW: Complications of wrist arthroscopy. Hand Clin 15:529–535, 1999.
5. Friedman SL, Palmar AK: The ulnar impaction syndrome. Hand Clin 7:296–310, 1991.
6. Geissler WB, Freeland AE: Arthroscopic management of intra-articular distal radius fractures. Hand Clin 15:455–465, 1999.
7. Hermansdorfer JD, Kleinman WB: Management of chronic peripheral tears of the triangular fibrocartilage complex. J Hand Surg 16A:340–346, 1991.
8. Koman LA, Poehling GG, Toby EB, et al: Chronic wrist pain: Indications for wrist arthroscopy. Arthroscopy 6:116–119, 1990.
9. Osterman AL: Basic wrist arthroscopy and endoscopy. Hand Clin 10:4, 1994.
10. Osterman AL: Advanced wrist arthroscopy and endoscopy. Hand Clin 11:1, 1994.
11. Osterman AL, Raphael J: Arthroscopic resection of dorsal ganglion of the wrist. Hand Clin 11:7–12, 1995.
12. Rizzo M, Berger RA, Steinmann SP, Bishop AT: Arthroscopic resection in the management of dorsal wrist ganglions: Results with a minimum 2-year follow-up. J Hand Surg 29A:59–62, 2004.
13. Ruch DS, Poehling GG: Arthroscopic management of partial scapholunate and lunotriquetral injuries of the wrist. J Hand Surg 21A:412–417, 1996.
14. Slade JF III, Jaskwich D: Percutaneous fixation of scaphoid fractures. Hand Clin 17(4):553–574, 2001.
15. Slutsky DJ: Volar portals in wrist arthroscopy. J Am Soc Hand Surg 2(4):225–232, 2002.
16. Trumble TE, Gilbert M, Vedder N: Isolated tears of the triangular fibrocartilage: Management by early arthroscopic repair. J Hand Surg 22A:57–65, 1997.
17. Whipple T: Arthroscopic surgery. In Whipple T (ed): The Wrist. Philadelphia, J. B. Lippincott, 1992.

CHAPTER 21

DISTAL RADIUS INJURIES
Adam Mirarchi, MD

1. **Which radiographic features/parameters are important to assess in the evaluation of distal radius fracture?**
   - The extent of comminution involving the dorsal cortical metaphysis of the distal radius
   - The status of the articular surfaces (scaphoid facet, lunate facet, and sigmoid notch)
   - Radial inclination (normal at approximately 22 degrees)
   - Sagittal tilt (normal at 11 degrees palmar)
   - Radial length (11 mm)
   - Status of the distal radioulnar joint (DRUJ) (Fig. 21-1)

2. **In which age group do distal radius fractures commonly occur?**
   Distal radius fractures have a bimodal incidence pattern. The two peak age ranges are childhood (6–10 years of age) and the transition between middle and early old age (60–69 years of age).

3. **How are distal radius fractures classified?**
   Various classification systems have been described for fractures of the distal radius. The most commonly used systems are Frykman's classification, Melone's classification, the Swiss Association for Study of Internal Fixation's (AO-ASIF's) classification, and the Mayo Clinic's Universal classification.
   - **Frykman's classification system** consists of eight types of distal radius fractures (I–VIII) that are classified by fracture location (extra-articular versus intra-articular) and the presence or absence of an associated ulnar styloid fracture. The more complex the fracture, the higher the Roman numeral. More complex fracture types have a poorer prognosis (Fig. 21-2).
   - **Melone's classification system** identifies five fracture types and four major fracture components: the shaft, the radial styloid, and the dorsal-medial and volar-medial fragments.
   - The **AO-ASIF's classification system** divides distal radius fractures into three major groups of group A (extra-articular), group B (simple intra-articular), and group C (complex). The major groups are subdivided into 27 different fracture patterns. The AO-ASIF classification system is the most complex.
   - The **Mayo Clinic's Universal classification system** divides distal radius fractures into four types. Type I is an extra-articular fracture, whereas types II, III, and IV are intra-articular with the fracture lines involving the radioscaphoid, radiolunate, and radioscapholunate articulations, respectively. Further emphasis is given to whether the fracture is reducible or irreducible and displaced or nondisplaced.

   Because of the relative subjectivity involved in classifying distal radius fractures and the lack of interobserver and intraobserver reliability, many surgeons classify distal radius fractures as displaced versus nondisplaced, extra-articular versus intra-articular, stable versus unstable, and open versus closed.

4. **Which factors influence the treatment approach or strategy for fractures of the distal radius?**
   - Associated injuries
   - Activity demands
   - Fracture pattern

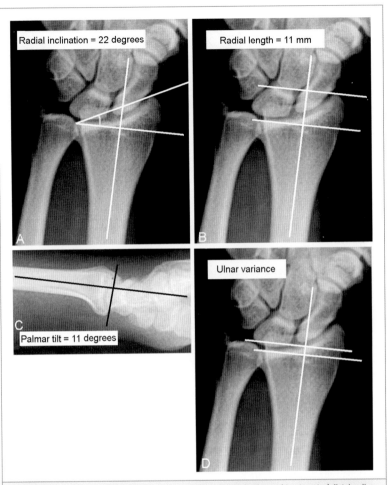

Figure 21-1. Important radiographic features to assess in the evaluation and treatment of distal radius fractures. **A**, Radial inclination of 22 degrees. **B**, Radial length of 11 mm. **C**, Palmar tilt of 11 degrees. **D**, Ulnar variance. In the sagittal plane, the normal palmar tilt averages 11 to 12 degrees. In the frontal plane, the average ulnar inclination is 22 to 23 degrees, and the radial length (height) averages 11 to 12 mm. (From Beredjiklian PK, Bozentka DJ: Review of Hand Surgery. Philadelphia, Saunders, 2004, p 119.)

- General medical condition
- Patient compliance
- Patient age

5. **Why should the entire upper extremity be examined in the patient with a distal radius fracture?**
   Associated injuries of the hand (carpal and metacarpal fractures, intrinsic ligament tears), elbow (radial head fracture, dislocation), arm (humeral shaft fracture), and shoulder regions (proximal humeral fracture, shoulder dislocation, clavicle fracture, acromioclavicular separation) are

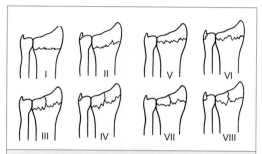

Figure 21-2. Frykman's classification of distal radius fractures. (From Brown DE, Randall RD [eds]: Orthopedic Secrets. Philadelphia, Hanley & Belfus, 1995.)

common. These areas should be assessed in the evaluation of *every* patient with a distal radius fracture.

6. **What is the nonoperative treatment strategy for an acceptably reduced, closed, and stable extra-articular distal radius fracture?**
   - The fracture is reduced, followed by immobilization in a well-molded sugar-tong plaster splint or short arm cast. Occasionally, a long arm cast may be necessary to prohibit forearm rotation and subsequent fracture displacement.
   - The patient is evaluated weekly for the first 3 weeks after the reduction, and radiographs are obtained in the splint or cast if it is fitting well. If the fracture redisplaces, repeat reduction or operative treatment is indicated. If satisfactory alignment has been maintained, the cast or splint is changed 3 weeks after the reduction and a new cast applied for an additional 3 weeks.
   - At 6 weeks after the injury, the cast is removed and is followed by the initiation of a therapy program to restore motion and strength.

7. **What is the Cotton-Loder position? Why should it be avoided?**
   To prevent displacement and to maintain the reduction of a distal radius fracture, the injured extremity may be splinted or casted in extreme palmar flexion, pronation, and ulnar deviation (Cotton-Loder position). There is an increased incidence of median nerve injury (carpal tunnel syndrome), reflex sympathetic dystrophy (RSD), and finger and wrist stiffness with immobilization of the upper extremity in this position.

8. **What is a Smith's fracture?**
   A distal radius fracture involving comminution of the volar cortex, volar displacement, and volar angulation (Fig. 21-3).

9. **What is a Barton's fracture?**
   An intra-articular fracture-dislocation of the *dorsal* or *posterior* margin of the radius. The fracture extends through the radiocarpal surface, with the carpus and metaphyseal fragment translating dorsally and subsequently resting on the dorsal aspect of the wrist.

   Barton also described a similar intra-articular fracture occurring volarly. The dorsal variant is the

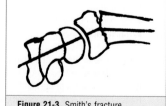

Figure 21-3. Smith's fracture.

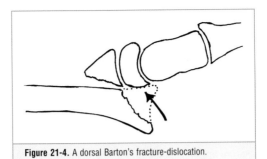

Figure 21-4. A dorsal Barton's fracture-dislocation.

one to which his name was originally applied although it is now understood that the volar variant is more common. These fractures are inherently unstable and require open reduction and internal fixation (ORIF) with buttress plate application because of the inability to obtain and maintain a satisfactory reduction by closed treatment methods (Fig. 21-4).

10. **What is a "die-punch fracture"?**
    An impacted displaced fracture of the lunate facet of the distal radial articular surface resulting from an axial load that drives the carpus into the distal radius. Such "lunate load" fracture fragments may be split into multiple pieces and are difficult to reduce by closed means (Fig. 21-5).

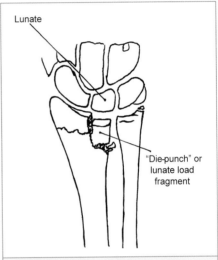

Figure 21-5. Fracture of the distal radius involving a "die-punch" fragment.

11. **What is a "chauffeur's fracture"?**
    A fracture of the radial styloid that was commonly seen in the 1900s. During this time period, cars commonly backfired when starting, causing the crank to strike or violently twist the hand. The injury was also referred to as a "backfire" fracture (Fig. 21-6).

12. **What injuries occur in association with a chauffeur's fracture?**
    A perilunate injury involving a tear of the intrinsic interosseous ligaments (scapholunate, lunotriquetral); tear of the triangular fibrocartilage complex (TFCC); or a fracture of the scaphoid, capitate, lunate, or ulnar styloid or a combination of these.

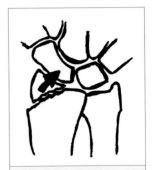

Figure 21-6. Chauffeur's fracture.

13. **What is the most common epiphyseal injury of the distal radius?**
    Distal radius epiphyseal injuries account for approximately 50% of all epiphyseal injuries. The most common is a Salter-Harris type II epiphyseal separation.

14. **What unique complication may occur after an epiphyseal fracture of the distal radius?**
    Premature physeal closure with a resultant deformity of the distal radius as growth in the remaining uninjured epiphysis continues. Incongruity of the distal radioulnar joint and ulnocarpal impaction syndrome may result from continued overgrowth of the ulna.

15. **How is a premature epiphyseal closure of the distal radius treated?**
    Treatment is individualized and based on the amount of physeal closure (partial or complete), the degree of deformity, the age of the patient, cosmetic appearance, and functional limitations. Treatment options include epiphyseal bar resection and insertion of a spacer such as fat, epiphysiodesis (surgical closure of the growth plate) of the distal ulna, ulnar shortening osteotomy, resection of the distal ulna, distal radial osteotomy, or a combination of these procedures.

16. **Is casting an acceptable treatment approach for a distal radius fracture that has undergone a satisfactory closed reduction?**
    Nonoperative treatment is preferred for distal radius fractures that are satisfactorily reduced and have no propensity for instability and subsequent collapse. It is imperative to follow all displaced distal radius fractures closely because a significant number of fractures will redisplace, particularly those that are comminuted or those that initially appeared satisfactorily reduced.

17. **Which radiographic parameters are considered acceptable following the reduction for a distal radius fracture?**
    - Radial length within 5 mm of the opposite side
    - Radial inclination >15 degrees
    - Sagittal tilt less than 10 degrees dorsal and up to 20 degrees volar
    - <1 mm of articular incongruity

18. **Which radiographic parameters are used to predict distal radius fracture instability?**
    - The extent of dorsal comminution (expressed as a percentage of the sagittal width of the radius)
    - The amount of initial fracture displacement and/or angulation

19. **Which intracarpal injuries are associated with a distal radius fracture?**
    - Intrinsic ligament tears (scapholunate, lunotriquetral)
    - Extrinsic ligament tears
    - Ulnar styloid fracture
    - Scaphoid fracture
    - Perilunate injury
    - Midcarpal instability

20. **Why is ligamentotaxis important in the treatment of distal radius fractures?**
    The capsule and extrinsic radiocarpal ligaments originate on the distal radial metaphyseal region and insert on the carpus. During traction or distraction, tension created in the ligaments and capsule indirectly elevates and reduces displaced or depressed fracture fragments.

KEY POINTS: DISTAL RADIUS INJURIES

1. Volar fixed-angle plating of unstable dorsally comminuted distal radius fractures is gaining in popularity because the plate may be covered by the thicker volar soft tissues, and the construct is sufficiently rigid to permit early functional recovery.
2. ORIF and bone grafting of a distal radius fracture nonunion are recommended only if a minimum of 5 mm of subchondral bone is present beneath the lunate facet.
3. The risk factors for a nonunion following a distal radius fracture are open fracture, diabetes mellitus, immunodeficiency disorder, peripheral vascular disease, and prolonged excessive overdistraction with external fixation.
4. Lactated Ringer's (LR's) solution is recommended for arthroscopic-assisted reduction and fixation (AARF) of a distal radius fracture because it is rapidly reabsorbed from the soft tissues following extravasation, thus minimizing the risk of a compartment syndrome.
5. DRUJ instability can occur with a fracture of the ulnar styloid that extends through the base and disrupts the deep DRUJ ligaments.
6. The Cotton-Loder position of immobilization (extreme palmar flexion, pronation, and ulnar deviation) should be avoided in the management of distal radius fractures because of the risk of median nerve injury and the development of a sympathetic mediated pain syndrome.

21. **What is the normal loading pattern of the carpus on the distal radius?**
    In the ulnar neutral wrist (ulnar and radial lengths are equal), approximately 80% of the axial load occurs through the distal radius and 20% through the ulna. The ulna is subjected to increasing load with an increase in ulnar length (radial shortening) or forearm pronation and is also subjected to increasing load during power grip.

22. **What degree of articular step-off is acceptable after reduction of a displaced intra-articular distal radius fracture?**
    The most important factor in determining the outcome of an intra-articular distal radius fracture is the amount of residual articular incongruity following treatment. Incongruity >1 mm is strongly associated with the subsequent development of symptomatic posttraumatic arthritis.

23. **What is the role of distraction in the diagnosis and treatment of a distal radius fracture?**
    Through ligamentotaxis, distraction may assist in obtaining an adequate reduction of displaced fracture fragments, and it assists the examiner in identifying additional injuries, particularly those involving the carpus and intrinsic interosseous ligaments.

24. **What are the treatment options for an unstable distal radius fracture?**
    - Closed reduction and percutaneous pinning (CRPP)
    - External fixation
    - ORIF
    - AARF
    - A combination of the previously mentioned methods

25. **What is "crossed pinning" of a distal radius fracture?**
    A technique of percutaneous pinning of a distal radius fracture that involves advancing Kirschner wires from the radial styloid and dorsal ulnar corner of the radius proximally across the fracture site at a 45-degree angle to the bone's long axis (Fig. 21-7).

26. **What is the Kapandji technique of intrafocal pinning?**

    The insertion of a series of Kirschner wires directly into the fracture site. The wires are then used as levers against the intact volar cortex to improve fracture alignment, particularly the volar tilt and radial inclination (Fig. 21-8). This technique may be combined with other percutaneous techniques or external fixation to improve surgical restoration of articular congruity and distal radius anatomy without actually opening the fracture site. It requires careful placement of the wires to avoid tenting the skin after levering the fracture into a reduced position and to avoid inadvertent injury to the extensor tendons and the superficial radial nerve. The technique is most effective when volar comminution is absent, allowing the wires to gain purchase in the volar cortex.

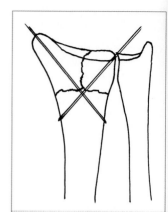

Figure 21-7. Three-part intra-articular fracture of the distal radius successfully reduced and stabilized by crossed Kirschner wires (also called K-wires).

27. **What is the primary indication for ORIF of a distal radius fracture?**

    Failure to obtain *or* maintain an acceptable reduction by closed methods, percutaneous or intrafocal pinning, external fixation, or limited open methods.

28. **What is limited open reduction?**

    Limited open reduction refers to exposure of the fracture through a small skin incision and limited dissection. The technique is particularly useful for ORIF of a dorsal lunate facet die-punch fracture fragment. This technique is typically combined with percutaneous pinning and/or external fixation. A 2-cm dorsal incision is used, with the dissection proceeding between the third and fourth extensor compartments. Some prefer to expose the die-punch fragment through the fourth compartment. An elevator is used to elevate the depressed fragment back into an anatomic position, where it is then maintained with some form of internal fixation with supplemental bone grafting of the metaphyseal defect.

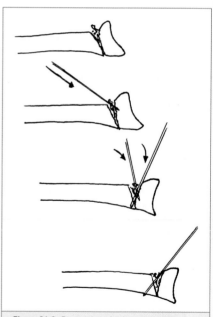

Figure 21-8. Restoration of palmar tilt via the Kapandji technique of intrafocal pinning.

29. **The displacement of which component of an intra-articular distal radius fracture is most likely to require ORIF?**

    The dorsal and/or volar die-punch fragments that often malrotate are difficult, if not impossible, to reduce by closed methods. The dorsal and volar Barton's fracture-dislocation injury patterns also require ORIF with application of a buttress plate to prevent displacement of these inherently unstable injuries.

30. **What are the most common complications associated with external fixation of a distal radius fracture?**
    - Stiffness
    - Sympathetic mediated pain syndrome
    - Pin tract infection

31. **What is a major concern of using dorsal buttress plating for the treatment of a distal radius fracture?**
    Irritation of the extensor tendons resulting in secondary tenosynovitis, pain, and tendon rupture. This has led to increasing popularity with volar fixed-angle plating.

32. **What are the advantages to plating a distal radius fracture through the volar approach?**
    - Because there is greater soft tissue coverage on the volar aspect of the wrist and distal forearm, a thicker, more rigid fixation device (plate) can be used.
    - The extensor tendons are not disturbed by extensive dissection and, hence, are less prone to rupture.
    - There is less disruption of the remaining intact dorsal periosteum attached to the dorsally comminuted fragments, which presumably enhances fracture union.
    - The volar scar is more cosmetic and better tolerated.

33. **Discuss the concept and design of volar fixed-angle plates.**
    Integral to the success of the volar approach of a dorsally comminuted and unstable distal radius fracture is the concept of fixed-angle plating. Volar fixed-angle plates consist of a precontoured plate with a T-shaped distal end that accommodates a series of screws or pegs that screw and lock into the plate. The screws or pegs are inserted in a spread pattern and are aimed slightly dorsally to create a supporting surface for the dorsal subchondral bone. This construct creates a fixed-angle device that can support the distal fracture fragments by transferring forces through the pegs/screws to the plate and ultimately into the intact proximal radius (Fig. 21-9).

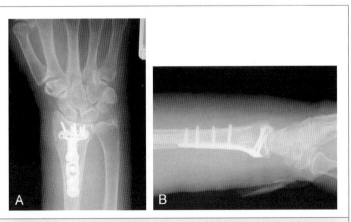

Figure 21-9. PA view *(A)* and laterl view *(B)* demonstrating ORIF of a distal radius fracture using a volar fixed-angle plate.

34. What is fragment-specific fixation?

A distal radius fracture may be divided into radial column, volar column, intra-articular depression, dorsal cortical, and dorsal ulnar split fragments. Specific implants have been developed to address each fracture components of these.

- The radial fragment is secured with a plate containing fixation pins.
- The volar column fracture is stabilized by a buttress plate that resists shear.
- Fixed-angle "wire-form" implants are used with articular depression and dorsal cortical fractures to support the subchondral bone and maintain radial length.
- Dorsal ulnar splint fragments are secured with implants that act as tension bands and function in an antiglide mechanism (Fig. 21-10).

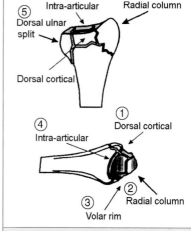

Figure 21-10. The five cortical components of the structural anatomy as depicted by Medoff. (From Trumble TE: Hand Surgery Update 3: Hand, Elbow, and Shoulder. Rosemont, IL, American Society for Surgery of the Hand, 2003, p 86.)

35. What is the importance of the superficial radial nerve relative to the treatment of a distal radius fracture?

The superficial radial sensory nerve and its branches are at risk for injury during the insertion of percutaneous Kirschner wires, external fixator pins, or screws because of its location on the dorsoradial aspect of the distal forearm and wrist/hand. The multiple branches are quite sensitive, and continued irritation may result in significant morbidity, including the precipitation of a sympathetic mediated pain syndrome in susceptible individuals.

36. What are the proposed advantages of AARF of a distal radius fracture?

- Less invasive than ORIF
- Enables the examiner to assess the accuracy of an articular reduction
- Permits identification and treatment of associated injuries of the TFCC, carpus, and intrinsic or extrinsic radiocarpal ligaments

37. Which nerve is most frequently compromised after an acute distal radius fracture?

The median nerve.

38. How is a patient with a distal radius fracture and an associated median nerve injury treated?

Examination of median nerve function should be accurate and expedient. The fracture should be reduced as quickly and as accurately as possible, followed by a repeat neurovascular examination. Most patients have a transient neuropraxia that resolves over 24–48 hours after fracture reduction, splinting, and elevation. The indications for surgical treatment (i.e., carpal tunnel release) vary. If the symptoms worsen or fail to improve within 24–48 hours after reduction, or if a reduction cannot be obtained or maintained, and median nerve compromise persists, a carpal tunnel release with concomitant surgical stabilization of the fracture is recommended. A carpal tunnel release is also recommended for a dense median motor and sensory palsy that persists after a closed reduction.

39. **What is the most common tendon to rupture following a distal radius fracture?**
    Extensor pollicis longus (EPL). Rupture typically occurs several weeks after the fracture and often occurs in patients with a nondisplaced or minimally displaced fracture, suggesting an ischemic etiology rather than an attritional rupture.

40. **What are the effects of distal radius deformity on DRUJ mechanics?**
    - Residual deformity of the distal radius after a fracture results in a significant disturbance of the distal radioulnar joint with a subsequent loss of forearm rotation.
    - Radial shortening, in particular, causes the greatest disturbance in DRUJ kinematics.

41. **How does an associated injury of the DRUJ affect the outcome of a distal radius fracture?**
    Failure to recognize and appropriately treat an associated injury of the DRUJ (usually instability) has been associated with a poor outcome following a distal radius fracture and is the focus of increased investigation. **DRUJ injuries are classified based on their stability:**
    - **Type 1:** Lesions are clinically stable with a congruent distal radioulnar joint.
    - **Type 2:** Characterized by a subluxation or dislocation injury. A tear of the triangular fibrocartilage (TFCC) or an ulnar styloid avulsion fracture is present.
    - **Type 3:** This is a potentially unstable injury. Disruption of the sigmoid notch or ulnar head is present.

42. **What effect does a distal radius malunion have on the carpus?**
    - Radiocarpal intra-articular incongruity results in the development of posttraumatic arthritis.
    - Increased dorsal tilt results in increased transmission of axial load through the ulnar side of the wrist.
    - Radial shortening disturbs DRUJ kinematics, leading to limited, painful forearm rotation.
    - Midcarpal instability may occur in those distal radius fractures that heal with excessive dorsal angulation. This pattern is typically a dorsal intercalated segmental instability (DISI) and is often associated with dorsal wrist pain and decreased grip strength.

43. **When should surgical treatment be undertaken for a malunited distal radius fracture?**
    The primary indication for surgical treatment of this type of fracture is unremitting pain and/or loss of function. Treatment is individualized, but the malunion may be corrected by a distal radius osteotomy combined with bone grafting and ORIF. Concomitant procedures are often necessary, including arthrodesis of the DRUJ (Sauvé-Kapandji procedure), resection or shortening of the distal ulna, or reconstruction of the DRUJ ligaments.

44. **Nonunion of a distal radius fracture occurs in which types of cases?**
    Nonunion of a distal radius fracture is extremely rare but has been reported in patients with:
    - Immunodeficiency disorders
    - Diabetes mellitus
    - Peripheral vascular disease
    - Infection following an open fracture
    - Prolonged and excessive distraction by an external fixator
    - Associated distal ulna fracture (Fig. 21-11)

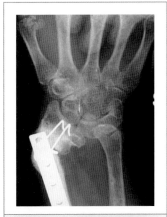

Figure 21-11. Distal radius fracture nonunion following previous ORIF.

45. A 37-year-old man presents with pain and deformity 7 months after treatment of an open distal radius fracture. His medical

history is unremarkable. He does not smoke. Physical examination demonstrates a 25-degree angle of dorsiflexion, a 25-degree angle of palmar flexion, a 15-degree angle of supination, and 70-degree angle of pronation. He has an obvious deformity of the wrist, and the hand appears radially deviated. Radiographs demonstrate a distal radius fracture nonunion. What is the recommended treatment?
- Any patient who presents with a distal radius nonunion should be encouraged to discontinue smoking because nicotine compromises bone and soft tissue healing.
- Preoperative radiographs should be studied carefully. It has been stated that ORIF is technically feasible only if a minimum of 5 mm of subchondral bone is present beneath the lunate facet of the distal radius. This amount of bone is necessary for screw purchase.
- The synovial pseudoarthrosis should be resected, with correction of the deformity and stabilization with internal fixation. Plates placed in orthogonal planes are preferred to enhance stability.
- The nonunion site should be augmented with autogenous cancellous bone graft, preferably from the iliac crest.
- Capsulotomies of the radiocarpal joint and DRUJ also may be necessary if significant stiffness is noted.
- Concomitant arthrosis at the DRUJ must be addressed with resection of the distal ulna, a hemiresection interpositional arthroplasty, or an arthrodesis of the DRUJ.
- If correction of the deformity and stable fixation of the nonunion are not feasible, total wrist arthrodesis that incorporates the nonunion site is recommended.

46. About 6 weeks after dorsal plate fixation of an unstable extra-articular distal radius fracture, a 23-year-old man presents with a loss of active thumb interphalangeal joint flexion. What is the most likely explanation?
Attritional rupture of the flexor pollicis longus caused by a prominent screw tip. There have been sporadic reports of spontaneous tendon ruptures with some of the newer implants that use self-tapping screws. The differential diagnosis should also include an anterior interosseous nerve palsy.

47. Why is it recommended that AARF be performed 4–7 days after injury?
Waiting 4–7 days permits stabilization of fracture site bleeding that otherwise obscures visualization of the articular surfaces and surrounding structures.

48. Why is LR solution used for AARF?
Although fluid extravasation is associated with AARF, LR is rapidly reabsorbed from the soft tissues, thus minimizing the risk of a compartment syndrome.

49. What are the most common intra-articular ligament injuries noted during AARF of a distal radius fracture?
- The most common injury pattern is a tear of the TFCC (40% incidence).
- Partial or complete tears of the scapholunate ligament occur in approximately 30% of patients.
- Lunotriquetral interosseous ligament tears occur less frequently (15% incidence).

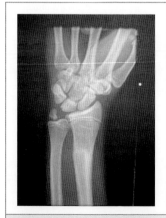

Figure 21-12. A fracture through the ulnar styloid base associated with DRUJ instability.

50. **A 24-year-old man sustained a fracture of the distal radius with an associated fracture through the base of the styloid, with 2 mm of displacement and dorsal subluxation of the DRUJ. How should treatment proceed?**
   - After reduction and stabilization of the distal radius fracture, DRUJ stability is assessed.
   - If the DRUJ is unstable, open reduction and internal fixation of the ulnar styloid are recommended to restore stability and avoid the complications of a painful nonunion. Fixation may be accomplished with tension band wiring, screw fixation, or multiple Kirschner wires (Fig. 21-12).

BIBLIOGRAPHY

1. Axelrod T, Paley D, Green J, McMurtry RY: Limited open reduction of the lunate facet in comminuted intra-articular fractures of the distal radius. J Hand Surg 13A:372–377, 1988.
2. Bailey DA, Wedge JH, McCulloch RG, et al: Epidemiology of fractures of the distal end of the radius in children as associated with growth. J Bone Joint Surg 71A:225–1231, 1989.
3. Bartosh RA, Saldana MJ: Intra-articular fractures of the distal radius: A cadaveric study to determine if ligamentotaxis restores radiopalmar tilt. J Hand Surg 15A:18–21, 1990.
4. Fernandez DL, Ring D, Jupiter JB: Surgical management of delayed union and nonunion of distal radius fractures. J Hand Surg 26A:201–209, 2001.
5. Geissler WB: Arthroscopically assisted reduction of intra-articular fractures of the distal radius. Hand Clin 11:19–29, 1995.
6. Hastings H, Leibovic SJ: Indications and techniques of open reduction internal fixation of distal radius fractures. Orthop Clin North Am 24:309–326, 1993.
7. Knirk JL, Jupiter JB: Intra-articular fractures of the distal end of the radius in young adults. J Bone Joint Surg 68A:647–659, 1986.
8. Konrath GA, Bahler S: Open reduction and internal fixation of unstable distal radius fractures: Results using the trimed fixation system. J Orthop Trauma 16(8):578–585, 2002.
9. McKay SD, MacDermid JC, Roth JH, Richards RS: Assessment of complications of distal radius fractures and development of a complication checklist. J Hand Surg (Am) 26(5):916–922, 2001.
10. Orbay JL, Fernandez DL: Volar fixation for dorsally displaced fractures of the distal radius: A preliminary report. J Hand Surg 27A:205–215, 2002.
11. Rogachefsky RA, Lipson SR, Appelgate B, et al: Treatment of severely comminuted intra-articular fractures of the distal end of the radius by open reduction and combined internal and external fixation. J Bone Joint Surg 83A:509–519, 2001.
12. Rozental TD, Branas CC, Bozentka DJ, Beredjiklian PK: Survival among elderly patients after fractures of the distal radius. J Hand Surg 27A:948–952, 2002.
13. Segalman KA, Clark GL: Un-united fractures of the distal radius: A report of 12 cases. J Hand Surg 23A:914–919, 1998.
14. Short WH, Palmer AK, Werner FW, Murphy DJ: A biomechanical study of distal radial fractures. J Hand Surg 12A:529–534, 1987.
15. Trumble TE, Schmitt SR, Vedder NB: Factors affecting functional outcome of displaced intra-articular distal radius fractures. J Hand Surg 19A:325–340, 1994.
16. Trumble TE, Wagner W, Hanel DP, et al: Intrafocal (Kapandji) pinning of distal radius fractures with and without external fixation. J Hand Surg 23A:381–394, 1998.

CHAPTER 22: DISTAL RADIOULNAR JOINT

Michael J. Moskal, MD

1. **What is the distal radioulnar joint (DRUJ)?**
   The DRUJ is the articulation between the sigmoid notch of the radius and the ulnar head of the ulna at the wrist.

2. **What is ulnar variance? How is it measured?**
   The variance is the relationship between the osseous end of the lunate fossa of the radius and the osseous end of the ulnar head of the ulna. It details relative ulnar length as compared to the radius. A "zero rotation" posteroanterior x-ray is taken with the wrist in neutral flexion-extension and neutral forearm pronation-supination. When the ulnar head is colinear with the radius, ulnar variance is zero; when the ulna is longer or shorter, ulnar variance is positive or negative, respectively (as measured in millimeters).

3. **How does the forearm rotate?**
   The radius rotates about the ulna, which is fixed and does not move.

4. **How does the relationship between the radius and ulna change during forearm rotation?**
   The position of the ulna relative to the radius becomes increasingly ulnar positive with pronation. The ulnar head translates dorsally upon the sigmoid notch with pronation.

5. **What percentage of axial (compressive) force is transferred across the radiocarpal and ulnocarpal joints of the wrist?**
   In general, 80% of axial force is transmitted through the radiocarpal joint and 20% through the ulnocarpal joint; the percentages vary based on the position of forearm rotation.

6. **Where is the triangular fibrocartilage complex (TFCC) located?**
   The TFCC is located just distal to the DRUJ. The TFCC arises from the lunate fossa of the radius and inserts into the fovea at the base of the ulnar styloid.

7. **What is the TFCC?**
   The TFCC comprises the dorsal and volar radioulnar ligaments (which originate from the dorsal and volar aspects of the radius and insert into the ulnar styloid base), the articular disk in the central portion, and the tendon sheath of the extensor carpi ulnaris. It is triangular in shape, with the base attached to the radius and the apex at the confluence of insertion onto the ulna.

8. **How does the TFCC attach to the ulna?**
   The TFCC attaches to the fovea near the axis of forearm rotation at the base of the ulnar styloid via the ligamentum subcuetum or the deep portion of the TFCC. The superficial portion inserts into the base of the styloid.

9. **Name the palmar (extrinsic) ligaments on the ulnar side of the wrist. What is their relationship to the TFCC and ulna?**
   The ulno-carpal ligaments are more commonly known as the disk-carpal ligaments (the disk-lunate, disk-triquetral, and ulnocapitate). The disk-lunate and disk-triquetral ligaments

originate at the palmar margin of the TFC and insert, in general, into the lunate, triquetrum, and lunotiquetral interosseous ligament. The disk-carpal ligaments play a critical role in the suspensory function of the TFCC. The ulnocapitate ligament originates from the ulna, lies just palmar to the disk-carpal ligaments, and inserts into the ulnar aspect of the carpus and capitate.

10. **Compare the vascularity of the peripheral TFC versus the central portion.**
    The central portion of the TFCC is relatively avascular as opposed to the periphery, which is well vascularized. The blood supply of the TFCC has implications regarding the treatment of TFCC disorders.

11. **How are TFCC tears classified?**
    Tears of the TFCC are traditionally organized by acuity, location, and associated damage to the ulnar carpal structures (Table 22-1). TFCC tears can be degenerative or traumatic, central or peripheral, and may involve the disk-carpal ligaments, lunotriquetral interosseous ligament, or articulation (Figs. 22-1, 22-2, and 22-3).

12. **How are central TFCC tears treated?**
    Tears (degenerative or traumatic) that involve the relatively avascular central portion of the TFCC that do not extend into the radioulnar ligaments are debrided with excision of the central disk

TABLE 22-1. PALMER'S CLASSIFICATION OF TFCC TEARS

Type of Tear	Location of Tear	Treatment Recommendation
Traumatic		
1A	Linear or horizontal	Debridement
1B	Distal ulna insertion or styloid fracture	Repair
1C	Tear with disk-carpal ligaments tear	Controversial: repair versus debride
1D	Tear from radius insertion	Repair if extends into radioulnar ligament, otherwise debridement
Degenerative		
2A	Wear without perforation	Wafer procedure or ulnar shortening
2B	Wear without perforation and chondromalacia of the lunate or ulna	Wafer procedure or ulnar shortening
2C	Central perforation and chondromalacia of lunate or ulnar head	TFC debridement and wafer procedure or ulnar shortening
2D	2-D with LT ligament tear	TFCC debridement, LT stabilization, and/or ulnar shortening
2E	2-D with ulnocarpal arthritis	Controversial: debridement/lavage, ulnar shortening

LT = lunotriquetral joint, TFC = triangular fibrocartilage, TFCC = triangular fibrocartilage complex.

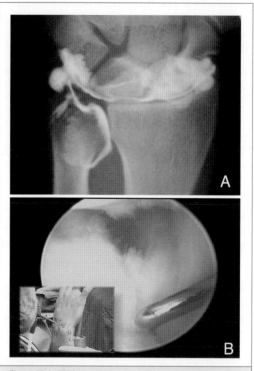

Figure 22-1. TFCC tear. *A*, An arthrographic example of a peripheral TFCC tear. Note the dye that is located near the tear and traverses the peripheral (ulnar) aspect of the TFCC *(arrow)*. *B*, An arthroscopic example of a peripheral TFCC tear. The tears begin dorsal to the prestyloid recess. The tear results in reduced tension within the TFC.

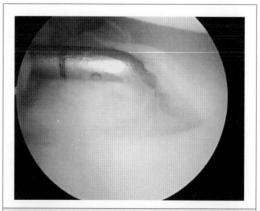

Figure 22-2. Wrist arthroscopy demonstrating a radial-sided TFC tear. The lunate fossa of the radius is in the foreground, and the dorsal wrist is to the left.

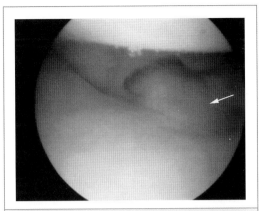

Figure 22-3. An arthroscopic view of a central TFCC degenerative tear due to ulnar impaction syndrome *(arrow)*. The lunate chondromalacia and cartilaginous fragments are not seen in this view. The dorsal aspect of the wrist is to the right of the picture.

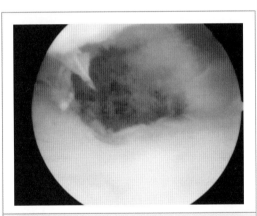

Figure 22-4. An arthroscopic example of a wafer procedure. The central TFCC tear has been debrided to a stable margin while maintaining the integrity of the radioulnar ligaments. The distal ulnar head has been resected approximately 2 mm through the debrided TFC tear.

portion to stable peripheral margins, preserving the dorsal and volar radioulnar ligaments. The debridement is thought to decrease the mechanical impingement by the flaps of torn tissue (Fig. 22-4).

13. **How are peripheral TFCC tears treated?**
    When peripheral TFCC tears are encountered acutely, one can consider a period of cast immobilization with the wrist and forearm in neutral position. Peripheral tears extending into the dorsal radioulnar ligament are repaired by arthroscopic-assisted suture techniques. Peripheral

tears along the radius extending into the radioulnar ligament(s) are typically repaired with sutures passed through the radius arthroscopically.

14. **How does a fracture of the ulnar styloid process affect the TFCC?**
    A fracture at the base of the ulnar styloid may lead to instability of the radioulnar joint due to disruption of the radioulnar ligaments that attach to the base of the styloid at the forea. A fracture of the base of the styloid may require internal fixation to restore stability.

15. **How does a fracture of the radius affect the TFCC?**
    Fractures of the distal radius, with or without an ulnar styloid fracture, may be associated with a TFCC tear, especially those radius fractures that are initially significantly displaced. The normal ligamentous attachments between the radius and ulna may be disrupted in association with a distal radius fracture.

16. **What is the relationship of the width of the TFC and ulnar variance?**
    With increasing amounts of negative ulnar variance, the thickness of the TFC increases.

17. **What is ulnocarpal impaction syndrome?**
    Ulnocarpal impaction syndrome refers to the increased repetitive contact force between the lunate articular cartilage, the TFC, and the head of the ulna. The condition is typically associated with positive ulnar variance (Fig. 22-5).

18. **How is ulnocarpal impaction treated?**
    The treatment is individualized. In Palmer's type 2A and 2B tears, wrist arthroscopy to evaluate and or treat the TFCC and lunate chondral surfaces followed by decompression of the ulnocarpal joint by ulnar shortening should be considered after failure of nonoperative treatment. For type 2C and 2D changes, the ulnar head is accessible through the torn TFCC. Many consider arthroscopic distal ulnar head excision as described by Feldon as a useful treatment (*see* Fig. 22-4). Alternatively, for advanced cases with significant positive ulnar variance or dynamic positive variance, TFCC debridement followed by ulnar shortening osteotomy can be performed.

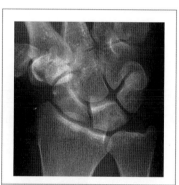

Figure 22-5. A posteroanterior neutral rotation view of the wrist. Note the loss of internal trabecular architecture of the lunate corresponding to cystic changes and chondromalacia. The ulnar head has a positive ulnar variance. This is the x-ray of the same wrist as seen in Figs. 22-3 and 22-4.

19. **When is an ulnar shortening procedure contraindicated?**
    When the morphology of the sigmoid notch is such that it is oriented with reverse inclination. If an ulnar shortening is performed with this morphology, joint reaction forces on the ulnar head at the distal radioulnar joint would likely be increased.

DISTAL RADIOULNAR JOINT

KEY POINTS: THE DRUJ

1. The DRUJ is affected by many traumatic and degenerative disorders.
2. Stability of the DRUJ is compromised in TFC tears, ulnar styloid base fractures, and fractures or malunions of the distal radius.
3. A supination contracture of the forearm is common following trauma to the DRUJ.

20. **How are DRUJ dislocations described?**
    The dislocations are described with respect to the direction of the ulna in relationship to the radius. With a palmar DRUJ dislocation, the ulna is palmar to the radius on a lateral x-ray. The dislocation is described as acute or chronic as well as simple or complex. A complex dislocation cannot be congruently reduced or is grossly unstable after reduction (Fig. 22-6)

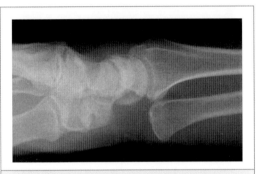

Figure 22-6. A true lateral view of the wrist demonstrating a palmar dislocation of the distal ulna with respect to the radius. This patient elected nonoperative treatment despite a significant loss of forearm rotation.

21. **How are DRUJ dislocations treated?**
    A simple acute dislocation is treated typically by closed reduction and immobilization of the forearm in near maximal supination for a dorsal dislocation and pronation for a palmar dislocation. Complex dislocations typically involve a fracture of the base of the ulnar styloid or tear of the TFCC. Occasionally, open reduction is required, especially in chronic cases or irreducible injuries.

22. **How does a distal radius fracture malunion affect the DRUJ?**
    A malunion that is excessively angulated in the palmar or dorsal direction leads to altered kinematic and DRUJ instability. A malunion that is excessively shortened results in ulnar positive variance and ulnocarpal impaction syndrome. Secondary arthritis of the ulnocarpal or radioulnar joints may result.

23. **Which imaging technique is useful in the diagnosis of subtle DRUJ instability?**
    A computed tomography (CT) scan. Both arms are positioned overhead, and the scans are taken through the sigmoid notch with the forearm in maximal supination, neutral forearm rotation, and maximal pronation (Fig. 22-7).

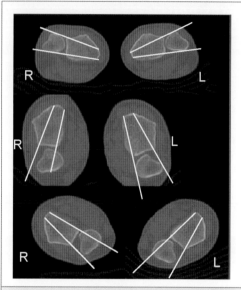

Figure 22-7. A CT scan of both DRUJs is demonstrated. Note the subtle palmar subluxation of the right ulna as seen on the lowest images when the forearms are in supination.

24. **What is Galeazzi's fracture?**
    A fracture of the radial (typically at the junction of the middle and distal third) shaft with associated DRUJ dislocation or subluxation. The DRUJ injury consists of disruption of the radioulnar ligaments from either the radius or ulna or an ulnar styloid fracture.

25. **How is Galeazzi's fracture treated?**
    Anatomic reduction and plate fixation of the radius fracture followed by closed reduction of the DRUJ. If the DRUJ is reduced and stable, closed treatment in supination or pronation typically suffices. If it is unstable, CRPP is preferred. Open reduction is used if the DRUJ dislocation is irreducible.

26. **What is an Essex-Lopresti injury?**
    This injury consists of a radial head and/or neck fracture with an accompanying interosseous membrane injury and disruption of the DRUJ.

27. **How is an Essex-Lopresti injury treated?**
    A high index of suspicion that the injury may exist is very important. When a radial head/neck fracture exists, the DRUJ should be routinely examined both clinically and radiographically. In the presence of combined proximal and distal radioulnar joint injuries, both joints must be treated appropriately to prevent late longitudinal instability of the forearm. The radial head fracture should be anatomically reduced and stabilized, or a radial head metallic prosthesis should be inserted followed by treatment of the distal radioulnar joint pathology, usually repair of the TFCC.

28. **What is the Darrach procedure?**
    The Darrach procedure is a resection of the distal end of the ulna and is usually indicated in those patients who place *very low* demands on the upper extremity.

29. **What is the Suavé-Kapandji procedure?**
   An arthrodesis of the distal radioulnar joint with an associated surgically created pseudoarthrosis of the ulna proximal to the arthrodesis. The benefit is that ulnocarpal support is preserved; however, impingement between the radius and the ulna proximally still may occur.

30. **In the absence of elbow pathology, what is a common cause of a loss of forearm rotation after trauma?**
   DRUJ capsular contracture. A loss of supination occurs frequently and can be improved or prevented by splinting the forearm in supination. If significant supination or pronation is lost, a palmar or dorsal DRUJ capsulectomy, respectively, followed by splinting and intensive therapy can be beneficial.

BIBLIOGRAPHY

1. Adams BD, Holley KA: Strains of the articular disk of the triangular fibrocartilage complex: A biomechanical study. J Hand Surg 18A:919–925, 1993.
2. Feldon P, Terrono AL, Belsky MR: Wafer distal ulna resection for triangular fibrocartilage tears and/or ulna impaction syndrome. J Hand Surg 17A:731–737, 1992.
3. Kalainov D, Culp R: Arthroscopic treatment of TFCC tears. Tech Hand Upper Extrem Surg 1:175–182, 1997.
4. Kleinman WB, Graham TJ: The distal radioulnar joint capsule: Clinical anatomy and the role in posttraumatic limitation of forearm rotation. J Hand Surg 23A:588–599, 1998.
5. Koppel M, Hargreaves IC, Herbert TJ: Ulnar shortening osteotomy for ulnar carpal instability and ulnar carpal impaction. J Hand Surg 22B:451–456, 1997.
6. Palmer AK, Werner FW: The triangular fibrocartilage complex of the wrist: Anatomy and function. J Hand Surg 6A:153–162, 1981.
7. Schurman AH, Bos KE: The ulnocarpal abutment syndrome: Follow-up of the wafer procedure. J Hand Surg 20B:171–177, 1995.
8. Shaaban H, Giakas G, Bolton M, et al: The distal radioulnar joint as a load-bearing mechanism—A biomechanical study. J Hand Surg 29A:85–95, 2004.
9. Tolat AR, Stanley JK, Trail IA: A cadaveric study of the anatomy and stability of the distal radioulnar joint in the coronal and transverse planes. J Hand Surg 21B:587–594, 1996.

CHAPTER 23
CARPAL FRACTURES AND INSTABILITY
Adam Mirachi, MD

CARPAL INSTABILITY

1. **What are the components of wrist stability?**
   The wrist is a sophisticated joint in which the complex integrated motion of the carpal bones depends on several factors, including;
   - Normal bony architecture of the carpal bones
   - The articular surfaces of the radius and ulna
   - The integrity of the intrinsic and extrinsic radiocarpal ligaments and triangular fibrocartilage complex (TFCC)
   - Balanced musculotendinous forces across the radiocarpal, intercarpal, and carpometacarpal (CMC) joints

2. **What are the key ligaments responsible for maintaining carpal stability?**
   1. Volar extrinsic radiocarpal ligaments
      - Radioscaphocapitate
      - Radiolunotriquetral
      - Radioscapholunate (ligament of Testut)
      - Ulnotriquetral ligament
      - Disk-carpal ligaments (disk-lunate and disk-triquetral). Together considered the *ulnocarpal volar ligament*, the disk-carpal ligaments originate from the volar distal radioulnar joint ligament and insert independently into the respective carpal bones. They are the primary stabilizers between the ulna and the volar carpus.
   2. Volar intrinsic ligaments
      - Arcuate or deltoid ligament
   3. Dorsal extrinsic radiocarpal ligaments
   4. Dorsal intrinsic ligaments
   5. Interosseous ligaments (Fig. 23-1)

3. **What are the stabilizing ligaments of the radiocarpal and midcarpal articulations?**
   - The radiocarpal joint is stabilized predominantly by the radioscaphocapitate, long radiolunate, dorsal radiolunotriquetral ligament, and ulnotriquetral ligaments.
   - The proximal row is stabilized by the scapholunate (SL) and lunotriquetral (LT) interosseous ligaments and the dorsal intercarpal ligament.
   - The midcarpal joint is stabilized by the congruent, closely matched articular surfaces and the compressive forces created by the extrinsic flexor and extensor tendons.
   - The distal carpal row is stabilized by the arcuate and interosseous ligaments.

4. **What is the space of Poirier?**
   A fenestration at the capitolunate (CL) articulation between the stout ligaments stabilizing the carpus. A lunate dislocation or perilunate fracture-dislocation is associated with a transverse capsular rent through this inherently weak region.

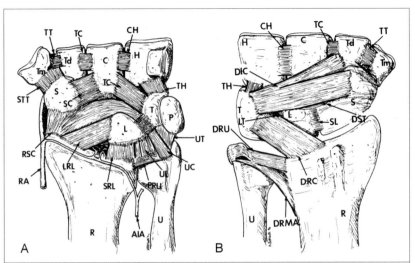

Figure 23-1. Extrinsic and intrinsic ligaments of the wrist. **A,** Palmar carpal ligaments from a palmar perspective. **B,** Dorsal carpal ligaments from a dorsal perspective. RSC = radioscaphocapitate, LRL = long radiolunate, SRL = short radiolunate, UL = ulnolunate, UT = ulnotriquetral, UC = ulnocapitate, PRU = palmar radioulnar, STT = scaphotrapeziotrapezoid, SC = scaphocapitate, TC = triquetrocapitate, CH = capitohamate, R = radius, U = ulna, S = scaphoid, L = lunate, T = triquetrum, P = pisiform, Tm = trapezium, Td = trapezoid, C = capitate, H = hamate, DRU = dorsal radioulnar, DRMA = dorsal radial metaphyseal arcuate, DRC = dorsal radiocarpal, DIC = dorsal intercarpal, DST = dorsal scaphotriquetral, SL = scapholunate, LT = lunotriquetral, TH = triquetrohamate, TT = trapeziotrapezoid, TC = trapeziocapitate, CH = capitohamate. (From Berger RA: The ligaments of the wrist: A current overview of anatomy with considerations of their potential functions. Hand Clin 13:63–82, 1997.)

5. **What is the function of the ligament of Testut?**
   The radioscapholunate ligament or ligament of Testut is a neurovascular conduit for the SL interosseous membrane and not a true extrinsic ligament of the wrist as previously thought.

6. **Describe the evaluation of a patient with suspected carpal instability.**
   - A detailed clinical history should be obtained, including the mechanism of injury, the character of the symptoms, and details about any prior wrist injury.
   - Physical examination is performed with the patient's elbow on an examining table to permit examination of the patient's right wrist with the examiner's right hand and vice versa. Palpation for areas of tenderness and passive motion to elicit instability should be performed. Specific provocative maneuvers should be performed based on the patient's complaints.
   - The findings are always compared with the contralateral extremity to improve diagnostic accuracy.

7. **What is the most common carpal instability pattern encountered?**
   SL instability.

8. **What are the common provocative tests used to evaluate for carpal instability?**
   - The **Watson scaphoid shift test** is a maneuver that assesses SL ligament competence. The wrist is positioned in ulnar deviation, and the scaphoid is stabilized in extension. The wrist is then brought into radial deviation, and scaphoid flexion is resisted by the examiner's thumb,

which applies a dorsally directed force. When the SL ligament is incompetent, the scaphoid shifts dorsally. An audible or palpable painful click or clunk may be elicited. Many patients will exhibit some mobility of the scaphoid, but unequivocal instability with reproduction of the patient's pain are the critical factors for a positive result.

- **Midcarpal instability** is assessed by applying axial compression of the wrist while simultaneously moving the wrist from extreme ulnar deviation to radial deviation. If the entire proximal row snaps from dorsiflexion into palmarflexion, then midcarpal instability is suspected.

Three tests have been described for assessing the competency of the LT ligament:

- The **Reagan ballottement test** relies on "trapping" the lunate between the thumb and index finger of one of the examiner's hands and the triquetrum in the other as the bones are moved independently and in an opposite direction from each other.
- In the **Kleinman shear test**, the examiner's thumbs are placed on the dorsal aspect of the pisiform and lunate. The bones are usually translated in an opposite direction with respect to each other.
- The **Linscheid ulnar snuffbox compression test** is performed with the wrist stabilized in slight radial deviation. Direct palpation of the triquetrum is performed. Pain at the LT interval suggests a ligamentous injury.

9. **Which radiographs are obtained in the evaluation of carpal instability?**
   A "carpal instability series" includes:
   - Posteroanterior (PA) view of the wrist (zero rotation preferred)
   - Lateral view of the wrist (zero rotation preferred)
   - Grip series (PA and lateral)
   - PA view of the wrist in radial and ulnar deviation
   - Comparison views of the opposite extremity

10. **When should a "clenched fist view" of the wrist be ordered? How is it obtained?**
    When SL dissociation is suspected. The view is obtained by having the patient make a fist while ulnarly deviating the wrist. This results in a compression load across the carpus that accentuates widening of the SL interval, which is normally 3–4 mm.

11. **Which additional imaging modalities are valuable in assessing carpal instability?**
    - **Wrist arthrography:** To detect ligament tears, TFCC lesions, and articular cartilage injuries
    - **Real-time fluoroscopy:** To assess for abnormal carpal motion
    - **Distraction PA and lateral radiographs of the wrist:** To assess for intercarpal ligament disruptions
    - **A technetium-99 bone scan:** To detect occult fractures or abnormalities of bone metabolism (i.e., degenerative arthritis)
    - **Ultrasound:** To localize suspected small or occult wrist ganglia
    - **CT scan:** To detect carpal fractures, malunions, and nonunions
    - **Magnetic resonance imaging (MRI):** To visualize the intrinsic and extrinsic radiocarpal ligaments and TFCC
    - **Diagnostic wrist arthroscopy:** To visualize the intrinsic/extrinsic ligaments, TFCC, and articular cartilage (the gold standard for detecting these conditions)

12. **Which radiographic features should be assessed in the evaluation of a patient with suspected carpal instability?**
    1. Presence of any fracture, malunion, or nonunion
    2. Ulnar variance

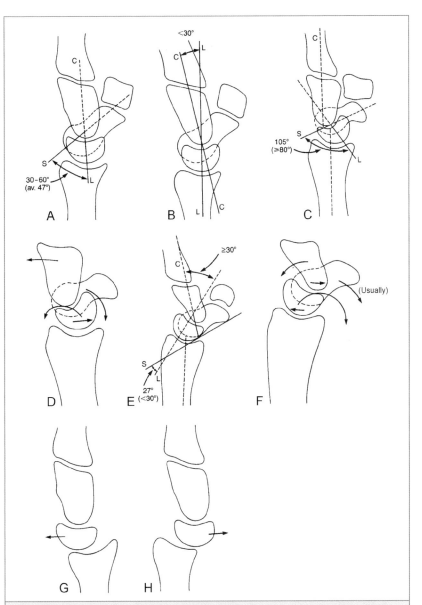

Figure 23-2. Radiographic measurement of the intercarpal angles analyzed in carpal instability. *A,* The SL angle. It is normally between 30 and 60 degrees and sometimes up to 80 degrees. *B,* The CL angle. It is normally <30 degrees. *C,* DISI is suspected when there is dorsal tilting of the lunate and ventral tilting of the scaphoid with a resultant increased SL angle, with or without an increased CL angle. *D,* Example of DISI. As the lunate tilts dorsally and the scaphoid tilts volarly, the lunate tends to move volarly and the capitate dorsally. *E,* Volar intercalated segmental instability (VISI) is suspected with an SL angle that is <30 degrees and/or a CL angle of ≥30 degrees. *F,* The scaphoid and lunate both tilting volarly. This is characteristic of VISI. *G,* With dorsal carpal subluxation, the center of the carpus is dorsal to the center of the mid-axis of the radius. *H,* Palmar carpal subluxation is recognized when the central axis of the carpus and lunate is volar to the mid-axis of the radius. (From Gilula LA, Weeks PM: Posttraumatic ligamentous instabilities of the wrist. Radiology 129:641–651, 1978.)

3. Intercarpal angles (*true* lateral view of the wrist)
    - **Scapholunate angle:** Normal = 45 to 60 degrees
    - **Capitolunate angle:** Normal = −20 to +10 degrees
    - **Radiolunate angle:** Normal = −20 to +10 degrees
4. Carpal height index (CHI), which is the carpal height divided by the length of the third metacarpal (normal 0.46–0.61)
5. SL or LT interosseous distance of ≥3–4 mm suggesting ligament disruption
6. Disruption of the carpal arcs (Fig. 23-2)

13. **Describe the triangulation method for determining midcarpal instability.**
    On a lateral view of the wrist, three landmarks are noted:
    - The dorsal pole of the lunate
    - The palmar pole of the lunate
    - A point between the dorsal and middle third of the capitate at the level of the third CMC joint.

    The points are connected to form a triangle with a dorsal limb (DL) and a palmar limb (PL). DL–PL ratios of greater than 1.0 indicate dorsal intercalated segmental instability (DISI). DL–PL ratios less than 0.5 are consistent with volar intercalated segment instability (VISI) deformities. If there is a borderline value, then the radiometacarpal (RM) angle is measured.

14. **Which specific radiographic findings are associated with SL instability?**
    - Widening of the SL interval ≥3 mm
    - Increased SL angle (>60 degrees) resulting from excessive scaphoid volar flexion and dorsiflexion of the lunate
    - A cortical ring sign in which the axial view of the volarflexed scaphoid forms a round projection on a PA view of the wrist
    - The V sign of Taleisnik on the lateral view of the wrist, with (the V referring to the volar silhouette of the palmar flexed scaphoid and radius)

15. **What is the "Terry Thomas" sign?**
    The finding of greater than 3 mm of widening between the scaphoid and lunate on a PA view of the wrist that is considered indicative of SL instability. The finding was named after Terry Thomas, a famous British comedian who had a large gap between his upper front teeth.

16. **How are carpal instabilities classified?**
    Classifying carpal instabilities is complex and challenging. There is no universally accepted system. Dobyns and Cooney have proposed the system in Table 23-1.

17. **What are DISI and VISI?**
    - **DISI**, which stands for dorsal intercalated segmental instability, involves a dorsiflexed (extended) posture of the lunate noted on a lateral view of the wrist. This is usually seen with rupture or degeneration of the SL ligament.
    - **VISI**, which stands for volar intercalated segmental instability, refers to the volar flexed lunate that commonly occurs following rupture of the LT ligament.

18. **Why does the lunate adopt a DISI posture when both the SL and LT ligaments have been ruptured?**
    The lunate adopts a DISI posture as a result of loss of the volarflexing influence of the scaphoid. In addition, the geometry of the distal radial articular surface and the relative dorsal position of the capitate axis on the smaller dorsal pole of the lunate also contribute.

TABLE 23-1. MAYO CLINIC'S CLASSIFICATION OF CARPAL INSTABILITY

Type, Site, and Name	Radiographic Pattern
I. Carpal Instability Dissociative (CID)	
1.1 Proximal carpal row CID	
a. Unstable scaphoid fracture	DISI
b. SL dissociation	DISI
c. LT dissociation	VISI
1.2 Distal carpal row CID	
a. Axial radial disruption	Radial translation or proximal translation
b. Axial ulnar disruption	Ulnar translation or proximal translation
c. Combined axial radial and axial ulnar disruption	
1.3 Combined proximal and distal CID	
II. Carpal Instability Nondissociative (CIND)	
2.1 Radiocarpal CIND	
a. Palmar ligament rupture	DISI, ulnar translation of entire proximal carpal row *or* ulnar translation with increased SL space; proximal translation (actually carpal instability complex)
b. Dorsal ligament rupture	VISI, dorsal translation
c. After "radius malunion"	Madelung's deformity, scaphoid malunion, lunate malunion (*see* section IV)
2.2 Midcarpal CIND	
a. Ulnar midcarpal instability from palmar ligament damage	VISI
b. Radial midcarpal instability from palmar ligament damage	VISI
c. Combined ulnar and radial midcarpal instability, palmar ligament damage	VISI
d. MCI from dorsal ligament damage	DISI
2.3 Combined radiocarpal-midcarpal CIND	
a. Capitolunate instability pattern (CLIP)	VISI, DISI, alternating
b. Disruption of radial and central ligaments	Ulnar translation with or without VISI or DISI
III. Carpal Instability Complex (CIC)	
a. Perilunate with radiocarpal instability	DISI and ulnar translation
b. Perilunate with axial instability	Axial ulnar instability and ulnar translation

Continued

TABLE 23-1. CONT'D	
Type, Site, and Name	Radiographic Pattern
c. Radiocarpal with axial instability	Axial radial instability and ulnar translation
d. SL dissociation with ulnar translation	DISI and ulnar translation
IV. Carpal Instability Adaptive ("Adaptive Carpus")	
a. Malposition of carpus with distal radius malunion	DISI or dorsal translation
b. Malposition of carpus with scaphoid nonunion	DISI
c. Malposition of carpus with lunate malunion	DISI or VISI
d. Malposition of carpus with Madelung's deformity	Ulnar translation, DISI, proximal translation

19. **What is the pattern of carpal force transmission during a perilunate injury?**
    The pattern of force transmission involves wrist dorsiflexion, ulnar deviation, and intercarpal supination. The injury begins at the radial aspect of the wrist and propagates toward the ulnar side through the midcarpal space.
    Mayfield described this concept as *progressive perilunar instability*, which may be divided into four sequential stages:
    - **Stage I** involves rupture of the SL ligament.
    - **Stage II** involves a midcarpal dislocation (usually dorsal).
    - **Stage III** involves rupture of the LT ligament.
    - **Stage IV** consists of complete dislocation of the lunate (usually volar) (Fig. 23-3).

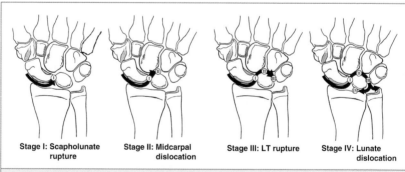

Figure 23-3. Progressive perilunar instability. (Adapted from Mayfield JK: Patterns of injury to carpal ligaments. A spectrum. Clin Orthop 187:36–42, 1984.)

20. **What is the "spilled tea cup" sign?**
    A radiographic finding noted on a lateral view of the wrist with stage IV perilunate instability. The lunate is usually rotated 90 degrees toward the palm on its axis, resembling a spilled tea cup.

21. **What is the most common form of carpal instability?**
    Carpal instability dissociative (CID), specifically an SL ligament disruption with a DISI deformity.

22. **What do the terms CID, CIND, and CIC mean?**
    - **CID**, carpal instability dissociative, refers to a disruption between the bones of the proximal row (SL and LT ligament disruption).
    - **CIND**, carpal instability nondissociative, is characterized by the bones of the proximal and distal carpal rows maintaining their relationship with respect to one another. However, the radiocarpal or midcarpal articulations are disrupted.
    - **CIC**, carpal instability complex, is a greater or lesser arc injury that results in a lunate dislocation or perilunate fracture-dislocation.

23. **Describe the difference between a greater, lesser, and inferior arc injury.**
    - A **greater arc** injury is a perilunate injury that involves a fracture of one or more carpal bones and disruption of the midcarpal articulation. The most common example is a trans-scaphoid dorsal perilunate fracture-dislocation.
    - A **lesser arc** injury involves the intercarpal and extrinsic ligaments without an associated carpal fracture.
    - An **inferior arc** injury is propagated through the radiocarpal joint instead of traversing the carpus (Fig. 23-4). The volar and dorsal extrinsic radiocarpal ligaments may be disrupted, or a

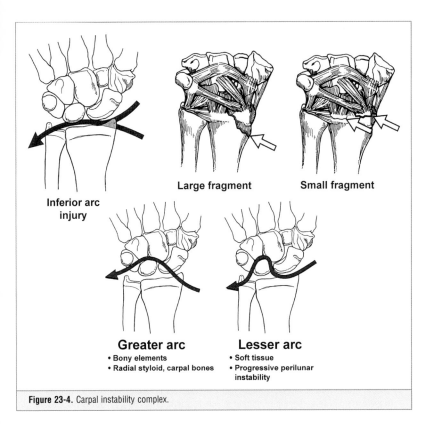

Figure 23-4. Carpal instability complex.

fracture of the radial styloid may occur. Associated fractures of the radial and ulnar styloid suggest an inferior arc injury. These injuries are potentially unstable. The classic inferior arc injury is a radiocarpal dislocation.

24. **How is SL instability treated?**
    Treatment of an SL ligament injury depends on:
    - Severity of the injury
    - Presence of associated injuries
    - Duration of instability
    - Absence or presence of carpal collapse
    - Integrity of the articular surfaces

    Treatment options include:
    - Casting or splinting
    - Primary SL ligament repair
    - Capsulodesis procedure to tether the distal scaphoid and prevent palmar flexion of the scaphoid
    - Limited intercarpal arthrodesis (scaphotrapeziotrapezoid, scaphocapitate, or capitate-hamate-lunate-triquetrum [four-corner] with scaphoid excision)
    - Proximal row carpectomy
    - Wrist arthrodesis
    - Wrist implant arthroplasty

25. **What are the signs and symptoms of LT instability?**
    - Pain in the area of the LT joint that is reproduced or exacerbated with provocative maneuvers
    - Dyskinesis of the lunate and triquetrum on physical examination and a palpable and/or visual clunk
    - Instability noted on real-time fluoroscopy of the injured patient
    - Arthrography possibly revealing dye leakage through the LT Interval
    - Static VISI deformity on a lateral view of the wrist

26. **How is LT instability treated?**
    Treatment options for an LT ligament injury include:
    - Primary ligament repair or LT joint pinning
    - Ligament reconstruction using a portion of the extensor carpi ulnaris or dorsal radiotriquetral ligament
    - LT arthrodesis
    - Limited intercarpal arthrodesis (four-corner)

27. **What are the causes and treatment options for midcarpal instability?**
    Midcarpal instability may be noted in young females with generalized ligamentous laxity. Trauma to the wrist may also destabilize the midcarpal joint, resulting in pain and dysfunction. Midcarpal instability has also been noted in some patients with a malunited distal radius fracture.

    Treatment is individualized and etiology specific. Exhaustive use of nonoperative modalities, such as splinting and activity modification, is recommended prior to proceeding with surgical reconstruction. If the patient's symptoms are recalcitrant to nonoperative management, surgical options include limited intercarpal arthrodesis, wrist fusion, dorsal and volar ligament reefing, or an osteotomy of the distal radius (if secondary to a malunion).

28. **Describe axial instability of the carpus (CIA), and discuss treatment.**
    Axial instability of the carpus is a rare injury that occurs following significant trauma to the extremity (e.g., fall injury, motor vehicle accident, crush injury). CIA is manifested

as a dissociation between the bones of the distal carpal row. *The characteristic finding is a longitudinal cleavage between the metacarpals and distal carpal row.* Various injury patterns have been described (Fig. 23-5). These injuries are often open. Compartment syndrome and/or an acute ulnar or median nerve compression neuropathy is common. Treatment is individualized but often involves thorough wound irrigation and debridement, fracture reduction and stabilization, fasciotomies, and nerve releases.

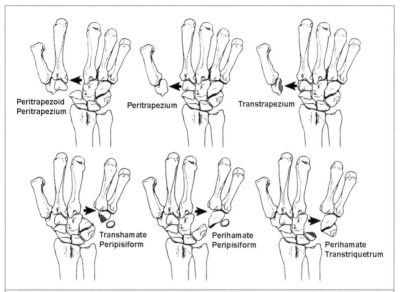

Figure 23-5. The various injury patterns of Carpal Instability Complex. (Courtesy of the Mayo Foundation.)

29. **Describe posttraumatic ulnar translocation of the carpus and its treatment.**
    Ulnar translocation of the carpus is a rare instability pattern that is the result of extensive injury to the extrinsic radiocarpal ligaments. The carpus subsequently "migrates" ulnarward with the lunate resting on the distal ulna. Two injury patterns have been described:
    - **Type 1:** The entire carpus translates ulnarly, and the radioscaphoid articulation is widened, often mimicking an SL dissociation.
    - **Type 2:** The radioscaphoid articulation is maintained, and the lunate and remaining carpus translate ulnarly.

    Treatment of this rare injury is challenging. Ligament repair (dorsal and volar extrinsic radiocarpal ligaments or interosseous ligaments) or reconstruction is unpredictable and unsatisfactory. Radiolunate or complete wrist arthrodesis may be indicated in difficult, late-presenting cases or following failed ligament repair.

CARPAL FRACTURES

30. **Which carpal bone is most frequently fractured?**
    - The scaphoid is commonly fractured.
    - Approximately 80% of such fractures in adults occur at the waist region of the scaphoid.
    - In children, the distal third is most commonly fractured.

31. **Describe the mechanism of injury for a scaphoid fracture.**
    Scaphoid fractures most commonly occur with a fall on an outstretched hand, resulting in wrist dorsiflexion usually greater than 95 degrees and greater than 10 degrees of radial deviation.

32. **How are scaphoid fractures classified?**
    - **Russe's classification** has three types based on the fracture orientation in relation to the longitudinal axis of the scaphoid. They are the horizontal oblique (HO), transverse (T), and vertical oblique (VO) (Fig. 23-6).
    - **Herbert's classification** is more extensive and consists of four types of fractures. Type A includes stable acute fractures; type B includes unstable acute fractures; type C includes delayed union, and type D includes established nonunion (Fig. 23-7).
    - The **simplest and most commonly used classification** is based on the location of the fracture within the scaphoid: distal pole or tuberosity, waist, or proximal pole.

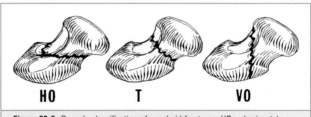

Figure 23-6. Russe's classification of scaphoid fractures. HO = horizontal oblique, T = transverse, VO = vertical oblique.

33. **Which scaphoid fracture types have the lowest rate of healing?**
    A fracture of the proximal pole has the propensity to develop a nonunion and/or avascular necrosis.

34. **How is a scaphoid fracture treated?**
    - **Nondisplaced** scaphoid fractures may be treated with cast immobilization or percutaneous or open screw fixation. Debate persists regarding the merits of immobilizing the elbow (long arm versus short arm cast). An initial long arm cast appears to reduce the time of healing compared with a short arm cast. An acceptable strategy involves placing a long arm cast for 6 weeks and then a short arm cast until there is radiographic evidence of healing. Screw fixation has been advocated to avoid the morbidity of prolonged cast immobilization.
    - **Displaced or unstable** fractures require open reduction and internal fixation (ORIF) (Fig. 23-8).

35. **Describe the blood supply to the scaphoid.**
    The scaphoid receives its primary blood supply through ligamentous and capsular attachments that enter via the dorsal ridge. The major blood supply is a branch of the radial artery. This branch supplies approximately 70% of the scaphoid, including the proximal third. The remaining 30% of the distal scaphoid is supplied by a second group of vessels that enter the tuberosity region. Additional vessels enter the lateral, volar, and distal scaphoid. It is widely agreed that the proximal pole is the region with the most tenuous blood supply, owing to the predominantly distal-to-proximal (retrograde) orientation of the vascular supply.

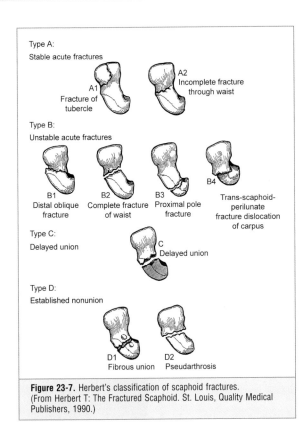

Figure 23-7. Herbert's classification of scaphoid fractures. (From Herbert T: The Fractured Scaphoid. St. Louis, Quality Medical Publishers, 1990.)

36. **What are the criteria for classifying a scaphoid fracture as displaced?**
    - 1 mm of displacement on any radiographic view
    - Angular displacement >10 degrees
    - Fracture comminution

37. **What other imaging modalities are used in the evaluation of a scaphoid fracture or nonunion?**
    - A technetium-99 bone scan may be used in the early postinjury period to identify an occult fracture of the scaphoid and can be performed as early as 48–72 hours after injury. The test is highly sensitive but has a relatively low specificity.
    - A CT scan is helpful for delineating an acute fracture but has a more significant role in the evaluation of a scaphoid malunion or nonunion. CT is the procedure of choice for evaluating a collapsed or angulated fracture malunion or nonunion and for confirming healing of a scaphoid fracture.

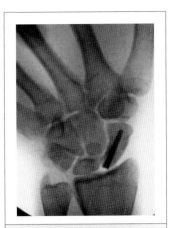

Figure 23-8. ORIF of a Herbert A2 scaphoid fracture with a headless cannulated screw.

- An MRI with gadolinium is helpful for identifying avascular necrosis (AVN) of the proximal pole.
- Trispiral tomography is effective in the evaluation of a suspected acute fracture and nonunion and to assess scaphoid healing following fracture or bone grafting of a nonunion.

38. **What is Preiser's disease?**
    Idiopathic avascular necrosis of the scaphoid that has been associated with repetitive trauma, collagen vascular disease, or corticosteroid use.

39. **Which factors are associated with the development of a scaphoid fracture nonunion?**
    - Delay in treatment
    - Inadequate immobilization
    - Proximal fracture location
    - Initial fracture displacement
    - Fracture comminution
    - Advanced patient age
    - Presence of an associated carpal fracture (Fig. 23-9)

40. **What is the scaphocapitate syndrome?**
    A rare carpal injury pattern consisting of a fracture of the scaphoid waist and capitate neck with rotation of the proximal capitate fragment such that the articular surface is directed distally. The recommended treatment is open reduction internal fixation ORIF.

41. **What are the different bone grafting techniques used to treat a scaphoid nonunion?**
    - The treatment of choice for a scaphoid nonunion without collapse, shortening, or angulation is bone grafting.

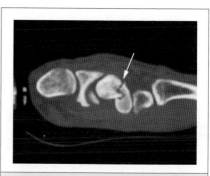

Figure 23-9. CT scan of a scaphoid fracture nonunion. Note the humpback deformity of the scaphoid.

- Cancellous grafting of the nonunion site may be performed through a volar approach, as described by Russe. The graft is usually harvested from the distal radius.
- A volar wedge graft, as described by Fernandez, is used to restore length and alignment in the collapsed, foreshortened scaphoid.
- Transferring a bone graft on its arterial pedicle from the radius to the scaphoid is becoming a popular procedure. The Zaidemberg graft relies on a septal vessel between the first and second dorsal extensor compartments (1,2 intercompartmental supraretinacular artery [ICSRA]) and the bone flap that accompanies it. A pedicled bone graft with a pronator quadratus attachment has also been described.

42. **What are the characteristics of the scaphoid nonunion advanced collapse (SNAC) pattern of wrist arthrosis?**
    The SNAC wrist can be described in four stages:
    - **Stage I:** Arthritis between the radial styloid and distal scaphoid
    - **Stage II:** Scaphoid fossa involvement
    - **Stage III:** Capitolunate (CT) arthrosis
    - **Stage IV:** Diffuse arthritis involving the entire wrist (Fig. 23-10)

43. **What is the relative incidence of carpal bone fractures?**
    - **Scaphoid:** Approximately 75–80% of all carpal fractures
    - **Triquetrum:** 10–15%
    - **Trapezium:** 2–5%
    - **Remaining carpal bones (hamate, lunate, pisiform, capitate, and trapezoid):** 1–2%
    - **Trapezoid:** Least common

44. **What are the different types of triquetral fractures?**
    Isolated triquetral fractures are relatively uncommon. They are more commonly associated with other carpal injuries, perilunate, and axial carpal fracture-dislocations. Cast or splint immobilization for 4–6 weeks is the preferred treatment approach
    There are two types of triquetral fractures:
    - The dorsal cortical fracture, which is produced by avulsion of the dorsal extrinsic ligaments or impaction against the ulnar styloid
    - The body fracture, which occurs less frequently

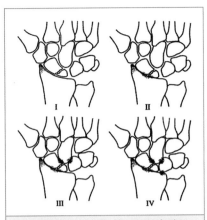

Figure 23-10. The predictable pattern of wrist arthritis associated with a scaphoid nonunion collapse pattern (SNAC) of wrist arthritis. Stage I is radial styloid arthritis; stage II is scaphoid fossa arthritis; stage III is CL arthritis, and stage IV is diffuse arthritis of the wrist. (From Trumble TE: Hand Surgery Update 3 Hand, Elbow, and Shoulder. Rosemont, IL, American Society for Surgery of the Hand, 2003.)

45. **What is the mechanism of injury for a pisiform fracture?**
    Direct trauma to the hypothenar eminence (Fig. 23-11).

46. **Why are pisiform fractures often initially not recognized?**
    - They commonly occur in association with other upper extremity injuries, which may be more obvious, thus delaying the diagnosis.
    - The fracture may not be detected on routine radiographs of the wrist or hand.
    - Special radiographic views, such as a supinated oblique or carpal tunnel view, are usually necessary to make the diagnosis.

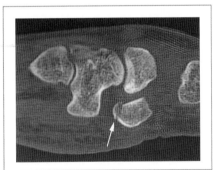

Figure 23-11. Pisiform fracture detected by CT scan.

47. **Which injuries occur in association with a trapezium fracture?**
    Fractures of the other carpal bones, the first metacarpal, and the distal radius.

48. **How are fractures of the trapezium classified?**
    There are three trapezial fracture patterns:
    - Body
    - Marginal
    - Volar ridge

49. **Which plain radiographic view should always be obtained in the evaluation of a suspected trapezial ridge fracture?**
Carpal tunnel view.

> ### KEY POINTS: CARPAL FRACTURES AND INSTABILITY
>
> 1. The ligament of Testut or radioscapholunate ligament primarily functions as a neurovascular conduit for the SL interosseous ligament.
> 2. The most common carpal instability pattern is an SL dissociation.
> 3. A scaphoid fracture occurs when the wrist is loaded and dorsiflexed more than 95 degrees.
> 4. A scaphoid fracture is considered displaced when there is more than 1 mm of gapping on any radiographic view or if there is more than 10 degrees of angulation between fragments.
> 5. Preiser's disease is idiopathic avascular necrosis of the scaphoid.
> 6. Excision is the preferred treatment for a symptomatic nonunion of the hamate hook.

50. **What are the three types of hamate fractures?**
    - **Body:** A fracture of the hamate body usually occurs following direct trauma to the ulnar aspect of the hand or wrist, or it may occur as part of an axial carpal instability injury.
    - **Hook of the hamate:** This fracture occurs following a direct injury to the hypothenar region, often in those who participate in racquet sports.
    - **Osteochondral or marginal fracture:** This fracture occurs in association with a fracture, dislocation, or fracture-dislocation of the ring and small finger metacarpals (CMC joint injuries).

51. **What complications occur following a hook of the hamate fracture?**
    - Nonunion (most common), which requires excision
    - Ulnar nerve compression
    - Rupture of the adjacent digital flexor tendons (usually flexor digitorum profundus [FDP] small)

52. **A 23-year-old man presents with severe wrist pain and hand numbness two hours after a motorcycle accident. He has no other injuries. Radiographs demonstrate a dorsal trans-scaphoid perilunate fracture-dislocation. What is the recommended treatment?**
    - Immediate closed reduction should be followed by observation to determine if the hand numbness (median nerve injury) resolves.
    - Closed reduction is performed with adequate analgesia and 10–15 lb of fingertrap countertraction.
    - If a satisfactory closed reduction cannot be achieved, open reduction is necessary.
    - If a satisfactory closed reduction of the associated midcarpal dislocation can be achieved and the numbness resolves, the patient may be scheduled electively for ORIF of the fracture and associated intercarpal ligament injury within 7–10 days after injury.

53. **What are the proposed advantages of performing percutaneous screw fixation versus cast immobilization for a nondisplaced scaphoid waist fracture?**
    - Percutaneous screw fixation avoids the morbidity of cast immobilization, specifically joint stiffness and muscle wasting.
    - Return to full activity occurs earlier in those fractures treated with internal fixation.
    - The time to union is shorter following screw fixation.

54. A 43-year-old avid golfer presents with a 2-day history of the sudden inability to actively flex the distal interphalangeal joint of the small finger. He denies any trauma or pain. What is the most likely cause?
Hook of the hamate fracture nonunion with an attritional rupture of the small finger FDP. Treatment consists of excision of the hook of the hamate nonunion and coverage of the exposed cancellous surface with a local periosteal flap combined with side-to-side transfer of the small finger FDP to the ring finger FDP.

55. A 28-year-old intoxicated orthopedic surgery resident falls on an outstretched hand. He presents with discomfort within the thenar region and the basilar aspect of the thumb. Physical examination demonstrates moderate swelling and ecchymosis throughout the thenar region. Tenderness is noted volarly over the basilar joint region, but no instability is detected. PA and lateral radiographs of the wrist and hand are unremarkable. How should the evaluation proceed?
The clinical history and physical findings are consistent with a fracture of the trapezial ridge, which is commonly missed on routine radiographic views. Diagnosis requires a carpal tunnel view radiograph or CT scan.

BIBLIOGRAPHY

1. Bond CD, Shin AY, McBride MT, Dao KD: Percutaneous screw fixation or cast immobilization for nondisplaced scaphoid fractures. J Bone Joint Surg 83A:483–488, 2001.
2. Cerezal L, Abascal F, Canga A, et al: Usefulness of gadolinium-enhanced MR imaging in the evaluation of the vascularity of scaphoid nonunions. Am J Roentgenol 174(1):141–149, 2000.
3. Dumontier C, Meyer zu Reckendorf G, Sautet A, et al: Radiocarpal dislocations: Classification and proposal for treatment. A review of twenty-seven cases. J Bone Joint Surg 83A(2):212–218, 2001.
4. Fleege MA, Jebson PJL, Renfrew DL, et al: Pisiform fractures. Skel Radiol 20:169–172, 1991.
5. Garcia-Elias M: Dorsal fractures of the triquetrum-avulsion or compression fractures? J Hand Surg 12A:266–268, 1987.
6. Garcia-Elias M, Dobyns JH, Cooney WP, Linscheid RL: Traumatic axial dislocations of the carpus. J Hand Surg 14A:336–357, 1989.
7. Gellman H, Caputo RJ, Carter V, et al: Comparison of short and long thumb-spica casts for nondisplaced fractures of the carpal scaphoid. J Bone Joint Surg 71A:354–357, 1989.
8. Kozin SH: Incidence, mechanism, and natural history of scaphoid fractures. Hand Clin 17(4):515–524, 2001.
9. Mayfield JK: Patterns of injury to carpal ligaments: A spectrum. Clin Orthop 187:36–42, 1984.
10. Mayfield JK: Wrist ligamentous anatomy and pathogenesis of carpal instability. Orthop Clin North Am 15:209–216, 1984.
11. Palmer AK: Trapezial ridge fractures. J Hand Surg 6:561–564, 1981.
12. Palmieri TJ: Pisiform area pain treated by pisiform excision. J Hand Surg 7A:477, 1982.
13. Pin PG, Young VL, Gilula LA, Weeks PM: Management of chronic lunotriquetral ligament tears. J Hand Surg 14A:77–83, 1989.
14. Pirela-Cruz MA, Hansen MF: Assessment of midcarpal deformity of the wrist using the triangulation method. J Hand Surg 28A(6):938–942, 2003.
15. Rand JA, Linscheid RL, Dobyns JH: Capitate fractures: A long-term follow-up. Clin Orthop 165:209–216, 1982.
16. Rayhack JM, Linscheid RL, Dobyns JH, Smith JH: Posttraumatic ulnar translation of the carpus. J Hand Surg 12A:180–189, 1987.
17. Shin AY, Weinstein LP, Berger RA, Bishop AT: Treatment of isolated injuries of the lunotriquetral ligament: A comparison of arthrodesis, ligament reconstruction, and ligament repair. J Bone Joint Surg 83B(7):1023–1028, 2002.
18. Steinmann SP, Bishop AT, Berger RA: Use of the 1,2 intercompartmental supraretinacular artery as a vascularized pedicle bone graft for difficult scaphoid nonunion. J Hand Surg 27(3):391–401, 2002.
19. Taleisnik J: Carpal Instability. J Bone Joint Surg 70A:1262–1267, 1988.

20. Trumble TE, Gilbert M, Murray LW, et al: Displaced scaphoid fractures treated with open reduction and internal fixation with a cannulated screw. J Bone Joint Surg 82A(5):633–641, 2000.
21. Vance RM, Gelberman RH, Evans EF: Scaphocapitate fractures: Patterns of dislocation, mechanism of injury, and preliminary results of treatment. J Bone Joint Surg 62A:271–276, 1980.
22. Watson HK, Ballet FL: The SLAC wrist: Scapholunate advanced collapse pattern of degenerative arthritis. J Hand Surg 9A:358–365, 1984.
23. Zaidemberg C, Siebert JW, Argrigiani C: A new vascularized bone graft for scaphoid nonunion. J Hand Surg 16A:474–478, 1991.

METACARPAL AND PHALANGEAL FRACTURES

Steven Shah, MD

CHAPTER 24

1. **Which radiographic views should be ordered to assess a fractured digit?**
   Three views—anteroposterior (AP), lateral, and oblique.

2. **What is the intrinsic plus position of immobilization?**
   The intrinsic plus position is used in the management of metacarpal or proximal phalangeal fractures. The wrist is immobilized in 30 degrees of extension, the metacarpophalangeal (MP) joints in 80–90 degrees of flexion, and the interphalangeal (IP) joints in full extension (Fig. 24-1).

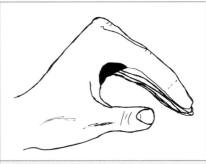

Figure 24-1. The "intrinsic plus" position of hand immobilization.

3. **What is the most common fracture in the hand?**
   Fracture of the distal phalanx.

4. **Which injury frequently occurs in association with a distal phalanx tuft fracture?**
   Laceration and/or avulsion of the nail matrix with a painful subungual hematoma.

5. **Describe the treatment of a displaced transverse fracture of the distal phalanx.**
   The fracture is occasionally open and often associated with laceration of the overlying nail matrix. The fracture should be irrigated followed by debridement of all devitalized tissue. The fracture is reduced and stabilized with a longitudinal Kirschner wire placed percutaneously. Failure to reduce and stabilize the fracture may result in a persistent deformity with displacement of the repaired nail bed.

6. **What is a mallet fracture?**
   An avulsion fracture of the dorsal base of the distal phalanx resulting in an extensor lag or droop at the distal interphalangeal (DIP) joint due to disruption of the extensor tendon mechanism (Fig. 24-2).

7. **How is a mallet fracture treated?**
   - Small, minimally displaced fractures without associated palmar subluxation of the remaining distal phalanx are amenable to splinting of the DIP joint in extension for 6 weeks.
   - Large displaced and/or malrotated fracture fragments or fractures that have persistent palmar subluxation of the distal phalanx despite closed reduction and splint immobilization require an open anatomic reduction of the articular surface and internal fixation.

8. **What are the most common complications associated with a fracture of the distal phalanx?**
   - Chronic fingertip pain
   - Cold sensitivity

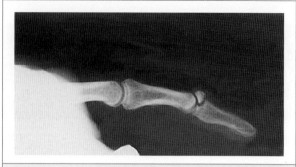

Figure 24-2. Mallet fracture.

- Hyperesthesia
- Loss of DIP joint motion
- Nail growth abnormalities

9. **How are condylar fractures of the phalangeal head classified?**
   - **Type I:** Unicondylar, nondisplaced, stable
   - **Type II:** Unicondylar, unstable, displaced
   - **Type III:** Bicondylar or comminuted, unstable (Fig. 24-3).

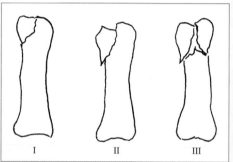

Figure 24-3. The three types of phalangeal condyle fractures.

10. **How are condylar fractures treated?**
    - **Type I:** Splint or cast immobilization with careful frequent radiographic follow-up to avoid fracture displacement and malunion
    - **Types II and III:** Closed reduction and percutaneous pinning (CRPP) or open reduction and internal fixation (ORIF)

11. **What is a pilon fracture of the proximal interphalangeal (PIP) joint?**
    A comminuted intra-articular fracture of the proximal articular surface of the middle phalanx with depression of the central articular surface and sagittal and coronal splaying of the remaining articular pieces (Fig 24-4).

12. **Describe the evaluation and treatment of a patient who presents with 30 degrees of active *and* passive PIP joint flexion after closed treatment of a proximal phalanx neck fracture. Clinical alignment is acceptable.**
    A proximal phalanx neck fracture has the potential to heal with excessive shortening and dorsal translation, resulting in a prominent, volar spike of bone that projects into the subcondylar space

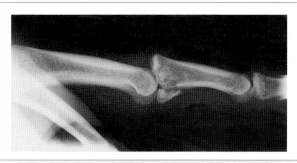

Figure 24-4. Pilon fracture of the PIP joint.

of the PIP joint subsequently blocking PIP flexion. The evaluation should consist of appropriate AP and true lateral radiographs of the involved digit. Surgical treatment involves excision of the bony block (exostectomy) through a mid-axial incision.

13. **What are the surgical treatment options for an unstable phalangeal shaft fracture?**
    - CRPP
    - ORIF
    - External fixation

14. **What are the indications for each of these techniques?**
    - CRPP is preferred for transverse, oblique, and spiral fractures.
    - ORIF is reserved for irreducible fractures, fractures that cannot be satisfactorily reduced and stabilized by closed methods, and displaced Salter-Harris type III and IV epiphyseal separations.
    - External fixation is reserved for contaminated open fractures, highly comminuted fractures, and fractures with associated bone loss or soft tissue compromise.

15. **About 5 weeks after closed reduction and splint immobilization of a stable transverse phalangeal shaft fracture, radiographs reveal a persistent fracture line. What is the appropriate treatment?**
    The radiographic and clinical signs of phalangeal fracture union do not correlate; radiographic fracture healing lags behind clinical healing. Clinical union is believed to occur by 5–6 weeks for phalangeal fractures. If the fracture is nontender and no motion is detected at the fracture site, immobilization may be discontinued and a range of motion program begun.

16. **Which metacarpal head is most frequently fractured?**
    The index finger.

17. **What are the treatment options for a comminuted metacarpal head fracture?**
    - A brief period (2–3 weeks) of splint immobilization followed by early motion
    - ORIF
    - External fixation
    - Soft tissue interpositional arthroplasty
    - Implant arthroplasty

18. **What is a "boxer's fracture"?**
    A fracture of the fifth metacarpal neck that occurs when a clenched MP joint strikes a solid object (e.g., wall). It is one of the most common hand fractures.

19. **Why do metacarpal neck fractures angulate in an apex dorsal manner?**
    The metacarpal head is flexed by the intrinsic tendons, which pass volar to the axis of MP joint motion, and the impact on striking a solid object results in comminution of the volar metacarpal neck.

20. **What amount of angulation is considered acceptable in the treatment of metacarpal neck fractures?**
    - **Index:** <10 degrees
    - **Long:** <15 degrees
    - **Ring:** <30 degrees
    - **Small:** <40 degrees

21. **How are metacarpal neck fractures treated?**
    - Most metacarpal neck fractures are amenable to closed manipulation and splint or cast immobilization for 3–4 weeks.
    - Unstable fractures usually are treated by CRPP.
    - Irreducible fractures or fractures that cannot be satisfactorily reduced by closed methods require ORIF (Fig. 24-5).

22. **How are metacarpal shaft fractures classified?**
    Transverse, oblique (short or long), spiral, or comminuted.

23. **Which metacarpal shaft fractures are associated with rotational malalignment?**
    Displaced spiral and oblique (long or short) fractures.

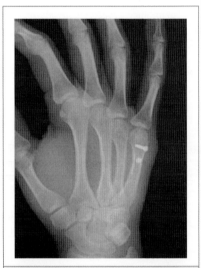

Figure 24-5. ORIF of a metacarpal shaft fracture.

KEY POINTS: METACARPAL AND PHALANGEAL FRACTURES

1. A tuft fracture of the distal phalanx is commonly associated with a nail bed laceration and subungual hematoma.
2. A metacarpal shaft fracture displaces with an apex dorsal pattern as a result of the proximal and volar pull of the intrinsics.
3. A transverse fracture of the proximal phalanx is amenable to fixation with flexible intramedullary wires.
4. The radiographic signs of healing of a hand fracture lag behind the clinical signs.

24. **How is rotational malalignment of a metacarpal or phalangeal fracture determined?**
    Ask the patient to flex the MP and PIP joints. In the injured hand, each digit should point toward the scaphoid tubercle; no digital overlap or divergence should be noted, and the planes of the nail plates should be identical to that in the uninvolved hand (Fig. 24-6).

Figure 24-6. Malrotation associated with a metacarpal shaft fracture. Note the divergence between the index and long fingers.

25. **What is the relationship between the amount of malrotation and digital overlap?**
    5 degrees of malrotation results in 1.5 cm of digital overlap in flexion.

26. **What is the typical displacement pattern of a metacarpal shaft fracture?**
    Metacarpal shaft fractures angulate in an apex-dorsal manner as a result of the proximal and volar pull of the intrinsics.

27. **How much angulation is acceptable in the treatment of a metacarpal shaft fracture?**
    - **Index:** <10 degrees
    - **Long:** <20 degrees
    - **Ring and small:** <30 degrees

28. **List the indications for ORIF of a metacarpal shaft fracture.**
    - Inability to obtain and/or maintain a satisfactory reduction with closed methods
    - Open fracture
    - Multiple metacarpal fractures
    - Fracture with segmental bone loss
    - Associated soft tissue loss

29. **What are the advantages and disadvantages of plate and screw fixation of a metacarpal or phalangeal fracture?**
    Advantages:
    - Rigid fixation, which permits early range of motion and return to function

    Disadvantages:
    - Technical difficulties in application
    - Extensive soft tissue dissection, resulting in adhesion formation and tendon irritation

- Plate interference with extensor tendon function
- Usually requires implant removal

30. **Which radiographic parameter is used to determine whether metacarpal length has been restored after closed or open treatment of a metacarpal shaft, neck, or base fracture?**
    On the AP view of a normal, uninjured hand, an oblique line may be drawn connecting the most distal aspects of the long, ring, and small finger metacarpal heads. This is referred to as the oblique metacarpal line (Fig. 24-7).

31. **What is the most commonly injured carpometacarpal (CMC) joint?**
    The small finger.

32. **Which injuries are associated with CMC joint injuries?**
    - Index and long—fractures of the trapezoid or trapezium and/or injury to the deep palmar arch, median, and ulnar nerves
    - Ring and small—injuries to the ulnar artery, ulnar nerve, or dorsal rim or body of the hamate

33. **Why is a fracture of the thumb proximal phalanx ulnar base important?**
    This fracture represents an avulsion injury of the ulnar collateral ligament complex of the thumb and may be associated with MP joint instability. Chronic instability can result in significant functional disability and the development of degenerative arthritis.

34. **How are ulnar collateral ligament avulsion fractures treated?**
    - Nondisplaced, or small and displaced <2 mm—thumb spica cast immobilization for 4–6 weeks
    - Displaced >2 mm, involvement >25% of the articular surface, and/or malrotated with associated clinical instability of the MP joint—ORIF (Fig. 24-8)

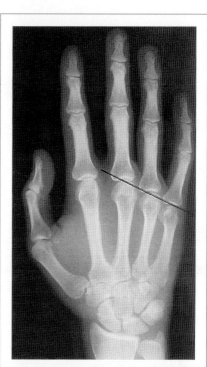

Figure 24-7. The heads of the long, ring, and small metacarpals aligned on the oblique metacarpal line.

35. **What is a Bennett's fracture?**
    - This injury is a fracture-dislocation of the first metacarpal base that occurs when the partially flexed thumb is axially loaded.
    - The volar ulnar aspect of the metacarpal base remains in anatomic position via the anterior oblique ligament. The remaining metacarpal shaft supinates, adducts, and displaces dorsally, proximally, and radially (Fig. 24-9).

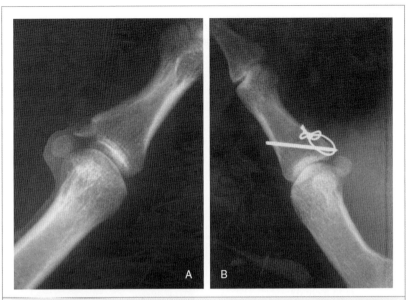

Figure 24-8. A displaced, malrotated avulsion fracture. **A,** The fracture involving the ulnar base of the thumb proximal phalanx ("bony skier's thumb"). **B,** The fracture reduced and stabilized with intraosseous wiring and a single Kirschner wire.

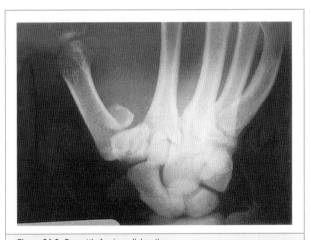

Figure 24-9. Bennett's fracture-dislocation.

36. **What is responsible for subluxation of the first metacarpal in a Bennett's fracture?**
    The first metacarpal base is displaced proximally, dorsally, and radially by the abductor pollicis longus (APL), which inserts on the first metacarpal base. Supination and adduction of the first metacarpal shaft are caused by the deforming force of the adductor pollicis.

METACARPAL AND PHALANGEAL FRACTURES

37. **How is a Bennett's fracture treated?**
    - If the volar ulnar fragment < 20% of the articular surface, CRPP is preferred.
    - If the fracture fragment > 25–30% of the articular surface, ORIF is recommended.

38. **What is a Rolando's fracture?**
    A Y- or T-shaped intra-articular fracture of the first metacarpal base was initially described by Silvio Rolando in 1910. The eponym now encompasses all comminuted intra-articular fractures of the first metacarpal base (Fig. 24-10).

39. **How is a Rolando's fracture treated?**
    The treatment approach depends on the degree of comminution. If there are two or three relatively large fracture fragments, ORIF may be feasible. If there is significant comminution, mini-external fixation with or without limited ORIF and bone grafting is recommended.

40. **What is the most common thumb fracture in a child?**
    Salter-Harris type II epiphyseal separation of the proximal phalanx.

41. **Why is residual deformity after a phalangeal fracture better tolerated in the thumb than in the fingers?**
    Because of compensatory motion at the adjacent CMC joint.

42. **What is a reverse Bennett's fracture?**
    A fracture-dislocation of the fifth metacarpal base. The radial base of the fifth metacarpal remains reduced via the carpometacarpal and intermetacarpal ligaments and joint capsule. The remaining metacarpal shaft is subluxated dorsally, proximally, and ulnarly by the deforming force of the extensor carpi ulnaris tendon (ECU).

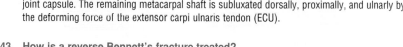

Figure 24-10. Rolando's fracture. This fracture has been reduced and stabilized using a combination of limited internal fixation with Kirschner wires and a mini-external fixator.

43. **How is a reverse Bennett's fracture treated?**
    A reverse Bennett's fracture is extremely unstable and usually not amenable to closed reduction and cast or splint immobilization. CRPP of the fifth metacarpal to the fourth metacarpal and hamate is recommended.

44. **Which specific type of proximal phalangeal fracture is amenable to fixation with flexible intramedullary Kirschner wires?**
    Transverse or short (minimally) oblique.

45. **Describe the relationship between metacarpal shortening after a fracture and metacarpophalangeal (MCP) joint motion.**
    There is a 7-degree extensor lag of the MCP joint for every 2 mm of metacarpal shortening.

46. **Which method of fixation for a metacarpal shaft fracture is best when tested in bending, torsion, and axial loading?**
 Dorsal plating with a lag screw.

47. **Which metacarpal fracture pattern is considered a contraindication to flexible intramedullary nailing?**
 Long oblique or comminuted. Rotational stability and maintenance of length are inadequate with flexible intramedullary nailing.

48. **Why should detachment of the collateral ligaments be avoided when performing ORIF of a displaced bicondylar fracture of the proximal or middle phalanx?**
 The condyles receive much of their blood supply via the collateral ligaments. Detachment may result in osteonecrosis and/or nonunion.

49. **A 19-year-old college student injures his right hand in a fight. He does not seek evaluation immediately but presents 2 months later complaining of overlapping of the small finger over the ring finger and an impaired grip. Physical examination demonstrates full active and passive range of motion of the MCP and IP joints. Scissoring of the small and ring fingers is noted on flexion. Radiographs demonstrate a healed spiral metacarpal shaft fracture. What is the recommended treatment?**
 The patient has a malunion due to neglect of a spiral metacarpal shaft fracture. If functional compromise is noted, a derotational osteotomy performed at the metacarpal base is recommended.

50. **Describe the typical pattern of displacement for a fracture of the proximal and middle phalanx.**
 Proximal phalanx:
 - **Apex volar**, flexion of the proximal fragment via the interosseous muscles, and extension of the distal fragment via the central slip

 Middle phalanx:
 - At or proximal to flexor digitorum sublimis (FDS) insertion, **apex dorsal**, extension of proximal fragment via central slip, and flexion of distal fragment by FDS
 - Distal to FDS insertion, **apex volar**, flexion of proximal fragment by FDS, and extension of distal fragment via terminal slip of extensor mechanism

BIBLIOGRAPHY

1. Buchler U, McCollam SM, Oppikofer C: Comminuted fractures of the basilar joint of the thumb: Combined treatment by external fixation, limited internal fixation, and bone grafting. J Hand Surg 16A:556–560, 1991.
2. Deitch MA, Kiefhaber TR, Comisar BR, Stern PJ: Dorsal fracture dislocations of the proximal interphalangeal joint: Surgical complications and long-term results. J Hand Surg (Am) 24(5):914–923, 1999.
3. Dinowitz M, Trumble T, Hanel D, et al: Failure of cast immobilization for thumb ulnar collateral ligament avulsion fractures. J Hand Surg 22A:1057–1063, 1997.
4. Dionysian E, Eaton RG: The long-term outcome of volar plate arthroplasty of the proximal interphalangeal joint. J Hand Surg (Am) 25(3):429–437, 2000.
5. Fusetti C, Meyer H, Borisch N, et al: Complications of plate fixation in metacarpal fractures. J Trauma 52(3):535–539, 2002.
6. Gonzalez MH, Hall RF: Intramedullary fixation of metacarpal and proximal phalangeal fractures of the hand. Clin Orthop 327:47–54, 1996.
7. Gonzalez MH, Igram CM, Hall RF: Intramedullary nailing of proximal phalangeal fractures. J Hand Surg 20A:808–812, 1995.

8. King HJ, Shin SJ, Kang ES: Complications of operative treatment for mallet fractures of the distal phalanx. J Hand Surg 26(1):28–31, 2001.
9. Kozin S, Bishop AT: Tension wire fixation of avulsion fractures at the thumb metacarpophalangeal joint. J Hand Surg 19A:1027–1031, 1994.
10. Krakauer JD, Stern PJ: Hinged device for fractures involving the proximal interphalangeal joint. Clin Orthop Rel Res 327:29–37, 1996.
11. Menon J: Correction of rotary malunion of the fingers by metacarpal rotational osteotomy. Orthopedics 13:197–200, 1990.
12. Page SM, Stern PJ: Complications and range of motion following plate fixation of metacarpal and phalangeal fractures. J Hand Surg (Am) 23(5):827–832, 1998.
13. Stern PJ, Roman RJ, Kiefhaber TR, McDonough JJ: Pilon fractures of the proximal interphalangeal joint. J Hand Surg 16A:844–850, 1991.
14. Strauch RJ, Rosenwasser MP, Lunt JG: Metacarpal shaft fractures: The effect of shortening on the extensor tendon mechanism. J Hand Surg 23A:519–523, 1998.
15. Trumble T, Gilbert M: *In situ* osteotomy for extra-articular malunion of the proximal phalanx. J Hand Surg 23A:821–826, 1998.
16. Wehbe MA, Schneider LH: Mallet fractures. J Bone Joint Surg 66A:658–669, 1984.
17. Weiss AP, Hastings H: Distal unicondylar fractures of the proximal phalanx. J Hand Surg 18A:594–599, 1993.

JOINT DISLOCATIONS AND LIGAMENT INJURIES

Peter J.L. Jebson, MD

1. **What are the clinical findings in the patient with a dorsally subluxed or dislocated distal ulna?**
   - The ulnar head is prominent dorsally when the forearm is pronated.
   - Supination of the forearm is markedly limited.
   - Positive piano key sign is noted (ballottement of the distal ulna from dorsal to palmar reduces it into the sigmoid notch of the distal radius).

2. **What is the usual mechanism of injury for a dorsal dislocation of the distal ulna?**
   A fall on an outstretched, hyperpronated forearm.

3. **What is the initial treatment of a dorsal dislocation of the distal radioulnar joint (DRUJ)?**
   Closed reduction with the forearm in supination and application of a long arm cast with the forearm maintained in maximal supination.

4. **What is Galeazzi's fracture? How is it treated?**
   Galeazzi's fracture is a specific injury of the distal forearm consisting of a fracture of the distal third of the radius with an associated dislocation of the DRUJ. Treatment consists of open reduction and internal fixation (ORIF) of the radial shaft fracture with closed reduction and immobilization of the DRUJ dislocation via long arm casting with the forearm in supination. If the DRUJ is unstable, temporary pinning or repair/reconstruction of the TFC may be performed.

5. **Which structure is most commonly responsible for preventing the closed reduction of a fracture-dislocation of the DRUJ?**
   The extensor carpi ulnaris (ECU) tendon, which can become entrapped.

6. **How is a palmar dislocation of the distal ulna treated?**
   A palmar or volar dislocation of the distal ulna is *not* as common as a dorsal dislocation. Treatment involves closed reduction with the forearm in pronation. If a satisfactory reduction cannot be obtained, an open reduction is necessary.

7. **How is a radiocarpal dislocation classified?**
   Radiocarpal dislocations usually are associated with a fracture of the ulnar and/or radial styloid (Fig. 25-1). They are classified into two types (Moneim's classification).
   - **Type 1** is an isolated radiocarpal dislocation.
   - **Type 2** consists of a dislocation of both the radiocarpal and intercarpal articulations.

8. **What is ulnar translocation of the carpus?**
   A rare carpal instability pattern that can occur following a radiocarpal dislocation and wrist trauma and is also encountered in patients with rheumatoid arthritis. Two types have been described by Taleisnik. In a type 1 injury, the entire carpus is displaced ulnarly, resulting in a widening of the radioscaphoid articulation. In a type 2 injury, the scapholunate interval is

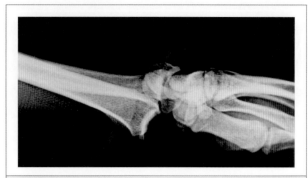

Figure 25-1. Lateral view of the wrist demonstrating a dorsal radiocarpal fracture-dislocation.

widened; the scaphoid maintains its relationship with the distal radius, and the remaining carpus translates ulnarly (Fig. 25-2).

9. **What is the most commonly injured carpometacarpal (CMC) joint?**
   The small finger CMC joint.

10. **What is the most common CMC joint injury pattern?**
    - Fracture-dislocations are more common than pure dislocations.
    - Dorsal dislocations are more common than volar or divergent dislocations.
    - Multiple dislocations are rare.

11. **Describe the treatment of a CMC dislocation.**
    After appropriate radiographs have been obtained and the injury pattern is identified, closed reduction should be attempted under regional or general anesthesia and 10 lb of longitudinal finger-trap traction. Once the CMC joint has been reduced, traction is discontinued, and the stability of the reduction is assessed. The CMC joints of the index and long fingers are usually stable. However, because of their mobility, a dislocation of the ring and small finger CMC joints is usually unstable, and a redislocation or subluxation usually occurs. If stable reduction is achieved, the hand is immobilized in a short arm cast for 4–6 weeks. Careful radiographic follow-up is necessary to document maintenance of the reduction. If the reduction is unstable, percutaneous pinning with Kirschner wires is recommended. An irreducible dislocation requires ORIF.

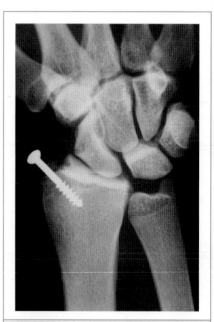

Figure 25-2. Type 1 ulnar translocation instability of the wrist.

12. **What key structure is injured after a dislocation of the thumb CMC joint?**
    The volar oblique trapeziometacarpal ligament (beak ligament).

13. **How is a thumb CMC joint dislocation treated?**
    Treatment is controversial. Previous treatment recommendations consisted of closed reduction with temporary percutaneous wire stabilization of the joint followed by cast immobilization. However, a recent study suggested that early surgical intervention, consisting of reconstruction of the volar oblique ligament using a slip of the flexor carpi radialis tendon, prevents the long-term complications of joint instability and chronic pain.

14. **A dorsal dislocation of the metacarpophalangeal (MCP) joint most commonly involves which finger?**
    The index finger.

15. **What is the difference between a complex and simple MCP dislocation?**
    A complex MCP dislocation is irreducible usually because of interposition of the volar plate and sesamoids (if present) (Fig. 25-3).

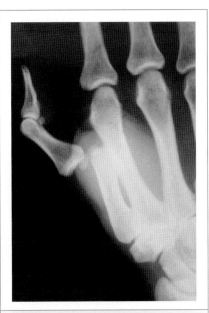

Figure 25-3. Complex dislocation of the thumb MCP joint. Note the sesamoids within the joint.

16. **Compare and contrast the clinical findings in a complex versus simple MCP joint dislocation.**

Complex Dislocation	Simple Dislocation
Closed reduction not possible	Closed reduction possible
Skin dimpling at the volar MCP crease	Skin dimpling absent
Slight hyperextension at the MP joint	Marked hyperextension of the MP joint
MP head prominent in palm	

17. **What are the radiographic features of a complex dorsal MCP joint dislocation?**
    - Widening of the MCP joint
    - Interposition of the sesamoids within the joint
    - Dorsally dislocated, slightly hyperextended proximal phalanx

18. **What is the reduction maneuver for a simple MCP joint dislocation?**
    Care must be taken during the reduction of an MCP joint dislocation to avoid converting a simple to a complex dislocation. *Longitudinal traction or hyperextension of the joint must be avoided*.

The reduction maneuver involves flexion of the wrist to relax the extrinsic flexor tendons, followed by direct digital pressure on the dorsal base of the proximal phalanx in a palmar direction to reduce the proximal phalanx into the MCP joint.

19. **What is a "gamekeeper's thumb"?**
    In 1955, Campbell described instability of the thumb MCP joint as a result of attritional changes in the ulnar collateral ligament (UCL). This condition originally was described as common in Scottish gamekeepers, whose technique of killing rabbits by twisting the head and neck resulted in progressive wear and degeneration of the UCL. The term *gamekeeper's thumb* has been inappropriately applied to all injuries of the UCL of the thumb MCP joint, including acute avulsion injuries that are more appropriately termed *skier's thumb* (Fig. 25-4).

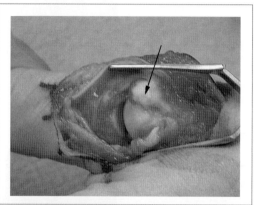

Figure 25-4. Skier's thumb. Note the avulsion of the proper collateral ligament *(arrow)* from the base of the proximal phalanx.

20. **What is a Stener lesion?**
    Interposition of the adductor pollicis aponeurosis between the distally avulsed UCL and its insertion at the base of the proximal phalanx (Fig. 25-5).

21. **What is the clinical significance of a Stener lesion?**
    Healing of the UCL to the base of the proximal phalanx is prevented. Treatment involves open reduction and reattachment of the UCL to the base of the proximal phalanx.

22. **How should you examine the thumb for a suspected injury to the UCL complex?**
    MCP joint stability is assessed with the joint in extension and flexion. Integrity of the UCL is assessed by radially deviating the proximal phalanx while simultaneously stabilizing the metacarpal shaft. This maneuver is performed with the MCP joint fully flexed and then fully extended. In the fully flexed position, the proper collateral ligament is the primary stabilizer of the MCP joint. The ability to radially deviate the thumb in this position >30–35 degrees implies complete ligament avulsion. Tenderness over the UCL, pain during stress testing and a radial deviation of the thumb <30 degrees imply incomplete injury to the UCL. Instability with the thumb MCP joint held in a fully extended position can only occur with avulsion of the UCL and volar plate complex.

23. **How is an acute UCL injury of the thumb MCP joint treated?**
    - **Partial tear:** Thumb spica splint or cast for 4–6 weeks
    - **Complete tear:** Surgical reattachment

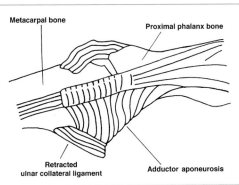

Figure 25-5. Schematic diagram of a Stener lesion. The torn UCL is displaced and trapped superficially to the adductor aponeurosis. It cannot return to its normal position beneath the adductor aponeurosis; therefore, surgical repair is required. (From Moschilla G, Briedahl W: Sonography of the finger. Am J Roentgenol 178[5]:1451–1457, 2002.)

- **Stener lesion:** Repair
- **Large displaced avulsion fracture:** ORIF

24. **How is a true gamekeeper's thumb treated?**
    - **Nonoperative:** Nonsteroidal anti-inflammatory medication and splinting.
    - **Operative:** Primary ligament repair is usually not feasible in chronic injuries. Options include ligament reconstruction using the adjacent joint capsule or a local or free tendon graft. Arthrodesis of the MCP joint is performed if posttraumatic arthritis is present.

25. **Which structures are responsible for stability of the proximal interphalangeal (PIP) joint?**
    The proper and accessory collateral ligaments in combination with the volar plate create a three-sided box around the PIP joint. Secondary stabilizers include the articular surfaces of the phalanges and the adjacent capsule and tendinous structures.

26. **How are PIP joint injuries classified?**
    PIP joint injuries are classified into three main types:
    - A **type 1** (hyperextension injury) is characterized by avulsion of the volar plate from the base of the middle phalanx with a reduced PIP joint and a minor split in the collateral ligament complex.
    - A **type 2** injury consists of a dorsal dislocation of the middle phalanx with a major split in the collateral ligament complex.
    - A **type 3 injury** (fracture-dislocation) consists of a comminuted fracture involving the articular surface of the middle phalanx and is associated with subluxation or dislocation of the middle phalanx.

27. **How is each type of injury treated?**
    Stable type 1 and type 2 injuries are treated by a brief period (2–3 days) of splint immobilization, followed by buddy taping and range-of-motion exercises. Treatment of a type 3 injury is based on the degree of fracture comminution. In a highly comminuted but reducible injury, treatment

involves extension block pinning or splinting. Additional treatment may include dynamic skeletal traction. If large fracture fragments are present, ORIF of the articular surface may be feasible. In highly comminuted injuries, a volar plate arthroplasty, primary arthrodesis, or implant arthroplasty may be indicated.

28. **What are the four types of PIP joint dislocations?**
    - Dorsal
    - Volar
    - Rotatory
    - Lateral (Fig. 25-6)

29. **Which structure is uniformly injured in a *volar* PIP joint dislocation?**
    The central slip of the extensor mechanism.

30. **Which structures are responsible for preventing the reduction of a *rotatory* PIP joint dislocation?**
    The central slip and lateral band.

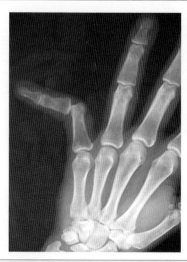

Figure 25-6. Lateral dislocation of the small finger PIP joint.

KEY POINTS: JOINT DISLOCATIONS AND LIGAMENT INJURIES

1. The ECU tendon is the most common structure responsible for preventing the closed reduction of a DRUJ fracture-dislocation.
2. A Stener lesion refers to interposition of the adductor aponeurosis between the distally avulsed UCL of the thumb MCP joint and its insertion on the base of the proximal phalanx.
3. The central slip is uniformly disrupted in a volar PIP joint dislocation.
4. A rotatory dislocation of the PIP joint may be irreducible because of "buttonholing" of the head of the proximal phalanx between the lateral band and central slip.
5. Joint stability following a PIP joint fracture-dislocation is dependent on the percentage of articular surface involved.

31. **Simultaneous dislocations of both the PIP and distal interphalangeal (DIP) joints most commonly involve which fingers?**
    The ring or small finger.

32. **Which structure is at risk during the open reduction of a complex MCP joint dislocation through a volar approach?**
    The neurovascular bundle is vulnerable and susceptible to inadvertent injury because of its displacement into the palm by the prominent metacarpal head.

33. A 24-year-old graduate student presents with swelling and discomfort at the MCP joint of his nondominant thumb 1 week after a fall from his bicycle. Radiographs are normal. Physical examination reveals that the joint is swollen with a 15-degree extensor lag. Tenderness is noted on palpation of the joint dorsally and radially. Ulnar deviation stress examination in both 30-degree of flexion and full extension elicits pain and increased laxity compared with the contralateral thumb. You suspect injury to which structures?
    The radial collateral ligament, joint capsule, and extensor pollicis brevis (EPB) tendon.

34. About 6 months after injuring her dominant thumb MCP joint, a 47-year-old woman presents with pain and a sensation of "looseness" of the joint during pinch activities. The only abnormalities noted during physical examination are pain and laxity with hyperextension stress testing. What is your treatment recommendation?
    The patient has sustained an injury to the volar plate complex involving either a sesamoid fracture or, more commonly, avulsion of the volar plate itself. Radiographs are thus indicated. Nonoperative treatment consists of a thumb spica splint or cast immobilization with the MCP joint in 20–30 degrees of flexion. Persistent pain or instability is an indication for surgical treatment, which involves reattachment of the volar plate. Sesamoid excision and volar plate repair are indicated if the instability is secondary to sesamoid nonunion.

35. A 20-year-old woman returns to the office 5 weeks after undergoing successful closed reduction of a dorsal radiocarpal fracture-dislocation. The wrist was believed to be stable after reduction, and she was placed in a long arm cast. Routine radiographs demonstrate a type 1 ulnar translocation Instability of the carpus. What is the recommended treatment?
    Treatment of posttraumatic ulnar translocation of the carpus is challenging and controversial. If the injury is recognized acutely, repair of all disrupted extrinsic radiocarpal ligaments has been advocated. This approach may be augmented with temporary pin fixation. Attempts to reconstruct the radiocarpal ligaments in subacute and chronic injuries have been unsuccessful and associated with a poor patient outcome. It has been recommended that such patients undergo a radiocarpal (radiolunate or radioscapholunate) fusion or total wrist fusion.

36. What are the treatment principles for an open interphalangeal joint dislocation?
    - Adequate anesthesia (usually digital block)
    - Intravenous antibiotics
    - Tetanus prophylaxis
    - Thorough wound irrigation and debridement
    - Joint reduction
    - Postreduction radiographs to confirm a concentric reduction

37. Which structures prevent a joint reduction in a complex DIP dislocation?
    - Interposed flexor digitorum profundus (FDP) tendon
    - Volar plate
    - Osteochondral fragments

38. What amount of PIP joint lateral instability is considered an indication for surgical repair?
    Greater than 20 degrees of lateral deviation with the PIP joint held in extension.

39. **What are the most common complications following the reduction of a dorsal PIP dislocation?**
    - Persistent pain
    - Persistent swelling
    - Flexion contracture
    - Stiffness

40. **What is the relationship between joint stability and the volar articular fracture in a PIP joint fracture-dislocation?**
    Stability is directly related to the percentage of articular surface involved.

Percent of Joint Surface Involved	PIP Joint Stability
<30%	Usually stable
30–50%	Potentially unstable
>50%	Unstable

41. **Why is the dorsal approach preferred for open reduction of a complex MCP dislocation?**
    To avoid potential injury to the displaced neurovascular structures.

42. **What is the key to obtaining a successful closed reduction of the dislocated DRUJ in the patient with a Galeazzi fracture-dislocation?**
    Anatomic reduction and stable fixation of the distal radius shaft fracture.

BIBLIOGRAPHY

1. Bilos ZJ, Vender MI, Bonavolenta M, Knutson K: Fracture subluxation of proximal interphalangeal joint treated by palmar plate advancement. J Hand Surg 19A:189–196, 1994.
2. Blazar PE, Steinberg DR: Fractures of the proximal interphalangeal joint. J Acad Orthop Surg 8:383–390, 2000.
3. Dinowitz M, Trumble TE, Hanel D, et al: Failure of cast immobilization for thumb ulnar collateral ligament avulsion fractures. J Hand Surg 22A:1057–1063, 1997.
4. Jebson PJL, Engber WD, Lange RH: Dislocations and fracture-dislocations of the carpometacarpal joints. Orthop Rev 23:19–28, 1994.
5. Kiefhaber TR, Stern PJ: Fracture dislocation of the PIP joint. J Hand Surg 23A:368–380, 1998.
6. Kozin SH, Bishop AT: Gamekeeper's thumb: Early diagnosis and treatment. Orthop Rev 23:797–804, 1994.
7. Moneim MS, Bolger JT, Omer GE: Radiocarpal dislocation—Classification and rationale for management. Clin Orthop 192:199–209, 1985.
8. Morgan JP, Gordon DA, Klug MS, et al: Dynamic digital traction for unstable comminuted intra-articular fracture-dislocations of the PIP joint. J Hand Surg 20B:565–573, 1995.
9. Schurman AH, Bos KE: Treatment of volar instability of the metacarpophalangeal joint of the thumb by volar capsulodesis. J Hand Surg 18B:346–349, 1993.
10. Simonian PT, Trumble TE: Traumatic dislocation of the thumb carpometacarpal joint: Early ligament reconstruction versus closed reduction and pinning. J Hand Surg 21A:802–806, 1996.
11. Stener B: Displacement of the ruptured ulnar collateral ligament of the metacarpophalangeal joint of the thumb. A clinical and anatomical study. J Bone Joint Surg 44B:869–879, 1962.
12. Wang KC, Hsu KY, Shih CH: Irreducible volar rotatory dislocation of the proximal interphalangeal joint. Orthop Rev 23:886–888, 1994.
13. Weiland AJ, Berner SH, Hotchkiss RN, et al: Repair of acute ulnar collateral ligament injuries of the thumb metacarpophalangeal joint with an intraossseous suture anchor. J Hand Surg 22A:585–591, 1997.

ARTHRODESIS AND ARTHROPLASTY OF THE HAND AND WRIST

Peter J.L. Jebson, MD, and Edwin Spencer, MD

1. **What are the indications and contraindications for total wrist arthroplasty?**
   - **Indications:** Inflammatory, posttraumatic, or degenerative arthritis of the wrist (Fig. 26-1)
   - **Contraindications:** Active infection, absence of motor tendons to control the wrist, and inadequate soft tissue coverage

2. **List the most common complications of total wrist arthroplasty.**
   - Implant loosening
   - Fracture
   - Dislocation
   - Infection

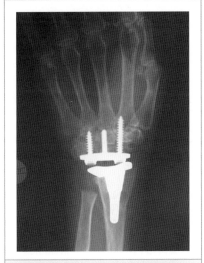

Figure 26-1. Total wrist arthroplasty in a patient with rheumatoid arthritis.

3. **What are the indications and contraindications for a wrist arthrodesis (fusion)?**
   - **Indications:**
     - Relieve pain secondary to primary degenerative, posttraumatic, inflammatory, or infectious arthritis
     - Reconstruction after tumor resection
     - Correction of deformity or instability after bone loss or paralytic condition
     - Total wrist arthroplasty, limited intercarpal arthrodesis, or other motion-preserving procedure
   - **Contraindications:**
     - Active infection
     - Lack of adequate soft tissue coverage

4. **List the general surgical goals when performing a wrist arthrodesis.**
   - Complete removal of all articular surfaces
   - Maintenance of carpal height and alignment
   - Use of internal fixation, bone graft, and postoperative external immobilization until radiographic evidence of union (Fig. 26-2)

5. **What is the preferred position for a wrist fusion?**
   Approximately 15–20 degrees of extension and 5–7 degrees of ulnar deviation.

6. **What activities are most difficult for the patient with a unilateral wrist arthrodesis?**
   - Forceful forearm supination and pronation with a simultaneous strong grip
   - Personal hygiene, particularly perineal care
   - Manipulating the hand in tight spaces

7. **Define proximal row carpectomy.**
   A proximal row carpectomy (PRC) is viewed as a "motion-preserving" procedure. A PRC consists of complete excision of the entire proximal row of the carpus (scaphoid, lunate, and triquetrum) to permit the proximal pole of the capitate to articulate with the lunate fossa of the distal radius. The procedure is performed in patients with radiocarpal or intercarpal arthrosis, spastic wrist contractures, or arthrosis with malalignment and/or deformity of the proximal carpus (Kienböck's disease, scapholunate advanced collapse [SLAC] pattern of wrist arthritis, scaphoid nonunion/malunion, and failed silastic implant).

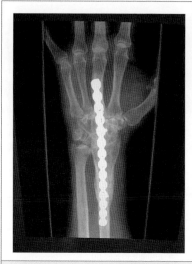

Figure 26-2. Total wrist arthrodesis.

8. **List the prerequisites for performing a PRC.**
   Preserved articular surfaces on the lunate fossa of the distal radius and proximal pole of the capitate.

9. **Why is a concomitant radial styloidectomy recommended for patients who undergo a PRC?**
   To prevent painful impingement of the trapezium against the distal radius.

10. **A PRC is relatively contraindicated in patients with which medical condition?**
    Rheumatoid arthritis. Contraindications for these patients exist because of the lack of predictable pain relief and the development of progressive deformities as a result of soft tissue imbalance.

11. **List the various intercarpal (limited wrist) fusions and their common indications.**
    - **Scaphotrapeziotrapezoid or triscaphe (STT)** arthrodesis is indicated for rotatory subluxation of the scaphoid, scaphoid nonunion, Kienböck's disease, triscaphe degenerative arthritis, and dorsal intercalated segmental instability (DISI) of the wrist.
    - **Lunotriquetral (LT)** arthrodesis is indicated for lunotriquetral ligament tear or instability.
    - **Capitolunate (CL)** arthrodesis is indicated for isolated midcarpal degenerative arthritis and articular fractures or bone loss.
    - **Scaphocapitate (SC)** arthrodesis is indicated for rotatory subluxation of the scaphoid, Kienböck's disease, and midcarpal instability.
    - **Capitate-lunate-hamate-triquetral** (four-corner) arthrodesis is indicated for ulnar midcarpal instability, the SLAC pattern of degenerative wrist arthritis, scaphoid nonunion, and isolated posttraumatic radioscaphoid arthritis after an intra-articular fracture.

12. **How much range of motion is lost in the wrist with the different types of intercarpal fusions?**
    - Radioscaphoid fusion causes approximately 60% loss of extension and flexion and 40% loss of radial and ulnar deviation.
    - Intercarpal fusions result in approximately 50% loss of wrist motion.
    - Midcarpal fusions cause a 40% loss of flexion and extension and a 10% loss of radial and ulnar deviation.

13. **What are the main indications for performing an arthrodesis of the finger or thumb joints?**
    - Pain
    - Deformity
    - Instability

14. **List the basic steps involved in arthrodesis of the thumb and finger joints.**
    - Remove all articular cartilage from the joint surfaces.
    - Establish adequate surface contact of the opposing bones.
    - Avoid excessive shortening.
    - Establish rigid internal or external fixation.

15. **What are the various techniques used to perform an arthrodesis of the thumb or finger joints?**
    - Single longitudinal or crossed Kirschner wires (Fig. 26-3)
    - Intraosseous wires (Fig. 26-3)
    - Plate and screws
    - Tension band wire
    - Compression screw
    - External fixation
    - Intramedullary bone graft

16. **Which joint in the hand is most likely to require an arthrodesis?**
    The proximal interphalangeal (PIP) joint.

17. **Which joint is least likely to require an arthrodesis?**
    The finger metacarpophalangeal (MCP) joint.

18. **What are the preferred positions for arthrodesis of the various joints of the fingers and thumb?**
    See Table 26-1.

19. **List the complications associated with arthrodesis of the thumb and finger joints.**
    - Infection
    - Skin slough
    - Nonunion
    - Delayed union
    - Painful/prominent hardware
    - Extensor tendon adhesions
    - Vascular insufficiency following correction of a severe flexion contracture

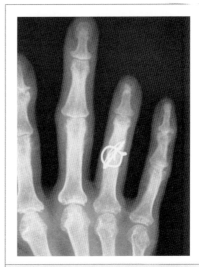

Figure 26-3. PIP arthrodesis using a combination of intraosseous and Kirschner wires.

TABLE 26-1.	PREFERRED POSITIONS FOR ARTHRODESIS OF FINGER AND THUMB JOINTS
Joint	Optimal Position
DIP	Between 10 and 20 degrees of flexion
PIP	Between 30 and 45 degrees of flexion, increasing in 5-degree increments from the index to the small fingers
Thumb MCP	About 15 degrees of flexion and 10 degrees of pronation (Fig. 26-4)
Thumb CMC	Between 30 and 40 degrees of palmar abduction, 35 degrees of radial abduction, and 15 degrees of pronation

DIP = distal interphalangeal, CMC = carpometacarpal.

20. **Why should every patient with a silicone implant in the hand or wrist be followed on an annual or biannual basis?**
    A severe foreign-body inflammatory response may occur as a consequence of silicone particulate debris from the implant. This response results in proliferative synovitis with cyst formation, bone resorption, and destruction with implant loosening as well as impending or actual pathologic fracture. Patients may be asymptomatic. Biannual follow-up with clinical *and* radiographic evaluation is recommended.

21. **Describe the treatment for a fractured hand or wrist silicone implant with significant particulate synovitis and cyst formation.**
    Treatment includes removal of the implant; synovectomy; curettage of reactive bone cyst with grafting of large cysts followed by reimplantation of a new prosthesis (rare); or a salvage procedure, such as resection arthroplasty, arthrodesis, proximal row carpectomy, or limited or total wrist arthrodesis.

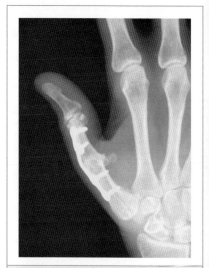

Figure 26-4. Thumb MCP arthrodesis using a custom-designed plate and screws.

22. **What are the indications for resecting the distal ulna?**
    Excision or resection of the distal ulna, also known as a Darrach procedure, is indicated for posttraumatic, degenerative, or inflammatory arthritis of the distal radioulnar joint (DRUJ) with limited and/or painful forearm rotation. The procedure is also indicated in symptomatic ulnocarpal impingement that results from excessive shortening of the radius after a distal radius fracture.

23. **List the most common complications associated with excision of the distal ulna.**
    - Bone regrowth
    - Impingement against the radius
    - Instability of the remaining proximal shaft with painful snapping and/or rupture of adjacent tendons

24. Approximately 10 years ago, a silicone interpositional wrist arthroplasty was performed on a 58-year-old woman. She had been doing well until the previous 6 months when she started noting progressive, diffuse wrist pain and swelling. She is afebrile, but all serological testing is normal. Radiographs reveal multiple cysts within the carpus and distal radius. What is the diagnosis?
Silicone synovitis.

25. List the histologic features of silicone synovitis.
Foreign body giant cells, fibroinflammatory tissue, and nonbirefringent crystals.

KEY POINTS: ARTHRODESIS AND ARTHROPLASTY OF THE HAND AND WRIST

1. Following a wrist arthrodesis, patients experience difficulty with perineal care and manipulating the hand in tight spaces.
2. A contraindication to a proximal row carpectomy (PRC) is damage to the articular surface of the lunate fossa of the distal radius or the proximal capitate.
3. Of all the intercarpal arthrodeses, the scapholunate has the highest nonunion rate.
4. The optimal position for a wrist arthrodesis is slight extension and ulnar deviation.

26. Which intercarpal arthrodesis has the highest nonunion rate?
Scapholunate.

27. What are the treatment options for patients who have undergone a Darrach procedure that has failed?
- If inadequate bone resection was initially performed or if bone has regrown within the defect, a repeat resection should be performed with excision of the adjacent periosteum.
- If instability of the ulnar shaft is encountered, the remaining ulnar shaft may be stabilized with a portion of the flexor and/or extensor carpi ulnaris tendons woven through the distal ulnar shaft.
- If symptomatic impingement between the distal radius and ulna is present, the pronator quadratus may be detached from the ulna and interposed between the radial and ulnar shafts. Alternatively, a distal ulna prosthesis may be used. Finally, salvage may be accomplished with the creation of a single or one-bone forearm.

28. What are the complications of using a headless cannulated screw for arthrodesis of the DIP joint?
- Distal phalanx fracture
- Nail bed injury (nail plate ridging)
- Fusion of the DIP joint in extension, rather than 10–20 degrees of flexion

29. Which recent development has been described in regard to interpositional arthroplasty of the osteoarthritic first CMC joint?
Satisfactory results have been reported with the use of a ceramic spherical implant as an interpositional arthroplasty in the patient with symptomatic osteoarthritis of the thumb CMC joint. However, a high rate of subsequent implant erosion or dislocation has also been noted.

30. A 72-year-old woman presents with advanced osteoarthritis of the thumb CMC joint. She is requesting surgical treatment because nonoperative treatment has failed to relieve her pain. A physical examination reveals that the thumb is severely adducted, and 45 degrees of hyperextension is noted at the MCP joint. Radiographs demonstrate significant subluxation of the first metacarpal base and degenerative changes within the MCP joint. What are your treatment recommendations?
    - Resection of the trapezium, soft tissue interpositional arthroplasty, and reconstruction of the volar trapeziometacarpal ligament (LRTI procedure)
    - Release of the adductor pollicis combined with a first web space deepening
    - MCP joint arthrodesis

31. A 34-year-old woman presents with wrist pain 1 year after closed treatment of a fracture that involved the sigmoid notch of the distal radius. Physical examination reveals that the DRUJ is stable, but pain is reproduced with palpation, compression, and translation. A three-compartment wrist arthrogram is normal. A bone scan demonstrates increased uptake at the DRUJ. Arthroscopy reveals an incongruous sigmoid notch and synovitis. The triangular fibrocartilage complex (TFCC) is intact. The patient does not want an arthrodesis, and she is not willing to have her distal ulna "cut off." What is your recommendation?
    A hemiresection interpositional arthroplasty consisting of excision of the radial portion of the distal ulna with preservation of the TFCC and ulnar styloid and interposition of a free tendon graft within the resected DRUJ.

32. What is a Sauvé-Kapandji or Lowenstein procedure?
    Arthrodesis of the DRUJ with resection of a segment of the distal ulnar shaft to create a pseudoarthrosis and permit forearm rotation.

33. About 6 months following a "successful scaphoid excision and four-corner arthrodesis" performed for a SLAC pattern of wrist arthritis, a 44-year-old male presents with a chief complaint of painful, limited wrist dorsiflexion. What is the most likely explanation?
    Impingement of an extended lunate against the dorsal distal radius. A DISI deformity with extension of the lunate occurs in an advanced SLAC pattern of wrist arthritis. Failure to correct the extended lunate at the time of the fusion results in painful impingement against the distal radius during wrist extension.

34. What is the most common functional complaint following an arthrodesis of the thumb CMC joint?
    Inability to lay the hand flat with the thumb in the plane of the hand.

BIBLIOGRAPHY

1. Alexander AH, Turner MA, Alexander CE, Lichtman DW: Lunate silicone replacement arthroplasty in Kienböck's disease: A long-term follow-up. J Hand Surg 15A:401–407, 1990.
2. Bamberger HB, Stern PJ, Kiefhaber TR, et al: Trapeziometacarpal joint arthrodesis: A functional evaluation. J Hand Surg 17A:605–611, 1992.
3. Carroll RE: Arthrodesis of the carpometacarpal joint of the thumb. A review of patients with a long postoperative period. Clin Orthop 220:106–110, 1987.
4. Cohen MS, Kozin SH: Degenerative arthritis of the wrist: Proximal row carpectomy versus scaphoid excision and four-corner arthrodesis. J Hand Surg 26A:94–104, 2001.

5. Divelbiss BJ, Sollerman C, Adams BD: Early results of the Universal total wrist arthroplasty in rheumatoid arthritis. J Hand Surg 27(2):195–204, 2002.
6. Fatti JF, Palmer AK, Greenky S, Mosher MF: Long-term results of Swanson interpositional arthroplasty. Part II. J Hand Surg 16A:432–437, 1991.
7. Ferlic DC, Clayton ML, Mills MF: Proximal row carpectomy: Review of rheumatoid and nonrheumatoid wrists. J Hand Surg 16A:420–423, 1991.
8. Fulton DB, Stern PJ: Trapeziometacarpal joint arthrodesis in primary osteoarthritis: A minimum two-year follow-up. J Hand Surg 26A(1):109–114, 2001.
9. Gellman H, Kauffman D, Lenihan M, et al: An *in vitro* analysis of wrist motion: The effect of limited intercarpal arthrodesis and the contributions of the radiocarpal and midcarpal joints. J Hand Surg 13A:378–383, 1988.
10. Hastings H, Weiss APC, Quenzer D, et al: Arthrodesis of the wrist for posttraumatic disorders. J Bone Joint Surg 78A:897–902, 1996.
11. Jebson PJL, Adams BD: Wrist arthrodesis: Review of current technique. J Am Acad Orthop Surg 9(1):53–60, 2001.
12. Jebson PJL, Hayes EP, Engber WD: Proximal row carpectomy: A minimum 10-year follow-up study. J Hand Surg 28A:561–569, 2003.
13. Kirschenbaum D, Schneider LH, Adams DC, Cody RP: Arthroplasty of the metacarpophalangeal joints with use of silicone-rubber implants in patients who have rheumatoid arthritis. Long-term results. J Bone Joint Surg 75A:3–12, 1993.
14. Kleinman WB, Carroll C: Scaphotrapeziotrapezoid arthrodesis for treatment of chronic static and dynamic scapholunate instability: A 10-year perspective on pitfalls and complications. J Hand Surg 15A:408–414, 1990.
15. Rettig ME, Dassa G, Raskin KB: Volar plate arthroplasty of the distal interphalangeal joint. J Hand Surg 26:940–944, 2001.
16. Smith RJ, Atkinson RE, Jupiter JB: Silicone synovitis of the wrist. J Hand Surg 10A:47–60, 1985.
17. Stahl S, Rozen N: Tension-band arthrodesis of the small joints of the hand. Orthopedics 24:981–983, 2001.
18. Tomaino MM, Pellegrini VD, Burton RI: Arthroplasty of the basal joint of the thumb. Long-term follow-up after ligament reconstruction and tendon interposition. J Bone Joint Surg 77A:346–355, 1995.
19. Watson HK, Ryu J, Akelman E: Limited triscaphoid intercarpal arthrodesis for rotary subluxation of the scaphoid. J Bone Joint Surg 68A:345–349, 1986.
20. Weiss APC, Hastings H: Wrist arthrodesis for traumatic conditions: A study of plate and local bone graft application. J Hand Surg 20A:50–56, 1995.
21. Weiss APC, Wiedemann G, Quenzer D, et al: Upper extremity function after wrist arthrodesis. J Hand Surg 20A:813–817, 1995.
22. Wilgis EFS: Distal interphalangeal joint silicone interpositional arthroplasty of the hand. Clin Orthop 342:38–41, 1997.
23. Zachary SV, Stern PJ: Complications following AO/ASIF wrist arthrodesis. J Hand Surg 20A:339–344, 1995.

CHAPTER 27
WRIST PAIN
Fred M. Hankin, MD

1. **Calcification of the articular surfaces of the carpal bones and the triangular cartilage are considered pathognomonic for which condition?**
   Pseudogout or calcium pyrophosphate dihydrate (CPPD) disease. The condition is characterized by recurrent attacks of pain and joint effusions. The wrist is involved in 50% of symptomatic patients. CPPD crystals can be deposited in hyaline cartilage (articular surface) and the triangular fibrocartilage.

2. **A 24-year-old woman in the third trimester of an otherwise normal pregnancy presents to her obstetrician with severe nocturnal wrist pain and numbness involving her thumb, index, and long fingers. What is the most likely diagnosis?**
   Carpal tunnel syndrome associated with pregnancy. Various symptoms may develop, including night pain, numbness, tingling in the median nerve distribution, weakness, and clumsiness. Occasionally, the pain can radiate proximally to the shoulder region. Fluid retention during pregnancy is thought to contribute. The symptoms may resolve after delivery.

3. **What is an effective first-time treatment for carpal tunnel syndrome?**
   Splinting of the wrist to avoid extreme flexion and extension, which is thought to exacerbate the symptoms. Splints are usually worn at nighttime.

4. **A tire repair technician presents with an acutely painful hypothenar region and numbness in his ring and small fingers. What is your presumptive diagnosis?**
   Ulnar artery thrombosis or hypothenar hammer syndrome due to repetitive and occasionally acute trauma to the hypothenar region of the hand. Secondary compression of the adjacent ulnar nerve can produce paresthesias. Other symptoms include pallor and coolness of the ring and small fingers.

5. **What is Preiser's disease?**
   Avascular necrosis of the carpal scaphoid bone.

6. **While walking home, a medical student trips on a crack in the sidewalk and falls onto his outstretched hand. He has acute pain in his wrist and anatomic snuffbox region. Radiographs obtained in the emergency department are normal. What is your working diagnosis?**
   Fracture of the carpal scaphoid bone. Despite normal radiographs, the history and clinical exam suggest the presence of a scaphoid fracture. The patient should be splinted, and follow-up evaluation should be arranged. Frequently, the oblique view or a dedicated scaphoid view will reveal the fracture. The precarious blood supply of the scaphoid can lead to nonunion or delayed healing.

7. **What is a carpometacarpal (CMC) boss?**
   A bony prominence usually arising at the base of the second or third carpometacarpal joint. The articulations of the index and long CMC joints can become hypertrophic, possibly as a result of an acute traumatic event or repetitive stress. A bursal sac may develop over the boss, leading the examiner to consider the mass a ganglion cyst. Occasionally, tendons or sensory nerves can subluxate over these bony prominences during wrist motion and become symptomatic (Fig. 27-1).

8. **A hand surgeon is clearing her gutters and falls from the roof. She has a comminuted radial head fracture. Her wrist is also very sore, but the initial radiographs are normal. What is your diagnosis?**

   An associated disruption of the interosseous membrane between the radial and ulnar shafts and ligamentous disruption of the distal radioulnar joint (DRUJ), also known as an Essex-Lopresti fracture-dislocation of the forearm. This injury can result in longitudinal translation of the ulna relative to the radius, which migrates proximally, resulting in ulnar-sided wrist pain (Fig. 27-2).

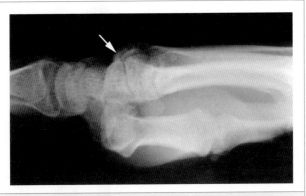

Figure 27-1. Lateral radiograph of the wrist demonstrating a CMC boss.

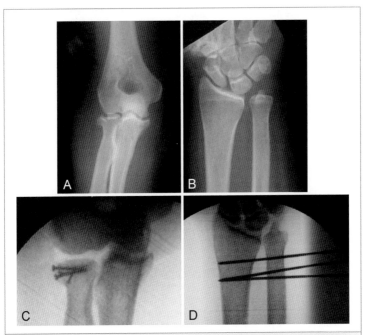

Figure 27-2. Essex-Lopresti injury of the forearm consisting of a radial head fracture with disruption of the interosseous membrane and DRUJ. The patient underwent open reduction internal fixation (ORIF) of the radial head fracture combined with closed reduction and percutaneous pinning of the DRUJ. ***A***, Displaced radial head fracture. ***B***, Disruption of the DRUJ. ***C***, ORIF of radial head fracture. ***D***, The DRUJ has undergone closed reduction and percutaneous pinning.

9. A 16-year-old girl presents with a 3-month history of wrist pain unresponsive to anti-inflammatory medications, physical therapy, steroid dose packs, vitamin supplement, and cervical spinal manipulations. What does this lesion represent (Fig. 27-3)?
   The radiographic features of well-demarcated sclerotic margins and sharp zone of transition between the lesion and normal bone are consistent with a benign process. The specific diagnosis in this case is a giant cell tumor of bone.

10. Asymmetric ossification of the carpal bones in a child with wrist stiffness and synovitis is consistent with what diagnosis?
    - Juvenile rheumatoid arthritis (JRA) is a possible diagnosis.
    - Increased peripheral circulation associated with synovitis can accelerate the ossification of carpal bones (i.e., more carpal bones are noted on the radiograph of the symptomatic wrist than on the asymptomatic side).

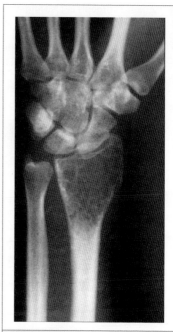

Figure 27-3. A giant cell tumor of bone involving the distal radius.

11. Define *ulnocarpal abutment*.
    When the distal ulna is longer than the radius (ulnar positive variance), it can impinge (abut) on the lunate and triquetrum. This impingement can result in inflammation and pain, restriction of motion, a degenerative triangular fibrocartilage (TFC) tear, cystic changes in the carpal bones, and disruption of the lunotriquetral ligament complex (Fig. 27-4).

12. What is a SLAC wrist?
    Scapholunate advanced collapse pattern of wrist arthritis. In a long-standing scapholunate (SL) ligament injury, disruption of the normal wrist mechanics can result in degenerative arthritis of the radioscaphoid, and capitolunate joints with preservation of the radiolunate articulation (Fig. 27-5).

13. The cortical ring, or "Terry Thomas sign," is associated with which wrist ligament injury?
    An SL ligament injury. The abnormal palmar flexion of the scaphoid results in a ring appearance in the distal scaphoid on a posteroanterior (PA) radiograph. The ring is the cortical margin of the distal pole of the scaphoid seen on profile. A diastasis between the

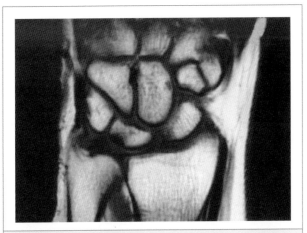

Figure 27-4. Magnetic resonance imaging (MRI) of the wrist demonstrating cystic changes in the ulnar aspect of the proximal lunate consistent with abutment by the distal ulna.

scaphoid and lunate (usually >3 mm) can be seen following an SL ligament injury and is termed the "Terry Thomas sign" in memory of a famous British comedian who had a large gap between his teeth.

14. **Explain the humpback deformity of the scaphoid.**
    When a fracture of the scaphoid waist region progresses to a malunion or nonunion, the distal pole palmar flexes relative to the proximal pole. A humpback outline of the dorsal border of the scaphoid is noted on a lateral wrist radiograph and can best be demonstrated on a sagittal computed tomography (CT) scan.

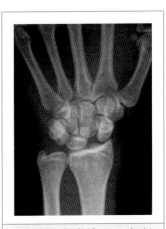

Figure 27-5. The SLAC pattern of wrist arthritis.

15. **The "spilled teacup sign" is seen with which wrist injury?**
    Volar lunate dislocation. The lunate is displaced toward the palm out of its fossa on the radius and into the carpal tunnel region, which resembles a spilled teacup.

16. **Negative ulnar variance is associated with which wrist condition?**
    Kienböck's disease or avascular necrosis of the lunate.

17. **What is a snapping extensor carpi ulnaris tendon?**
    Trauma to the extensor retinaculum of the sixth dorsal extensor compartment or, more commonly, synovitis of the extensor carpi ulnaris tendon can result in tendon subluxation out of its groove on the distal ulna, resulting in ulnar-sided wrist pain and a snapping sensation with resisted pronation and supination maneuvers.

18. **Define *intersection syndrome*.**
    Stenosing tenosynovitis of the second dorsal extensor compartment involving the radial wrist extensors (extensor carpi radialis longus and brevis) as they pass beneath the muscle bellies of the extensor pollicis brevis and abductor pollicis longus.

19. **Where does flexor carpi radialis tendinitis occur?**
    At the volar radial wrist region where the flexor carpi radialis (FCR) tendon passes through the fibrosseous tunnel adjacent to the scaphoid and trapezium.

20. **Your grandfather has had chronic pain and swelling along the volar radial aspect of his wrist at the base of his thenar eminence. One day, after picking up a gallon of milk, he feels a "pop" in this area, and reports that his pain has now vanished. What is the diagnosis?**
    An FCR tendon rupture following chronic tendinitis. The FCR can become irritated and inflamed as it passes through the fibrosseous tunnel along the volar surface of the trapezium.

KEY POINTS: WRIST PAIN

1. Preiser's disease is avascular necrosis of the carpal scaphoid that is not associated with a fracture.
2. Hypothenar hammer syndrome is thrombosis of the ulnar artery at Guyon's canal in the wrist.
3. Acute calcific tendonitis most commonly involves the flexor carpi ulnaris (FCU).
4. The classic radiographic findings of lead toxicity include broad bands of markedly increased density in the metaphyseal region adjacent to the growth plate.
5. Asymmetric ossification of the carpus in the child with wrist stiffness and synovitis is characteristic of JRA.

21. **What is the space of Parona?**
    A potential retrotendinous space in the distal forearm between the pronator quadratus and the flexor tendons. The synovial sheath enclosing the flexor pollicis longus can connect to the flexor tendon sheath of the small finger through this space. A horseshoe abscess occurs when all three spaces are involved in suppurative flexor tenosynovitis.

22. **A 35-year-old trial attorney presents with acute wrist pain along the course of the flexor carpi ulnaris (FCU). A lateral view of the wrist reveals linear calcification in the soft tissues. What is the diagnosis?**
    Acute calcific tendinitis involving the FCU tendon.

23. A plastic surgery resident is late for a conference and falls onto her outstretched hand as she sneaks in the back door of the lecture hall. Persistent pain and ecchymosis are noted over the base of the hypothenar region 2 weeks later. Routine plain radiographs are normal. What diagnostic test should you request?
    A carpal tunnel view and a 30-degree supinated oblique view will profile the pisotriquetral articulation. If the radiographs are normal and a hook of the hamate fracture is suspected, a CT scan is indicated (Fig. 27-6).

    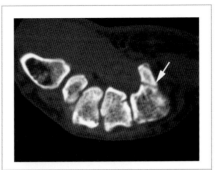

    Figure 27-6. CT scan demonstrating a fracture of the hook of the hamate.

24. Which type of injury occurs following an industrial microwave exposure to the wrist?
    Although the skin may demonstrate only minor cutaneous changes, such as erythema, the underlying muscle, nerves, and blood vessels can be severely damaged after microwave exposure. The patient can present with a benign-appearing limb and acute neurologic deficits.

25. A 65-year-old hand surgeon is an avid squash player. He presents with significant wrist pain associated with activity. What is the most likely diagnosis?
    Osteoarthritis. All of the joints in the hand can be involved in osteoarthritis, particularly those associated with repetitive joint loading and wear. The radiographs of this patient demonstrate involvement between the capitate and lunate (Fig. 27-7).

    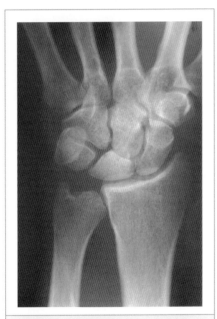

    Figure 27-7. Osteoarthritis is noted between the capitate and lunate. Radiographic changes include joint space narrowing and sclerosis.

26. A medical student falls while playing volleyball during the mid-afternoon break. Tenderness is noted over the distal radius. The initial plain radiographs are normal. What is your treatment approach?
    Immobilization in a splint or cast with reexamination in 7–10 days. If the patient is tender and swollen, immobilization is recommended for the treatment of a presumed occult fracture. A bone scan can help to clarify the diagnosis. An occult fracture of the distal radius or scaphoid

may not become apparent on radiographs for several weeks. The healing process consisting of new bone formation can usually be seen on subsequent follow-up radiographs.

27. **The small bone "chip" noted on the dorsum of the wrist on a lateral wrist radiograph following a fall on the outstretched hand usually originates from which structure?**
   The "chip" usually represents an avulsion fragment from the triquetrum. The clinical exam should rule out the presence of other associated carpal bone injuries (Fig. 27-8).

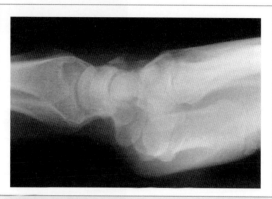

Figure 27-8. Dorsal avulsion fragment from the triquetrum is noted on a lateral wrist radiograph.

28. **About 2 weeks after a nurse's arm was placed in a cast for a nondisplaced distal radius fracture, she notices the sudden inability to extend her ipsilateral thumb. What is the cause?**
   Rupture of the extensor pollicis longus tendon. This infrequent complication usually occurs with a nondisplaced or minimally displaced fracture of the distal radius. Proposed causes include attritional wear at the fracture site and poor blood supply to the tendon within the narrow third dorsal extensor compartment. The keys to recognition are patient education and physician awareness, including careful physical examination.

29. **You are asked to evaluate the infant son of a malpractice lawyer in the emergency department. Plain radiographs demonstrate an acute nondisplaced distal radius fracture and a healing, spiral fracture of the humeral shaft. What should you suspect?**
   Child abuse. Multiple fractures in various stages of healing are the classic radiographic findings that should lead to a high index of suspicion.

30. **An intoxicated college student slams his hand through a glass window and sustains multiple lacerations. His neurovascular system is intact, and tendon function is normal. The lacerations are repaired in the emergency department. When pain persists, he is referred to your office. A radiograph shows a large piece of glass in the soft tissues. What advice do you give to the medical student rotating on your service?**
   Not all glass is visible on plain radiographs. Multiplanar plain radiographs should be routinely ordered as part of the evaluation of a suspected foreign body injury involving glass. Those

foreign bodies not visualized on plain radiographs may be detected with ultrasound, a CT scan, or MRI (Fig. 27-9).

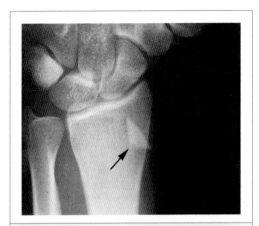

Figure 27-9. A radiograph revealing a large piece of glass *(arrow)* in soft tissue.

31. A physician assistant sustains a laceration to the dorsal forearm while cleaning a fish tank. During the next several weeks, he develops a warm and swollen wrist. Surgical debridement is performed. Routine aerobic and anaerobic wound cultures are normal. What specific microbiology test should be performed at the repeat surgical debridement?

The patient most likely has an infection involving *Mycobacterium marinum*. *M. marinum* is a fastidious organism found in both fresh and salt water. Such an infection commonly results from an injury sustained on a dock or pier or during the cleaning of a fish tank or preparing shellfish. When the clinical course is atypical, a high index of suspicion is necessary. Appropriate wound cultures should be obtained, including fungal cultures and specific cultures for *M. marinum*, which involves growing the organism at 31°C on Lowenstein-Jensen media (Fig. 27-10).

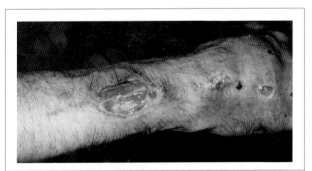

Figure 27-10. Surgical debridement of the distal forearm in a patient with a *Mycobacterium marinum* infection.

32. A 4-year-old girl injures her nondominant hand. The accident was not witnessed, but reports at the daycare center indicate that a traction mechanism may have been involved. Hand and wrist radiographs are normal. What further evaluation is necessary?

   Comparison radiographs of the opposite extremity are frequently helpful, particularly in children. Subtle differences in cortical alignment and growth plate width often are better appreciated with comparison films of the uninjured extremity. In addition, a physical examination and radiographs of the ipsilateral elbow are needed. An occult fracture of the radial head or distal humerus or a radial head dislocation can present with wrist or forearm pain and may go unrecognized. If an abnormality is not detected on radiographs, the child should still be protected with an appropriate splint and carefully reexamined several days later.

33. A 2-year-old child is admitted to the pediatric service with encephalopathy and abdominal pain. Wrist radiographs are obtained as part of a skeletal survey. What is the diagnosis?

   Lead poisoning. Classic radiographic findings include broad bands of markedly increased density in the metaphyseal region adjacent to the growth plate. Occasionally, the child may have bone or joint pain, but the presenting symptoms are usually encephalopathy, abdominal pain, colic, attention deficit disorder, or behavioral problems (Fig. 27-11).

34. About 2 months following a nondisplaced distal radius fracture that was treated with long arm cast immobilization, your spouse's grandmother presents to you with continued pain, stiff fingers, a stiff wrist, a lack of forearm rotation, and a "frozen" shoulder. What is the recommended treatment approach?

   The presence of a complex regional pain syndrome should be suspected particularly if there is pain out of proportion to the injury. A formal supervised therapy program is indicated to enhance joint mobility, avoid contracture formation, and improve function.

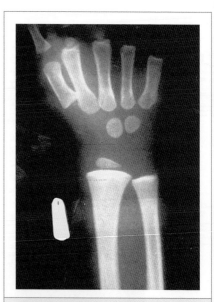

Figure 27-11. Classic radiographic findings of lead toxicity. These include broad bands of markedly increased density in the metaphyseal region adjacent to the growth plate.

35. A surgery intern places an arterial line into the radial artery of an intubated patient. The nurse calls him 4 hours later and reports that the thumb and index finger are blue. What crucial information did the senior resident fail to provide his peer?

   An incomplete palmar vascular arch exists in approximately 3–7% of the population. An Allen's test should have been performed prior to insertion of the arterial line to determine the status of the radial and ulnar arteries and their contribution to the arterial supply of the hand.

36. **Following a fall onto a glass bottle, a medical student has several lacerations on his hand. He reports numbness along the thenar eminence but normal sensation on the pads of his digits. What is the basis for this finding?**
   The patient has likely suffered a laceration of the palmar cutaneous nerve with preservation of the median nerve. The palmar cutaneous branch of the median nerve, which is responsible for thenar sensation, is a separate branch of the median nerve that originates proximal to the wrist crease.

37. **Upset that his favorite collegiate football team lost the championship game, an undergraduate is arrested at the campus riot. Following the removal of the handcuffs, he experiences numbness on the dorsal radial aspect of both hands and the dorsal aspect of both thumbs. What is your diagnosis?**
   Neuropraxic injury to the dorsal sensory branches of the radial nerve or the lateral antebrachial nerve. These fine nerve branches, as they course over the radial styloid, can be compressed by straps, splints, casts, garments, bands, and handcuffs.

BIBLIOGRAPHY

1. Albanese SA, Palmer AK, Kerr DR, et al: Wrist pain and distal growth plate closure of the radius in gymnasts. J Pediatr Orthop 9:23–28, 1989.
2. Chun S, Palmer AE: The ulnar impaction syndrome: Follow-up of ulnar shortening osteotomy. J Hand Surg 18A:46–53, 1993.
3. Cohen MS, Kozin SH: Degenerative arthritis of the wrist: Proximal row carpectomy versus scaphoid excision and four-corner arthrodesis. J Hand Surg 26A:94–104, 2001.
4. Cooney WP: Evaluation of chronic wrist pain by arthrography, arthroscopy, and arthrotomy. J Hand Surg 26A:815–822, 1993.
5. Hauck RM, Palmer AK: Classification and treatment of ulnar styloid nonunion. J Hand Surg 21A:418–422, 1996.
6. Jupiter JB, Ring D: A comparison of early and late reconstruction of malunited fractures of the distal end of the radius. J Bone Joint Surg 78A:739–748, 1996.
7. Terrill RA: Use of arthroscopy in the evaluation and treatment of chronic wrist pain. Hand Clin 10:593–603, 1994.
8. Weiss AP, Weiland AJ, Moore JR, et al: Radial shortening for Kienböck's disease. J Bone Joint Surg 73A:384, 1991.
9. Wintman BI, Gelberman RH, Katz JN: Dynamic scapholunate instability: Results of operative treatment with dorsal capsulodesis. J Hand Surg 20A:971–979, 1995.

CHAPTER 28
FLEXOR TENDON INJURIES
Christopher C. Schmidt, MD, and Samir M. Patel, MD

1. **How do you determine if the flexor digitorum profundus (FDP) is intact following a finger laceration?**
   - Active flexion of the distal interphalangeal (DIP) joint should be preserved.
   - Assessment involves holding the other fingers and the proximal interphalangeal (PIP) joint of the injured finger in full extension. If the DIP joint of the injured finger can be actively flexed, the FDP tendon is intact (Fig. 28-1).

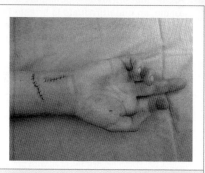

Figure 28-1. The resting posture of the hand seen in the patient with a flexor tendon laceration. Note the loss of the normal cascade posture of the hand.

2. **How do you determine if the flexor digitorum superficialis (FDS) is intact?**
   - Active flexion of the PIP joint should be preserved.
   - Assessment involves holding the metacarpophalangeal (MCP) and PIP joints of the other fingers in extension. Full flexion of the PIP joint of the injured finger indicates an intact FDS.

3. **Which clinical finding suggests a partial flexor tendon injury?**
   During testing, if pain is present with resisted PIP or DIP joint flexion, a partial flexor tendon is suspected.

4. **Discuss the zones used to describe flexor tendon injuries in the hand and wrist.**
   - **Zone I** extends from the fingertip to the center portion of the middle phalanx.
   - **Zone II** extends from the center portion of the middle phalanx to the distal palmar crease.
   - **Zone III** extends from the distal palmar crease to the distal portion of the transverse carpal ligament.
   - **Zone IV** overlies the transverse carpal ligament.
   - **Zone V** extends from the wrist crease to the level of the musculotendinous junction of the flexor tendons (Fig. 28-2).

5. **Discuss the zones used to describe flexor tendon injuries in the thumb.**
   - **Zone I** extends from the interphalangeal (IP) joint crease to the thumb tip.
   - **Zone II** extends from the MCP joint crease to the IP joint crease.
   - **Zone III** is beneath the thenar muscles.

6. **The lumbrical muscles are found in which zone? Where do the lumbricals originate?**
   Zone III. They originate on the radial side of the FDP tendons.

7. **Which zone is referred to as "no man's land"? Why?**
   Zone II. Sterling Bunnell termed zone II "no man's land" because of the high incidence of poor results after attempted primary repair of a tendon laceration. The main complication is adhesion formation because of the volume contained within the tendon sheath (two FDS slips and the FDP).

8. **What topographic landmark is used to indicate the origin of the digital flexor tendon sheath?**
   The distal palmar crease.

9. **Describe the pulley system of the fibro-osseous sheath in the fingers.**
   The pulleys are well-defined, thickened areas within the fibro-osseous sheath. There is a palmar aponeurosis pulley, and in the fingers, five annular (A1–5) and three cruciform (C1–3) pulleys. The palmar aponeurosis pulley is composed of the transverse fibers of the palmar fascia, which is attached to the deep transverse metacarpal ligament via vertical septae. The annular pulleys are thicker and more defined than the thin cruciform pulleys. The odd-numbered annular pulleys are located over the joints of the finger: MCP—first annular pulley (A1), PIP—third annular pulley (A3), and DIP—fifth annular pulley (A5).

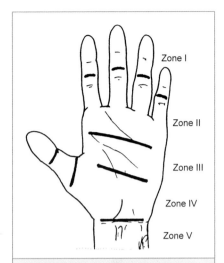

Figure 28-2. The zones of the hand used to describe flexor tendon injuries. (From Brown De, Neumann Rd [eds]: Orthopedic Secrets. Philadelphia, Hanley & Belfus, 1995.)

10. **What is the function of the annular pulleys?**
    To prevent tendon bowstringing and subsequent loss of digital motion during flexion (Fig. 28-3).

11. **What is the function of the cruciform pulleys?**
    The cruciate pulleys are thin and collapsible. They permit the annular pulleys to approximate one another during digital flexion.

12. **Which pulleys are the most important in preventing tendon bowstringing?**
    A2 and A4.

13. **Describe the pulley system of the thumb.**
    The thumb has three pulleys: A1, A2, and oblique. The A1 pulley is located at the level of the MCP joint,

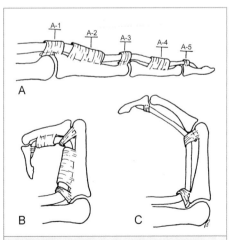

Figure 28-3. *A*, Annular pulley, with FDS omitted. *B* and *C*, Demonstration of function of annular pulleys and tendon bowstringing with absent A2 and A4. (From Brown DE, Newmann RD [eds]: Orthopedic Secrets. Philadelphia, Hanley & Belfus, 1995.)

the A2 pulley at the IP joint, and the oblique pulley in the center portion of the proximal phalanx.

14. **Which thumb pulley is most important to prevent bowstringing of the flexor pollicis longus (FPL) tendon?**
    Oblique pulley.

15. **Define *vincula*.**
    Folds of mesotenon that carry blood supply to the FDS and the FDP tendons. The arterial supply is a series of transverse communicating branches of the common digital artery. Normally, there is a short (vinculum brevis) and long vinculum (vinculum longus) for each FDS and FDP tendon. They enter the *dorsal* portion of each tendon. The vincula are often responsible for preventing proximal retraction of the tendon following injury.

16. **Are the vincula the only source of flexor tendon nutrition?**
    No. Flexor tendons are nourished by vascular perfusion (vincula system) and by diffusion of nutrients from synovial fluid within the sheath. Diffusion appears to be a more effective pathway than perfusion within the digital sheath.

17. **How do flexor tendons within the digital sheath heal after repair?**
    The exact nature of tendon healing remains somewhat controversial. It is believed that healing occurs via the fibroblastic response of the sheath and surrounding tissues and that adhesion formation is essential. However, several studies have demonstrated that flexor tendons have an intrinsic ability to heal via nutrients supplied by diffusion from the synovial fluid.

18. **What are the histologic phases of flexor tendon healing?**
    Flexor tendons heal through three overlapping phases of inflammation, proliferation, and remodeling.
    - The **inflammation phase** starts soon after injury with the invasion of white blood cells, which is followed by the formation of granulation tissue.
    - Fibroblast proliferation and matrix synthesis mark the onset of the **proliferation phase**. The endotenon cells migrate across closely approximated areas of the laceration, whereas epitenon cells tend to proliferate and migrate into the tendon gaps.
    - The **remodeling phase** begins around the sixth week, when the fibroblasts decrease in number and size, and their nuclei align along the long access of the tendon. The replacement tissue that bridges the cut tendon ends continues to mature, but the quality of the tissue remains inferior to normal tendon tissue.

19. **What is "jersey finger"?**
    Avulsion of the FDP tendon from its insertion at the distal phalanx, resulting in a loss of DIP joint flexion. The injury is the result of forcible extension of the DIP joint during grasp.

20. **Which finger is most commonly involved in a jersey finger injury?**
    The ring finger is involved in 75% of cases.

21. **How are FDP tendon ruptures classified?**
    - **Type 1:** Tendon retraction into the palm requiring reattachment within 7–10 days after injury
    - **Type 2:** Tendon retraction to the PIP joint level (most common) with reattachment possible up to 6 weeks
    - **Type 3:** Minimal tendon retraction with A4 pulley preventing the large avulsed fragment from proximal retraction, treatment involves ORIF

22. **Which technique is used to retrieve an avulsed or lacerated flexor tendon that has retracted into the palm?**
    To avoid passing a tendon retriever blindly through the fibro-osseous sheath or incising the entire finger and palm for retrieval of a retracted flexor tendon, a small incision may be made at the distal palmar crease. The retracted tendon is identified, and a small catheter or pediatric gastrostomy feeding tube may be placed retrograde through the finger incision down the sheath and out the palmar incision. The retracted tendon is then sutured to the catheter or tube, which is pulled back out the finger incision, thus retrieving the tendon into the opening in the sheath.

23. **What are the preferred surgical approaches for repair of a flexor tendon laceration in the fibro-osseous sheath?**
    The original wound is extended proximally and distally to allow incorporation into a volar zigzag incision, as advocated by Bunnell, or a mid-axial incision, as popularized by Strickland.

24. **What are "retinacular windows"?**
    The areas of the fibro-osseous sheath between the annular pulleys that may be incised to facilitate flexor tendon repair, thus preserving the important annular pulleys.

25. **Does closure of the sheath following a flexor tendon repair improve the outcome?**
    Many surgeons advocate repair of the sheath because of the role of synovial diffusion as a source of nutrition for flexor tendon healing. However, no study has provided undisputed statistical evidence that closure of the sheath improves the outcome of flexor tendon repair.

26. **What is the preferred suture technique for flexor tendon repair?**
    The preferred suture technique consists of:
    - Atraumatic handling of the tendon ends
    - Placement of a core suture
    - Augmentation with a running epitendinous suture

    The type and technique of suture placement are controversial. The ideal suture material should be nonreactive, strong, easy to use, of small size, pliable, and capable of maintaining a good knot. Most surgeons use a suture composed of a synthetic braided material (3-0 or 4-0 nonabsorbable suture, such as Ethibond, Prolene, Supramid, Mersilene, Tevdek) for the core suture. There are several methods for core suture placement. The most commonly used techniques are the modified Kessler, Bunnell, and Strickland techniques (Fig. 28-4).

27. **To what is the strength of a flexor tendon repair proportional?**
    The number of core suture strands crossing the repair site.

28. **What are the proposed advantages of an epitendinous suture?**
    - Tidies the repair edges, thus avoiding adhesion formation or triggering
    - Adds 10–50% strength to the repair
    - Reduces gap formation at the tendon ends

29. **Is placement of the core sutures important with respect to the tensile strength of the repair?**
    Yes. Dorsal placement of the core suture yields greater tensile strength than volar placement.

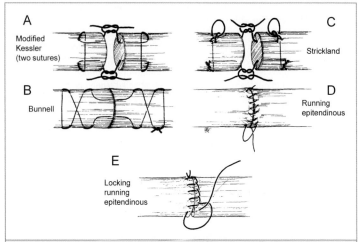

Figure 28-4. The techniques for core suture replacement. *A*, Modified Kessler technique. *B*, Bunnell technique. *C*, Strickland technique. *D*, Augmentation with a running epitendinous suture. *E*, Augmentation with a locking running epitendinous suture. (From Green DP [ed]: Operative Hand Surgery, 3rd ed. New York, Churchill Livingstone, 1993.)

30. **Does time from injury to repair affect function after an intrasynovial flexor tendon repair?**
    An *in vivo* canine study determined that delayed suture (>7 days after injury) leads to an increased incidence of adhesion formation.

KEY POINTS: FLEXOR TENDON INJURIES

1. The fibro-osseous flexor tendon sheath in the fingers includes a series of distinct thickenings known as *annular* and *cruciform* pulleys. There are five annular (A1–5) and three cruciform (C1–3) pulleys.
2. The annular pulleys improve the biomechanical efficiency of tendon excursion and prevent tendon bowstringing during flexion with a subsequent loss of motion.
3. Vincula are folds of mesotenon within the flexor tendon sheath that provide blood supply to the flexor tendons via transverse communicating branches of the common digital artery.
4. Early mobilization following flexor tendon repair promotes tendon gliding and reduces adhesion formation.
5. Flexor tendon injuries have a better overall outcome following repair in children than in adults.
6. Current techniques for flexor tendon repair are focusing on multistrand and multigrasp methods that permit early *active* range of motion exercises.

31. **Why should delayed primary flexor tendon repair be performed within 3 weeks of when the injury occurred?**
    Delaying flexor tendon repair more than 3 weeks postinjury may prevent end-to-end repair because of myostatic contraction of the injured muscle.

32. **How is the hand immobilized immediately following a flexor tendon repair?**
    The wrist is immobilized in 30 degrees of palmar flexion, the MCP joints in 60 degrees of flexion, and the PIP and DIP joints in full extension. In noncompliant patients or young children, extending the splint above the elbow or placement of a long arm cast in the aforementioned position is indicated.

33. **Describe the rehabilitation of a repaired flexor tendon laceration.**
    Postoperative rehabilitation involves *early motion* to prevent adhesion formation and improve tendon healing. Several early motion protocols have been described. A popular program involves immediate-controlled passive immobilization for 4 weeks and then a combination of passive and gentle active exercise for 2 weeks followed by active flexion and extension. Various modifications of this program have been described, depending on patient cooperation and the quality of the tendon repair. New programs involve the earlier use of active motion exercises and discontinuation of splinting. Regardless of which protocol is used, close supervision by both the operating surgeon and hand therapist is essential.

34. **What are the differences between repaired flexor tendons that are totally immobilized postoperatively and those treated with varying degrees of protected passive mobilization?**
    Tensile strength and gliding function are greater in mobilized tendons, whereas adhesion formation is greater in immobilized tendons.

35. **Why should both the FDS and FDP be repaired in a zone II injury?**
    Proposed advantages of repairing both tendons include preservation of the FDP blood supply via an intact vincular system, stronger flexion power, greater independent finger motion, avoidance of PIP joint hyperextension, and a better overall outcome.

36. **What are the most common complications following a flexor tendon repair?**
    - Adhesion formation between the FDP and FDS tendons and/or flexor sheath with a resultant loss of joint motion
    - Rupture of the repair
    - Contracture of the PIP and/or DIP joints

37. **When is a tendon repair the weakest?**
    Approximately 10–12 days postoperatively.

38. **When does a tendon rupture after primary repair most commonly occur?**
    Postoperative day 10.

39. **What is the recommended treatment of an acute flexor tendon rupture following repair?**
    Prompt reexploration and repair if feasible.

40. **How is bowstringing of the flexor tendons and median nerve avoided after flexor tendon repair in zone IV?**
    The transverse carpal ligament (TCL) is incised in zone IV repairs. To prevent postoperative bowstringing of the carpal tunnel contents, the TCL may be lengthened and repaired and/or the wrist immobilized in a neutral position instead of palmar flexion.

41. **How are the results of flexor tendon repair analyzed?**
    The outcome of a flexor tendon repair is determined by measuring the amount of active flexion of the DIP and PIP joints, then subtracting the amount of extension deficit at these joints during active extension. The results are graded as excellent (>132 degrees total motion), good (88–131 degrees), fair (44–87 degrees), or poor (<44 degrees).

42. **What are the physical findings in a patient with a partial flexor tendon laceration?**
    - Weakness or pain with resisted flexor tendon function
    - Evidence of catching or triggering of the partially lacerated site within the tendon sheath

43. **When should a partial flexor tendon laceration be repaired?**
    Although it is controversial, in general, a laceration involving more than 50% of the cross section of the tendon should be repaired because of the risk of rupture.

44. **A 25-year-old man underwent staged flexor tendon reconstruction of the ring finger with a free tendon graft. When he attempts to make a fist, flexion of the adjacent ring and small fingers is weak and incomplete. What is the most likely cause?**
    The tendon graft is most likely too tight. This is known as the *quadrigia effect*. Because the FDP muscles of the small, ring, and long fingers arise from a single muscle belly, "overtightening" one of the tendons results in laxity and incomplete excursion of the others, with a resultant loss of digital flexion.

45. **How are flexor tendon injuries in children treated?**
    Primary repair of a flexor tendon injury in children is associated with a better result following repair in an adult. The child younger than 6 years of age also requires immobilization of the limb above the elbow. In the cooperative child over the age of 6, the dorsal splint with rubber band traction as described by Kleinert is recommended. Making the diagnosis of a flexor tendon injury in the young child is difficult. If there is any doubt concerning the type of injury after carefully observing the child, the wound must be explored and tendons and nerves identified to assess their status. Even though the results of flexor tendon repair in children are usually very good, the tendons are small, the surgical technique must be meticulous, and loupe magnification is typically required.

46. **A 38-year-old woman underwent staged flexor tendon reconstruction of the long finger using an autologous free tendon graft. When she attempts to flex the MCP joint, the PIP joint extends. What is the most likely cause?**
    The phenomenon of paradoxical extension of the PIP joint during MCP flexion is known as *lumbrical plus*. The most likely cause in this patient is a long tendon graft. As the lumbrical contracts via the contracting tendon graft, the PIP joint extends as a consequence of the lumbrical's contribution to the extensor mechanism. Treatment involves release of the involved lumbrical.

47. **Which of the following autologous tendon grafts has the least gliding resistance between the graft itself and the A2 pulley: the palmaris longus, flexor digitorum longus, extensor hallicis longus, or peroneus longus?**
    The flexor digitorum longus. The ideal donor tendon for reconstruction in the digits should have low gliding resistance between the graft and the pulley system. An intrasynovial tendon graft demonstrates superior gliding function compared with an extrasynovial graft after 6 weeks of healing. Of the donor sites listed, only the flexor digitorum longus is *intra*synovial.

48. **A 23-year-old man lacerates his small finger volarly in zone II. He is unable to actively flex the PIP joint. What do you suspect?**
    A laceration of the FDS tendon or a normal variant since the FDS to the small finger is absent in approximately one third of all individuals. Comparison with the other hand should always be performed.

49. **What complication occurs if the FDP is advanced more than 1 cm during the repair of a zone I laceration?**
    The quadrigia effect.

BIBLIOGRAPHY

1. Al-Qattan MM: Conservative management of zone II partial flexor tendon lacerations greater than half the width of the tendon. J Hand Surg 25A:1118–1121, 2000.
2. Britton EN, Kleinert JM: Acute flexor tendon injury: Repair and rehabilitation. In Piemer CA (ed): Surgery of the Hand and Upper Extremity. New York, McGraw-Hill, 1996.
3. Coyle MP, Leddy TP, Leddy JP: Staged flexor tendon reconstruction fingertip to plam. J Hand Surg 27A:581–585, 2002.
4. Gelberman RH, Siegel DB, Woo SL-Y, et al: Healing of digital flexor tendons: Importance of interval from injury to repair. J Bone Joint Surg 734A:66–75, 1991.
5. Gerard F, Garbuio P, Obert L, Tropet Y: Immediate active mobilization after flexor tendon repairs in Verdan's zones I and II. Ann Chir Main:127–132, 1998.
6. Matloub HS, Dzwierzynski WW, Erickson S, et al: Magnetic resonance imaging scanning in the diagnosis of zone II flexor tendon rupture. J Hand Surg 21A:451–455, 1996.
7. Seiler JG, Uchiyama S, Ellis F: Reconstruction of the flexor pulley. The effect of tension and source of graft in an *in vitro* dog model. J Bone Joint Surg 80A:699–703, 1998.
8. Strickland JW: Development of flexor tendon surgery: Twenty-five years of progress. J Hand Surg 25:214–235, 2000.
9. Strickland JW, Schmidt C: Repair of flexor digitorum profundus lacerations: The Indiana Method. Oper Techn Orthop 8:73–80, 2001.
10. Uchiyama S, Amadio PC, Coert JH: Gliding resistance of extrasynovial and intrasynovial tendons through the A2 pulley. J Bone Joint Surg 79A:219–224, 1997.

EXTENSOR TENDON INJURIES

Mary Lynn Newport, MD

1. **Why are extensor tendons so susceptible to traumatic injury?**
   - Superficial location
   - Size (thin cross-sectional area and less tendon substance)

2. **Which zones are used to describe extensor tendon injuries of the fingers, hand, and arm?**
   There are **nine zones** identified by Roman numerals (I–IX). Zones I, III, and V are located over the distal interphalangeal (DIP), proximal interphalangeal (PIP), and metacarpophalangeal (MCP) joints, respectively. Zones II and IV are located over the distal and proximal phalanges, respectively. Zone VI is located over the metacarpals and zone VII beneath the retinaculum. Zone VIII involves the distal forearm, and zone IX includes the muscles in the proximal forearm (Fig. 29-1).

3. **Which zones are used to describe extensor tendon injuries of the thumb?**
   The **five zones** are identified by Roman numerals (I–V). Zone I is over the interphalangeal (IP) joint, zone II over the proximal phalanx, zone III over the MCP joint, zone IV over the first metacarpal, and zone V over the wrist (*see* Fig. 29-1).

4. **Extensor tendon injuries most commonly occur in which digit?**
   The long finger.

5. **Which extensor zone of *the finger* is most commonly injured?**
   Zone VI, directly over the metacarpals.

6. **Which extensor tendon zone of *the thumb* is most commonly injured?**
   Zone IV, over the first metacarpal.

7. **What is the strongest repair technique for an extensor tendon laceration?**
   Modified Bunnell or modified Kessler technique.

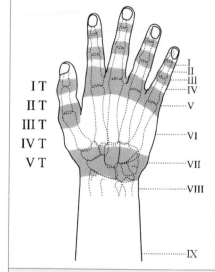

Figure 29-1. Zones identifying injuries to extensor tendon injuries in fingers and thumbs. Nine zones (listed on the right) are used to identify extensor tendon injuries in the fingers, hand, and forearm. Five zones (listed on the left) are used to identify extensor tendon injuries in the thumb.

8. **Which factors are associated with a poor outcome following an extensor tendon repair?**
   - Associated injury (e.g., fracture, joint dislocation, flexor tendon injury)
   - Distal zone injuries (zones I–IV)
   - Infection, soft tissue loss, or crush injury
   - Unreliable, poorly motivated patient

9. **After repair of a simple extensor tendon laceration and static splinting, which is more common—loss of finger flexion or extension? Why?**
   Loss of finger flexion due to adhesion formation.

10. **What is the most common complication following an extensor tendon repair at the level of the wrist (zone VII)?**
    Adhesion of the repaired tendon to the extensor retinaculum. Excising a portion of or lengthening the retinaculum and beginning early motion may decrease the incidence of adhesion formation and should be done if the repair site impinges on the retinaculum. Excision of the entire retinaculum should be avoided because bowstringing of the extensor tendons may occur (Fig. 29-2).

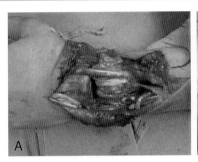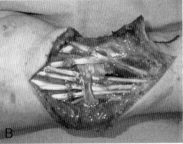

Figure 29-2. *A,* Multiple extensor tendon lacerations over the wrist joint with an associated traumatic arthrotomy. *B,* Tendons repaired with a modified Kessler technique.

11. **What is the preferred treatment for a late presenting, unrecognized extensor pollicis longus (EPL) laceration in zone IV?**
    Primary repair is usually not feasible because of myostatic contracture of the muscle belly. Tendon grafting may be performed, but the preferred procedure is a tendon transfer, such as an extensor indicis proprius (EIP) to EPL tendon transfer.

12. **Which specific injury must be ruled out in *every* patient who presents with an extensor tendon laceration over the MCP joint?**
    A human bite wound, or "fight bite."

13. **Which particular injuries occur in association with a human bite wound over the MCP joint (zone V)?**
    - Extensor tendon laceration
    - Traumatic MCP joint arthrotomy
    - Metacarpal head or neck fracture
    - MCP joint chondral injury
    - Foreign body introduction (e.g., tooth)

    A fight bite is a contaminated wound that can result in significant infection, dysfunction, or possible amputation if neglected or inappropriately treated.

14. **Describe the management of a complete extensor tendon laceration in zone V with an associated human bite wound.**
    - Tetanus immunization is provided as indicated.
    - Intravenous antibiotics that provide coverage of gram-positive bacteria (staphylococci and streptococci) are begun and maintained if an active infection is present. A short course of prophylactic oral antibiotics is usually prescribed in the absence of infection.
    - Radiographs of the hand should be obtained to note the presence or absence of a foreign body and/or fracture.
    - The wound is extended proximally and distally, and the underlying extensor tendon and MCP joint are explored. The articular surfaces and joint recesses should be inspected with the joint in **both** flexion and extension. If the injury is acute and infection is not present, the tendon may be repaired after thorough irrigation and debridement of the wound. Repair is not necessary if the laceration is <50%. If the wound is infected, the tendon is usually repaired secondarily. ***The wound is always left open***. Under no circumstances should a human bite wound be closed.

15. **What is the key to successful repair of a zone VIII extensor tendon laceration?**
    Repair at the myotendinous junction is difficult. To improve the quality of the repair, the tendon should be sutured to the intramuscular raphe.

16. **What is a "mallet finger"?**
    Disruption of the terminal slip of the extensor mechanism resulting in a characteristic flexion deformity or drooping of the DIP joint. The mechanism of injury (MOI) is sudden forceful flexion of the extended fingertip. The injury also can result in a variable-sized avulsion of bone from the distal phalanx or fracture through an open epiphysis.

17. **How are mallet finger injuries classified?**
    - **Type 1:** Closed rupture of tendon insertion
    - **Type 2:** Tendon laceration
    - **Type 3:** Deep abrasion with loss of skin, subcutaneous tissue, and tendon substance
    - **Type 4A:** Transepiphyseal plate fracture in children
    - **Type 4B:** Avulsion fracture of the distal phalanx involving 20–50% of articular surface
    - **Type 4C:** Fracture >50% with palmar subluxation of the distal phalanx

18. **How are type 1, 2, and 3 injuries treated?**
    - **Type 1:** DIP joint immobilization in full extension for 6 weeks followed by nighttime splinting for an additional 4–6 weeks
    - **Type 2:** Tendon repair with temporary immobilization of the DIP joint in full extension, consider transarticular K-wire
    - **Type 3:** Soft tissue coverage (reversed cross-finger flap) and tendon repair or grafting versus DIP arthrodesis

19. **Describe the treatment of an adult type 4 mallet finger.**
    - If the fracture is not associated with palmar subluxation of the distal phalanx, splinting of the DIP joint is preferred.
    - If the fracture is >40–50% of the articular surface and distal phalanx subluxation is noted, open reduction and internal fixation (ORIF) is recommended.

    Open treatment of a type 4 mallet finger should be undertaken carefully since a 50% complication rate has been reported.

20. **What are the most common complications associated with splint treatment of a type 1 mallet finger?**
    Skin maceration and noncompliance.

21. **Which characteristic deformity occurs following a laceration or rupture of the central slip?**
    Boutonnière (buttonhole) deformity (Fig. 29-3).

22. **What are the three components of an acute boutonnière deformity?**
    - Central slip disruption
    - Attenuation of the triangular ligament connecting the lateral bands
    - Volar migration of the lateral bands

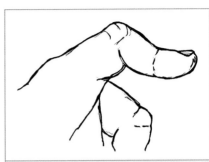

23. **How is an acute central slip rupture treated?**
    Immobilization of the PIP joint in full extension for 6–8 weeks. DIP joint motion is preserved to prevent contracture of the oblique retinacular ligaments and to keep the lateral bands mobile.

    Figure 29-3. A boutonnière deformity following traumatic rupture of the central slip insertion.

24. **Why is passive range of motion of the DIP joint performed during splint treatment of an acute boutonnière deformity?**
    Flexion of the DIP joint draws the lateral bands distally and dorsally, thus rebalancing the extensor mechanism.

25. **What is the MOI for a traumatic boutonnière deformity?**
    Acute forceful flexion of the PIP joint resulting in avulsion of the central slip or volar dislocation of the PIP joint. Immediately after injury, PIP joint extension is maintained by the lateral bands. The diagnosis is suspected on the basis of a swollen PIP joint, tenderness at the dorsal joint line, and weak PIP extension against resistance. *The development of a boutonnière deformity may not become apparent until 10–21 days after the time of injury.*

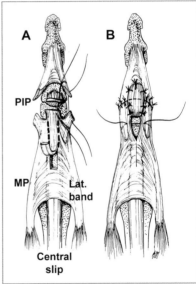

Figure 29-4. *A*, Reconstruction of the central slip following traumatic loss of the tendon. This procedure may be performed using a retrograde flap of the remaining central slip. *B*, Reconstruction performed by reapproximating the split lateral bands at the base of the middle phalanx.

26. **Which reconstructive techniques are useful for an acute boutonnière deformity with associated loss of the central slip?**
    A retrograde (distally based) flap from the remaining central slip reinforced over the central slip repair or longitudinally split lateral bands reapproximated in the midline over the base of the middle phalanx (Fig. 29-4).

27. **What are the clinical stages of a boutonnière deformity?**
    - **Stage 1:** Dynamic deformity that is passively supple
    - **Stage 2:** Established deformity that cannot be passively corrected
    - **Stage 3:** Established deformity with secondary PIP joint changes (fibrosis, arthrosis) and capsular or collateral ligament contracture

28. **What must be present before proceeding with surgical reconstruction of a chronic boutonnière deformity?**
    Full passive mobility of the PIP joint.

29. **A sagittal band tear with a resultant traumatic dislocation of the extensor tendon most commonly involves which finger?**
    The middle (long) finger.

30. **Which sagittal band (radial or ulnar) is most commonly torn?**
    The radial sagittal band.

31. **Describe the MOI for a sagittal band tear.**
    A direct blow to the MCP joint or forceful MCP joint flexion or extension.

32. **What are the typical clinical findings noted in a patient with a sagittal band tear?**
    - Swelling over the MCP joint
    - Localized tenderness
    - Pain with resisted finger extension
    - Incomplete MCP joint extension
    - Ulnar (radial sagittal band tear) or radial (ulnar sagittal band tear) deviation of the involved digit

33. **How is a sagittal band tear with an extensor tendon subluxation or dislocation treated?**
    Acute injuries (less than 2 weeks old) may be casted with the MCP joint in full extension and the IP joints free. Repair of the sagittal band or reconstruction with a distally based slip of the extensor mechanism or juncturae tendinum is indicated if cast immobilization fails or if the injury is chronic.

KEY POINTS: EXTENSOR TENDON INJURIES

1. The modified Bunnell or Kessler techniques are the most effective for repair of an extensor tendon laceration.
2. Loss of digital flexion is common following repair and static immobilization of an extensor tendon laceration.
3. A human bite wound should never be closed.
4. Full passive PIP joint mobility must be present before proceeding with reconstruction of a chronic boutonnière deformity.
5. A sagittal band tear with a secondary dislocation of the extensor tendon most commonly involves the radial sagittal band of the long finger.

34. **Describe the MOI for a dislocation of the extensor carpi ulnaris (ECU) tendon.**
    Forceful forearm supination combined with palmar flexion and ulnar deviation of the wrist.

35. **How is an acute ECU dislocation treated?**
    An acute injury is treated by long arm cast immobilization in full supination for 4–6 weeks. Chronic injuries usually require tendon relocation and stabilization with a soft tissue sling.

36. **Define *extrinsic extensor tendon tightness*.**
    Extrinsic extensor tendon tightness occurs when the extensor tendon becomes adherent over the metacarpal or foreshortened after repair. Dorsal tenodesis during MCP joint flexion limits PIP flexion when the MCP joint is flexed.

37. **How do you test for extrinsic tightness?**
    PIP joint flexion is first assessed with the MCP joint held in extension. PIP joint flexion is then determined with the MCP joint fully flexed. If extrinsic tightness is present, active and passive PIP joint flexion is greater with the MCP joint in extension than in full flexion.

38. **Which common clinical scenarios result in extrinsic extensor tendon tightness?**
    - Metacarpal fracture, particularly if the fracture was treated by ORIF
    - Soft tissue injury over the dorsum of the hand
    - Crush injury to the dorsum of the hand
    - Extensor tendon repair or transfer that was inadvertently performed with excessive tension or shortening

39. **What is the indication for surgical treatment of extrinsic extensor tendon tightness?**
    Functionally limiting loss of simultaneous MCP and PIP joint flexion despite a minimum of 6 months of appropriate hand therapy.

40. **Which surgical procedures are advocated for extrinsic extensor tendon tightness?**
    Extensor tenolysis, release, or lengthening. Tenolysis is preferred if appropriate tendon length is available and if the loss of motion results from tendon adherence or scarring. If extrinsic tightness is secondary to inadequate tendon length, an extrinsic extensor tendon release or lengthening is advocated.

41. **Define *swan-neck deformity*.**
    The swan-neck deformity refers to the characteristic appearance of the finger with PIP hyperextension and DIP flexion.

42. **What are the potential causes of a swan-neck deformity?**
    - PIP volar plate injury with hyperextension, dorsal displacement of the lateral bands dorsal to the axis of PIP joint rotation
    - Mallet finger injury in the patient with ligamentous laxity, with all extensor force secondarily transferred to the PIP joint and the joint pulled into hyperextension because of laxity

43. **Which surgical procedures are advocated for the neglected mallet finger with a secondary swan-neck deformity?**
    - Oblique retinacular ligament reconstruction
    - Flexor digitorum superficialis (FDS) tenodesis
    - Central slip tenotomy (Fowler procedure)

44. **When is *staged* extensor tendon reconstruction most commonly indicated?**
    Following an avulsion or degloving injury to the dorsum of the hand or wrist with associated extensor tendon and soft tissue loss. Initial treatment consists of irrigation and debridement of

all devitalized and contaminated tissue followed by reduction and stabilization of all associated fractures. Soft tissue coverage should be done as soon as is safely possible to prevent infection. Primary tendon repair, grafting, or tendon transfer is usually not feasible. Silicone rods attached to the distal extensor tendon stumps are placed at the time of soft tissue coverage if the soft tissue bed is appropriate. Secondary tendon grafting is typically performed 3 months later (Fig. 29-5).

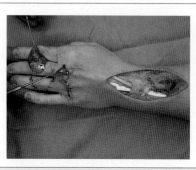

Figure 29-5. Staged reconstruction of the long finger extensor tendon following a chronic postoperative infection and tendon debridement.

45. **Which donor sites are most commonly used for staged extensor tendon grafting?**
    - Palmaris longus
    - Extensor digitorum longus (foot)
    - The FDS, extensor indicis proprius, and extensor digiti minimi (used when necessary)

BIBLIOGRAPHY

1. Brown EZ Jr, Ribik CA: Early dynamic splinting for extensor tendon injuries. J Hand Surg 14A:72–76, 1989.
2. Chester DL, Beale S, Beveridge L, et al: A prospective, controlled, randomized trial comparing early active extension with passive extension using a dynamic splint in the rehabilitation of repaired extensor tendons. J Hand Surg (Br) 27(3):283–288, 2002.
3. Crosby CA, Wehbe MA: Early protected motion after extensor tendon repair. J Hand Surg (Am) 24(5):1061–1070, 1999.
4. Hung LK, Chan A, Chang J, et al: Early controlled active mobilization with dynamic splintage for treatment of extensor tendon injuries. J Hand Surg 15A:251–257, 1990.
5. Ishizuki M: Traumatic and spontaneous dislocation of extensor tendon of the long finger. J Hand Surg 15:967–972, 1990.
6. Newport ML, Blair WF, Steyers CM Jr: Long-term results of extensor tendon repair. J Hand Surg 15A:961–966, 1990.
7. Newport ML, Pollack GR, Williams CD: Biomechanical characteristics of suture techniques in extensor zone IV. J Hand Surg 20A(4):650–656, 1995.
8. Newport ML, Williams CD: Biomechanical characteristics of extensor tendon suture techniques. J Hand Surg 17A:1117–1123, 1992.
9. Newport ML: Extensor tendons injuries in the hand. J Am Acad Ortho Surg 5(2):59–66, 1997.
10. Okafor B, Mbubaegbu C, Munshi L, Williams DJ: Mallet deformity of the finger: Five-year follow-up of conservative treatment. J Bone Joint Surg (Br) 79A(4):544–547, 1997.
11. Ritts GD, Wood MB, Engber WD: Nonoperative treatment of traumatic dislocations of the extensor digitorum tendons in patients without rheumatoid disorders. J Hand Surg 10:714–716, 1985.

NAIL BED AND FINGERTIP INJURIES
Jeffrey C. King, MD, and Keith G. Wolter, MD

1. **Describe the anatomy of the fingernail.**
   - The **perionychium** is composed of the nail bed (composed of the sterile and germinal matrix) and surrounding soft tissues.
   - The **hyponychium** is the junction of the nail bed and the skin at the distal aspect of the finger.
   - The **eponychium** or cuticle extends from the nail wall to the dorsal nail plate.
   - The **paronychium** extends along the lateral edge of the nail and the eponychium.
   - The **germinal matrix** produces the bulk (90%) of the nail plate, which is composed of stacked sheets of anucleate epithelial cells.
   - The **sterile matrix** extends from the lunula to the hyponychium. It contributes keratin, which thickens the nail and enables it to adhere to the nail bed as it migrates distally.
   - The **lunula**, a crescent-shaped white area in the proximal nail, is the visible portion of the germinal matrix. The remainder of the germinal matrix lies beneath the eponychium and nail wall, adjacent to the extensor tendon insertion on the distal phalanx (Fig. 30-1).

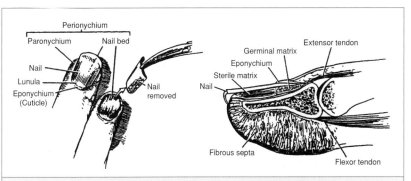

Figure 30-1. Anatomy of the fingernail and fingertip. The fingernail originates from the germinal matrix in the base of the nail fold. The deep surface of the nail bed is closely adherent to the periosteum of the distal phalanx. Numerous fibrous septae connect the volar digital skin to the underlying bone. (From Russell RC, Casas LA: Management of fingertip injuries. Clin Plast Surg 16:405–425, 1989.)

2. **Describe the anatomy of the fingertip.**
   The volar surface of the fingertip is highly specialized. The glabrous skin, with its unique pattern of stratified cells, results in the distinctive fingerprint. Fibrous septae anchor the skin to the bone, providing stability during grasp (*see* Fig. 30-1). Dense nerve receptors enable highly discriminatory sensation. The digital vessels arborize into a multitude of small branches surrounding the distal phalanx.

3. **What is the nail fold? Why is it important?**
   - The nail fold houses the proximal nail plate and is composed of a **dorsal roof** and **ventral floor**.

- The ventral floor is part of the germinal matrix; the dorsal roof tapers to the nail wall and the thin eponychium distally.
- The patency of the nail fold is crucial for normal nail growth; if patency is not maintained after injury, adhesions between the roof and floor (**synechia**) can result in nail grooving or splitting. The roof of the nail fold also produces keratin, which gives healthy nails their shine. Injury to the nail fold may cause the nail to appear dull.

4. **What is the significance of the hyponychium?**
   The hyponychium is the junction of the skin and nail bed just under the distal edge of the nail plate. A keratinous plug that prevents debris from getting under the nail plate demarcates the hyponychium.

5. **Which clinical history questions should be asked when evaluating a patient with a fingertip injury?**
   - When did the injury occur?
   - What was the mechanism of injury (i.e., crush, laceration, or avulsion)?
   - What is the patient's age, sex, occupation, and hand dominance?
   - Are there any associated injuries to the injured extremity?
   - What is the patient's tetanus immunization status?

6. **How long does it take for a new nail to grow?**
   The nail grows at an average rate of 0.1 mm/day. After the initial injury, there is typically a delay in growth, followed by an accelerated growth phase that subsequently diminishes to the normal growth rate. Complete longitudinal growth takes between 70 and 160 days.
   Nail growth follows a trimodal pattern with regard to age. Children who are less than 3 years of age have a low growth rate. The rate increases between ages 3 and 30 years, then slows after 30 years.

7. **What is the significance of a subungual hematoma?**
   A subungual hematoma indicates an injury to the nail bed. For a hematoma that involves up to 50% of the subungual region, a hole can be made in the nail using a cautery or heated needle to drain the hematoma directly. If the hematoma is larger than 50%, the nail plate should be removed, the hematoma evacuated, and the associated nailbed laceration repaired.

8. **How does a *splinter* hemorrhage differ from a *subungual* hematoma?**
   A splinter hemorrhage is a nontraumatic lesion beneath the nail plate of the patient with bacterial endocarditis and valvular vegetations. Splinter hemorrhages are often multiple and are caused by the embolization of bacteria or blood clots to the terminal arterioles of the fingertips. Associated hand findings include Osler's nodes, which are small painful lesions in the fingertip pad, and Janeway's lesions, which are larger, painless, discolored lesions on the palms and soles of the feet.

9. **Discuss the epidemiology of nail bed injuries.**
   - These injuries are most common in children and young adults.
   - They most commonly involve the long finger.
   - Most injuries are simple lacerations.
   - Approximately 50% of injuries are associated with a distal phalanx fracture.
   - Crush and avulsion injuries have the worst prognosis.

10. **Which injury has the highest predictive value for a nail bed laceration?**
    Fracture of the distal phalanx.

NAIL BED AND FINGERTIP INJURIES

11. **When is the best time to repair a nail bed injury?**
    The results of immediate repair are superior to those of delayed repair or reconstruction.

12. **How is the digit prepared for a nail bed repair?**
    - Adequate anesthesia is obtained via a regional block using a local anesthetic agent without epinephrine.
    - The entire hand should be sterilely prepped, draped, and comfortably positioned on an arm board or hand table.
    - Tourniquet ischemia is required for optimal visualization. A penrose drain around the base of the finger is useful for this purpose.
    - Loupe magnification is particularly helpful.

13. **What are the key steps for repairing a nail bed?**
    1. Remove the nail plate to visualize the nail bed.
    2. Irrigate and cleanse the nail bed.
    3. Remove any grossly contaminated and/or clearly nonviable tissue.
    4. Repair the nail margin first to establish anatomic alignment.
    5. Repair the nail bed and nail fold using a fine absorbable suture, such as 6-0 chromic.
    6. Repair any associated fingerpad laceration.
    7. Keep the nail fold *open* with replacement of the nail plate (preferred); alternatively, a substitute, such as silastic or Xeroform gauze, may be used.

14. **Which structure lies between the sterile matrix and distal phalanx?**
    Periosteum. Thus, a nail bed injury or avulsion with an associated distal phalanx fracture is by definition an open fracture. The nail plate should be removed and the wound copiously irrigated before nail bed repair. If the fracture is unstable and displaced, prohibiting a tension-free repair of the nail bed, the fracture should be reduced and stabilized, if necessary, with a Kirschner wire inserted percutaneously through the fingertip.

15. **Describe the treatment of a nail plate avulsion with an associated deep nail matrix avulsion.**
    - Inspect the undersurface of the nail plate for attached matrix.
    - Carefully dissect free the matrix from the nail plate to create a free graft if large enough.
    - For a small matrix avulsion, replace the nail without separating the matrix.
    - Cleanse the nail bed defect.
    - Place the free matrix graft into the defect, and repair with absorbable suture.
    - Replace the avulsed nail plate beneath the eponychial fold (Fig. 30-2).

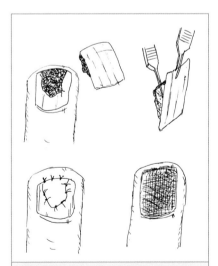

Figure 30-2. Nail matrix attached to an avulsed nail plate. The matrix is carefully excised and then inserted as a free graft. The matrix is then covered with a sterile nonadherent gauze or other nail substitute.

16. **What is the recommended treatment for a nail avulsion with loss of the matrix?**
    Replacement with a free full or partial-thickness graft from an adjacent finger or toe (usually the second toe).

NAIL BED AND FINGERTIP INJURIES

17. **What causes nail plate ridging?**
    A loss of sterile matrix following an injury or inadequate repair with a resulting scar tissue that prohibits normal growth and maintenance of the nail plate. Sterile matrix loss results in a permanent ridge, a split nail, or a nonadherent nail. Temporary nail ridging may be a sign of a systemic illness or hypoxic insult to the finger (**Beau's lines**).

18. **How is nail ridging treated?**
    - If the deformity is mild, the nail plate is trimmed appropriately.
    - For a moderate or severe deformity, the scar can be resected and the nail bed reconstructed with a split-thickness nail bed graft from the same nail bed or another finger or toe.

19. **What is the significance of a hook nail?**
    A hook nail occurs following loss of bone support for the sterile matrix, which requires a skeletal buttress along its entire length for normal longitudinal nail growth. If this support is lost, or if the nail bed is advanced over the tip of the distal phalanx, a hook nail occurs. A hook nail commonly occurs following a fingertip amputation.

20. **Incomplete removal of the germinal matrix after a fingertip amputation results in which condition?**
    A nail cyst or spike (Fig. 30-3). Treatment consists of complete excision of the germinal matrix.

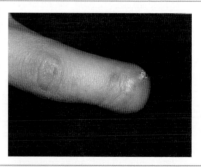

Figure 30-3. Nail spike.

21. **What are the treatment options for a fingertip amputation with a dorsally angulated defect?**
    - Volar V-Y advancement (Atasoy-Kleinert) flap
    - Lateral double V-Y advancement (Kutler) flap
    - Moberg volar neurovascular advancement flap (thumb only)
    - Neurovascular island flap

22. **Describe the volar (Atasoy-Kleinert) V-Y advancement flap.**
    - The flap involves a full-thickness V-shaped incision of the remaining volar fingertip, with the apex placed at the distal interphalangeal (DIP) crease.
    - The flap is elevated off of the distal phalanx and flexor tendon sheath, with division of the septae and preservation of the nerves and vessels.
    - The flap is advanced distally to cover the tip defect.
    - The flap may be advanced up to 1 cm.
    - The proximal incision is closed, resulting in a Y-shaped wound (Fig. 30-4).

23. **Describe the Kutler V-Y flap.**
    - The flap involves two V-shaped incisions at the lateral aspects of the remaining finger.
    - The flaps are mobilized and advanced distally to meet in the midline (Fig. 30-4).
    - The vascular supply of the lateral flaps is less reliable than that of the volar flap.

24. **Describe the Moberg volar advancement flap for thumb tip defects.**
    - The flap includes the entire volar surface of the remaining digit.
    - It is developed by making two longitudinal skin incisions just dorsal to the interphalangeal flexion creases.

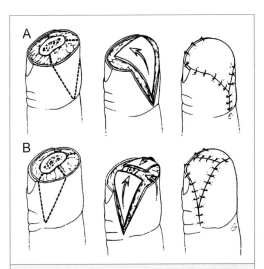

Figure 30-4. Two methods of V-Y advancement of tissue onto the distal part of the fingertip. *A,* The Atasoy-Kleinert volar V-Y flap. This flap works well for more dorsal losses of tissue, with sparing of volar tissue. *B,* The Kutler lateral V-Y flap. This flap works well for nearly transverse amputations. (From Chick L, Lister G: Trauma. In Ruberg RL, Smith DJ [eds]: Plastic Surgery: A Core Curriculum. St. Louis, Mosby, 1994.)

- The digit is flexed and the flap advanced to close the distal defect, leaving a proximal base.
- Alternatively, the proximal end of the flap can be managed with a V-to-Y closure (Fig. 30-5).

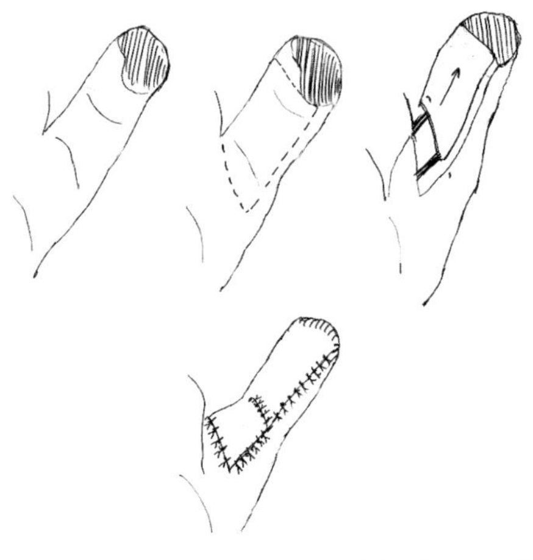

25. **What is the primary indication for the Moberg volar advancement flap?**
    Coverage of an amputated thumb tip, for which preservation of length is critical.

26. **What are the most common complications associated with the Moberg flap?**
    - Flexion contracture of the thumb
    - Flap necrosis

27. **What is the maximal amount of advancement for a Moberg flap?**
    The maximal gain in length is 1 cm.

Figure 30-5. The Moberg advancement flap. This flap is useful to cover a thumb tip amputation.

28. **What is a cross-finger flap?**
    - A vascularized, random pedicle flap is based on the dorsal aspect of the middle phalangeal region of a finger that is adjacent to the injured finger.

- The flap is designed to cover a volar defect.
- To minimize the donor defect, a template should be made of the area to be covered, and the flap should be 2 mm larger in all directions than the defect to be covered.
- Care is taken to preserve the paratenon of the extensor mechanism during flap elevation.
- Takedown of the flap usually occurs 10–14 days later.
- The donor defect is covered with a full or split-thickness skin graft (Fig. 30-6).

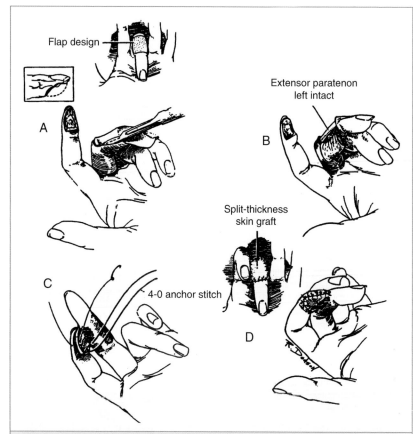

Figure 30-6. Standard cross-finger flap. *A,* A volar fingertip pulp amputation. This type of amputation is best covered with a standard cross-finger flap designed over the dorsal surface of the middle phalanx of an adjoining digit. *B,* The flap elevated superficial to the extensor tendon paratenon. The flap is turned over like the page of a book. *C,* The flexed injured finger. With the finger flexed, the flap is sutured over the tip defect. *D,* The donor site. The site is covered with a skin graft. (From Russell RC, Casas LA: Management of fingertip injuries. Clin Plast Surg 16:405–425, 1989.)

29. **Describe the thenar flap.**
    - The thenar flap is a vascularized, random pedicle flap that is harvested from the proximal radial aspect of the thumb in the thenar region (Fig. 30-7).
    - The flap is developed with its base perpendicular to the thumb metacarpophalangeal (MCP) flexion crease.

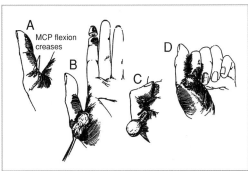

Figure 30-7. Thenar flap. ***A,*** The metacarpophalangeal (MCP) joint flexion crease. This crease is a useful donor site for the thenar flap. ***B,*** The radially based flap that is elevated at the level of the flexor tendon without injuring the neurovascular bundle. ***C,*** The flexed thumb. When it is flexed, the donor site is closed primarily, if possible, or covered with a skin graft. ***D,*** The injured fingertip flexed into the palm. The flap is inset. (From Russell RC, Casas LA: Management of fingertip injuries. Clin Plast Surg 16:405–425, 1989.)

30. **What are the indications for a thenar flap?**
    - Coverage of a transverse or volar oblique soft tissue defect should involve the index or long fingers.
    - Stiffness in the recipient finger is the major drawback to this flap. Therefore, its use is reserved for young patients who have no evidence of preexisting joint stiffness or arthrosis.

KEY POINTS: NAIL BED AND FINGERTIP INJURIES

1. A fracture of the distal phalanx has the highest predictive value for a nail bed laceration.
2. A nail spike occurs following incomplete removal of the germinal matrix during a revision amputation.
3. A hook nail occurs because of loss of bone support for the sterile matrix.
4. A Moberg advancement flap is used to treat a thumb tip amputation so that critical length and sensibility can be preserved.
5. The most common complications associated with the Moberg advancement flap are flap necrosis and a flexion contracture.

31. **List four important technical points to consider with the thenar flap.**
    1. Keep the flap as proximal as possible on the thenar eminence.
    2. Keep the flap as radial as possible.
    3. Minimize the amount of proximal interphalangeal joint flexion in the recipient finger.
    4. Release the pedicle, and initiate active motion as soon as possible.

32. **What are common complications of local pedicle flaps in the digits?**
    Pedicle flaps require 10–14 days of relative immobility of the injured digit and donor area, which most commonly results in joint stiffness and tendon adhesions.

33. **List three common patient complaints following a fingertip amputation, regardless of the treatment approach.**
    - Cold intolerance
    - Diminished sensation
    - Hypersensitivity

34. **What are the alternative sources for donor tissue following a fingertip amputation?**
    - The best option is the amputated segment. An appropriately sized skin graft may be fashioned from the amputated tissue, trimmed and defatted, and sutured over the defect.
    - A local advancement flap may be used if the amputated tissue is not available.
    - As a last resort, a split- or full-thickness skin graft can be used. However, the durability of a skin graft on the fingertip is not ideal. Potential donor sites for a skin graft include the hypothenar eminence and inner aspect of the upper arm. The hairless hypothenar skin closely matches that of the fingertip.

35. **What are the advantages of using a local flap versus a split-thickness skin graft for coverage of a fingertip injury without exposed bone?**
    - Sensation is reportedly better with a local flap.
    - The contour of a local flap more closely approximates that of the natural anatomy.
    - Split-thickness skin grafts are prone to scar contracture and breakdown, which are particularly undesirable in the fingertip.

36. **Name the treatment options for a fingertip amputation with exposed bone.**
    - Skeletal shortening and primary closure
    - Skeletal shortening and healing by secondary intention (Fig. 30-8)

37. **What are the advantages and disadvantages of the two options listed in question 36?**
    In most cases, healing by secondary intention gives a better final functional result. Results are surprisingly good, even in adults, and the return of sensibility is generally better. However, one of the concerns is the delay in return to function. Healing by secondary intention can take up to 3–6 weeks or more, and the wound requires meticulous care and daily dressing changes. For poorly compliant patients or patients who require rapid return to function, skeletal shortening and closure may be the best choice.

38. **What is the maximal tissue defect that will reliably heal by secondary intention?**
    A defect of up to 1 cm^2 should predictably heal by secondary intention.

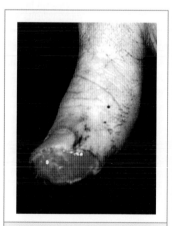

Figure 30-8. An amputation of the fingertip permitted to heal by secondary intention.

39. **What are the most common complications after revision and closure of a fingertip amputation?**
    - Nail remnant (spike) and neuroma formation are common.
    - Care must be exercised to remove all of the germinal matrix from both sides of the eponychial fold in attempting to ablate the nail. Otherwise, the remaining matrix will continue to produce a nail.

- A severely injured, unrepaired nerve can form a neuroma (Fig. 30-9). To minimize the risk of developing this complication, traction is applied to the severed nerve, and the nerve end is sharply transected as proximally as possible. The cut end retracts away from the fingertip.

40. A 42-year-old man sustains a ring finger amputation through the DIP joint. Shortening and primary closure are performed. The flexor digitorum profundus (FDP) is sutured to the extensor mechanism over the tip to provide padding. About 2 months later, he complains of the inability to make a tight fist. What caused loss of motion in the uninjured digits?
    - The loss of motion is the result of the *quadriga effect*, which refers to a Roman chariot that was pulled by four horses but controlled by one driver.
    - The FDP tendons of the long, ring, and small fingers share a common muscle belly. If the FDP tendon to one of these fingers is pulled too far distally, weakness of the other FDP tendons results, and digital flexion is impaired. The FDP to the index finger typically has a separate muscle belly and therefore is not affected.

Figure 30-9. A neuroma of the digital nerve following revision amputation of a ring finger.

41. Which other abnormal pattern of finger motion may occur after transection of the FDP tendon proximal to its insertion?
    A **lumbrical plus finger,** which consists of paradoxical extension of the proximal interphalangeal (PIP) joint of the affected finger during digital flexion. When the FDP tendon is released from its insertion, it shortens and thus places the lumbrical, which inserts on the lateral band of the extensor mechanism, on increased stretch. When grasp is attempted, active shortening of the FDP results in extension of the PIP joint via the lumbrical.

42. How do fingertip injuries differ between children and adults?
    Children heal faster and better. Spontaneous re-epithelialization after local wound care alone results in an excellent outcome in most children.

43. Why is it especially advisable to encourage patients with a fingertip injury to stop smoking (and *not* by using nicotine patches)?
    Nicotine inhibits the inflammatory phase of wound healing and increases the wound complication rate by:
    1. Decreasing local blood flow
    2. Elevating vasopressin and norepinephrine levels that inhibit epithelialization
    3. Stimulating the sympathetic ganglia and adrenal medulla
    4. Inhibiting the maturation and proliferation capacity of erythrocyte precursors

BIBLIOGRAPHY

1. Brown RE, Zook EG, Russell RC: Fingertip reconstruction with flaps and nail bed grafts. J Hand Surg 24:345–351, 1999.
2. Browne EZ: Complications of fingertip injuries. Hand Clin 10:125–137, 1994.

3. Goitz R, Westkaemper J, Tomaino M, Sotereanos D: Soft-tissue defects of the digits. Hand Clin 13:189–205, 1997.
4. Foucher G, Delare O, Citron N, Molderez A: Long-term outcome of neurovascular palmar advancement flaps for distal thumb injuries. Br J Plast Surg 52:64–68, 1999.
5. Martin C, Gonzalea Del Pino J: Controversies in the treatment of fingertip amputations: Conservative versus surgical reconstruction. Clin Orthop 353:63–73, 1998.
6. Russell RC, Casas LA: Management of fingertip injuries. Clin Plast Surg 16:405–425, 1989.
7. Zook EG, Neumeister MW: Fingernails. In Russell RC (ed): Plastic Surgery: Indications, Operations, and Outcomes, Vol. IV. St Louis, Mosby, 2000, p 1751.
8. Zook EG: Reconstruction of a functional and esthetic nail. Hand Clin 6:59–68, 1990.

SOFT TISSUE COVERAGE OF THE HAND

Donald P. Condit, MD, and Danielle A. Conaway, MD

1. **What are the five phases of wound healing?**
   1. Hemostasis
   2. Inflammation
   3. Cellular migration and proliferation
   4. Protein synthesis and wound contraction
   5. Remodeling

2. **During the early phases of wound healing, what is the primary component of the provisional wound matrix and what are its major functions?**
   Fibrin is the primary component. Its functions include:
   - Contributes to hemostasis
   - Binds with fibronectins (glycoproteins that facilitate attachment of migrating cells) to modulate migration of cells into the wound
   - Binds various cytokines at the time of injury and serves as a reservoir for these factors in later stages of wound healing
   - Stimulates inflammatory cell migration and angiogenesis

3. **What are the physical signs of inflammation?**
   Erythema, edema, pain, heat, and decrease in functional use of the involved area.

4. **What causes these physical signs?**
   - 10–15 minutes after the initial injury (marked by vasoconstriction), vasodilation occurs, which generates erythema and heat.
   - Capillaries develop gaps between the endothelial cells that line them. These gaps allow plasma (including inflammatory cells) to leak from the intravascular space to the extravascular compartment leading to edema.
   - Alterations in tissue pH resulting from breakdown products of tissue and bacteria, along with swelling and decreased tissue oxygenation from damage to the blood supply, produce pain in the injured area.

5. **In normal healing, after hemostasis, which cells predominate in the wound?**
   - **Initially:** Inflammatory cells with neutrophils predominate.
   - **48–72 hours:** Macrophages begin to outnumber and remain in wound for several days.
   - **After 5–7 days:** Few inflammatory cells remain, and fibroblasts become the predominant cell type.

6. **What cell is responsible for later stages of wound healing and why?**
   The fibroblast is responsible for collagen synthesis and other proteins generated during the repair process.

7. **What type of collagen is present in normal skin and scar tissue and in what amount?**
   - Collagen makes up 25% of protein in normal skin and more than 50% in scar tissue.

- Collagen type I predominates and makes up 80–90% seen in intact dermis; the remaining 10–20% is type III.
- In wound granulation tissue type III, collagen increases to approximately 30%.

8. **Why does scar tissue have increased stiffness versus normal skin?**
   - Elastin, a normal component of connective tissue matrix (dermis), is not synthesized in response to injury, and subsequently, is not found in scar tissue.

9. **When does wound contraction begin, what is its typical rate, and what affects it?**
   - Wound contraction begins 4–5 days after initial injury and actively continues for approximately 2 weeks. The process will continue for a longer period of time in an open wound.
   - The rate of contraction varies between locations but averages 0.6–0.7 mm/day.
   - The rate of contraction can often be predicted by the degree of skin laxity. A wound on the scalp or pretibial area will contract significantly more slowly than a buttock wound.
   - Wound shape affects the rate of contraction, with square wounds contracting more quickly than circular wounds.
   - Splints can temporarily slow wound contraction, but it will proceed at an accelerated rate after splint removal.

10. **What cell provides the mechanism for wound contraction?**
    Two general theories exist:
    - **Myofibroblast mediated:** Myofibroblasts at the periphery of the wound pull the wound edges together in a picture-frame fashion via actin–myosin interaction.
    - **Fibroblast mediated:** Contraction of the wound occurs as fibroblasts elongate and migrate through the surrounding matrix, effectively retracting collagen fibrils.

11. **Compare wound strength at different time intervals?**
    - Wound strength 1 week after injury is 3% of normal dermis.
    - After 3 weeks, when remodeling phase begins to predominate, the wound will have approximately 20% the strength of normal dermis.
    - By 6 weeks, the wound has reached 80–90% of its eventual strength.
    - The wound strength of scar never reaches that of normal skin. However, it reaches its maximum of about 80% of skin breaking strength at 6 months (continues to remodel up to 12 months).

12. **What are the principles for providing soft tissue coverage of the hand?**
    - Soft tissue coverage differs for dorsal and palmar skin.
    - The major principles for dorsal skin are that it should be thin, elastic, and loose enough to not limit flexion.
    - Palmar skin should be thicker and tougher, yet loose enough to allow motion. Most importantly, it must retain its function of sensibility.
    - Skin on each side must serve as a barrier to cover tendons and joints.

13. **What is the "reconstructive ladder" for coverage of a soft tissue defect in the hand?**
    - An algorithm developed to guide the choice of reconstructive options for a soft tissue defect in the hand or upper extremity.
    - The "ladder" progresses from simple to complex procedures as follows:
      1. Primary wound closure
      2. Healing by secondary intention
      3. Skin graft
      4. Flap (local, distant, or free)

An additional alternative available now is the vacuum assisted closure (VAC) sponge. This can increase the rate at which healing by secondary intention occurs. It can also be used as a temporizing measure for any of these steps (except that it changes a primary wound closure into a delayed wound closure).

14. **Which factors are important in assessing a wound that requires some form of soft tissue coverage?**
    - Wound location
    - Wound size and depth
    - Zone of injury
    - Absence or presence of wound contamination and/or infection
    - The key structures (nerve, tendon, artery, or bone) requiring coverage

15. **What are the three phases of wound healing by secondary intention?**
    - **Phase I:** Granulation tissue formation
    - **Phase II:** Wound contracture
    - **Phase III:** Epithelialization

16. **Where are cells derived from during epithelialization?**
    - Wound edges
    - Epithelial appendages, such as hair follicles, sweat glands, and sebaceous glands
    These appendages extend into the dermis and subcutaneous tissues, and these cells persist after the epidermal layer has been removed after an injury.

17. **Why are wound debridement and soft tissue coverage preferred within the first week after an injury?**
    To minimize the risk of infection and lessen the overall morbidity to the patient.

18. **What is the preferred wound status before proceeding with soft tissue coverage?**
    The wound should be clean (no contamination), free of infection, stable, and flat and should appear healthy and viable.

19. **What are the three types of skin grafts?**
    - Autograft (from the patient)
    - Allograft (from another human being)
    - Xenograft (usually porcine skin)

20. **Why is contact between a skin graft and the recipient site important?**
    Contact between the tissue bed and the graft is important to allow for *inosculation*, which is the ingrowth of vessels into existing vascular channels, and *neovascularization*, which is new vessel formation. This usually occurs after 24–36 hours.

21. **How does a skin graft survive prior to inosculation and neovascularization?**
    Via diffusion or serum imbibition.

22. **Which factors affect split-thickness skin graft survival?**
    - The most important single factor is the blood supply of the recipient site. Split-thickness skin grafting will survive on periosteum and paratenon; however, bone denuded of periosteum and tendons stripped of paratenon will not support a split-thickness skin graft.
    - Infection at the recipient site will usually prevent graft survival. The presence of 10^5 bacteria per gram of tissue is associated with a high incidence of failure.
    - Hematoma formation or seroma accumulation in the early postoperative period can result in a hydraulic separation of the graft and subsequent failure.

23. **How does the bacterial count of tissue affect healing?**
    - In normal skin, 1,000 organisms/g of tissue are present.
    - These are present in hair follicles, crevices, and skin recesses and have little effect on wound healing.
    - Contamination is an elevated bacterial count, generally greater than 10^5 organisms/g.
    - Bacteria count is less important when dealing with surface bacteria as these can be removed easily with a careful skin prep at the time of surgery.
    - Penetration of the wound bed is of great significance (10^5 organisms/g). Less than this amount has a high rate of graft take (>90%); higher counts have only a 20% success rate.

24. **How can this bacterial count be determined?**
    Perform a wound biopsy.

25. **What are the advantages of split-thickness skin grafting for wound coverage?**
    - Multiple potential donor sites
    - Variable graft thickness and color
    - Reliable graft take
    - Ease of graft harvest
    - Low donor site morbidity
    - Ease of graft application
    - Uncomplicated postoperative wound management
    - Graft shrinkage that reduces defect size

26. **List the disadvantages of split-thickness skin grafting.**
    - Hypersensitivity at the recipient site
    - Donor-site scarring
    - Lack of durability or stability in palmar hand and finger-tip regions
    - Pigmentation and color mismatch
    - Unattractive appearance following mesh expansion of the graft

27. **How does graft thickness affect the characteristics of graft healing?**
    - Wounds mature much more rapidly after full-thickness grafts compared to split-thickness grafts.
    - Graft thickness has a significant influence on the amount of wound contraction following skin grafting.
    - A thin graft contracts more. The thicker the graft, the less matrix protein is present, which correlates with less wound contraction.

28. **How long does wound maturation and contraction occur following split-thickness skin grafting?**
    - Following split-thickness skin grafting, scar tissue maturation and contraction continues for at least 12 months after wound closure and coverage.
    - Wound contracture can be reduced by graft compression and joint splinting. Compression garments may be worn for up to 18 months for maximum benefit.

29. **What is the best timing for a skin graft?**
    - Early grafting has replaced the concept of grafting over a healthy, rich bed of granulation tissue (which takes time to develop).
    - Timing should be based on an achievement of a clean wound with development of a network of fine vessels established (or present) in the wound bed.
    - Often, grafting can be achieved immediately after debridement if these conditions are met.
    - If not, no more than 2–3 days should be allowed to pass to establish these conditions and allow for grafting.
    - VAC sponges can serve as a temporizing measure until a skin graft can be applied.

30. **How and when does a split-thickness skin graft usually adhere?**
    - Split-thickness skin grafting is usually held to the wound bed by a carefully applied dressing.
    - Within hours of the operation, a fibrinous exudate develops and holds the graft to the bed.
    - This tenuous biologic bond is not strong enough to withstand any significant shearing force.
    - In the early postoperative period, graft survival is dependent on minimizing mechanical forces exerted by the dressing that is placed over the graft. **Lack of attention to this critical dressing is a common cause of graft loss.**
    - If the graft position relative to the wound bed has been maintained, the blood supply was adequate, and the bacterial count is low, graft adherence should be significant at 72–96 hours.
    - The graft–wound interface should be immobilized and undisturbed for the first 96 hours.
    - Concave recipient sites should have "tie-over" or "bolster" of moist cotton over nonadherent gauze.

31. **What are the contraindications to split-thickness skin grafting?**
    - Lack of an adequate blood supply
    - Significant contamination of the recipient bed
    - Exposed bone, tendon, or cartilage lacking periosteum or paratenon

32. **What are the advantages and disadvantages of a full-thickness graft?**
    Advantages:
    - Better protection, because more durable, especially at withstanding shear forces
    - Better sensibility (more epidermal appendages)
    - Less contraction

    Disadvantages:
    - Take less readily than a split-thickness graft
    - Requires a better wound bed for application
    - More prone to infection

33. **How does the location of the wound in the hand influence the skin graft type?**
    A full-thickness skin graft will contract less, is more durable and flexible, and has more sensation. Therefore, a full-thickness skin graft is preferred for areas prone to shear, such as the palm, areas over joints, and the fingertips.

34. **How are grafts harvested?**
    - Full-thickness grafts are harvested by freehand technique with a knife and the full layer of skin is taken down to the level of the subcutaneous fat. (These wounds must be closed.) They are generally excised in an ovoid fashion with the axis in the direction of minimal tension and the defect is closed in a layered fashion.
    - Split-thickness grafts are harvested in most cases with a dermatome, and the thickness can be set by using a depth gauge (because dermis remains with its associated appendages, healing occurs by epithelialization). The skin should be lubricated so that the dermatome will slide easily and not skip.

35. **What are the recommendations for the depth of split-thickness skin grafts of the hand?**
    - **Most wounds (adults) are best covered by a graft of intermediate thickness of 0.015 inches.**
    - **Wounds where graft survival may be at risk should be thinner, at 0.010–0.012 inches.**
    - **Infants:** Never over 0.008 inches.
    - **Prepubertal children:** If >0.010 inches necessary, remove from lower abdomen or buttock.
    - **Adult males:** 0.015 inches from thighs; 0.018 inches from abdomen/buttocks.

- **Adult females:** If >0.015 inches, use lower abdomen; try never to use inner thigh.
- **Elderly adults:** Treat similar to a child's skin (dermal thickness increases with age in youth, greatest in fourth decade and then decreases with age after).

36. **When should a split-thickness skin graft be meshed?**
    The potential improvement in graft survival must be balanced against the diamond-shaped scar pattern that is inherent with a meshed graft. The interstices created by the meshing device allow the escape of blood and serum from the interface between the recipient bed and the graft during the first few hours after the dressing has been applied, thus reducing the risk of seroma or hematoma formation and improving graft survival rates. Where cosmesis is of paramount concern to the patient, the use of unmeshed skin is recommended.

KEY POINTS: SOFT TISSUE COVERAGE OF THE HAND

1. A 60-degree Z-plasty results in a 75% increase in length of the long axis.
2. Meshing a split-thickness skin graft extends the surface area and allows drainage, thus reducing the risk of hematoma and seroma formation.
3. A split-thickness skin graft in an adult should be harvested at a thickness of 0.010 to 0.012 in.
4. Split-thickness skin grafting should not be performed if the wound involves cartilage, tendon, or bone that is devoid of paratenon or periosteum.

37. **What is the recommended mesh ratio?**
    A mesh ratio of 1:1.5 is recommended for use in the hand.

38. **How long does it take the skin graft *donor site* to epithelialize?**
    - Epithelialization will usually take 7–10 days.
    - Donor site healing will take place by epithelial migration from the hair follicles and glandular structures in the donor site.
    - The donor site can be covered with either a nonadherent gauze compressive dressing or a completely occlusive dressing.

39. **What are the commonly used donor sites for a full-thickness skin graft to the hand or forearm?**
    - The groin crease (lateral to the hair)
    - Hypothenar region
    - Abdomen
    - Medial arm
    - Antecubital fossa

    Note: If possible, the donor and recipient morbidity should be localized to the same extremity.

40. **What are the contraindications for full-thickness skin grafting?**
    Since full-thickness skin grafts require the wound bed to provide nutrition, the contraindications are similar to those for split-thickness grafting. Contaminated wounds or those with exposed tendon, bone, or nerve should not be covered permanently with a skin graft. Extensive burn wounds are generally better covered with a split-thickness skin graft.

41. **What are the two most common errors and causes of failure in full-thickness skin grafting?**
    - **Improper immobilization:** Following application of a compressive dressing, the recipient site is completely immobilized and undisturbed for 10–14 days to allow secure graft adherence.

- **Inadequate defatting of the graft:** Defatting must be meticulous to allow the ingrowth of capillaries for graft viability.

42. **What are the most important technical features of skin grafting?**
    - Selection of appropriate donor site to meet the size, cosmetic, and mechanical requirements of the recipient site
    - Harvesting of a graft with the desired size and thickness with minimal donor site morbidity
    - Appropriate graft preparation with defatting of a full-thickness skin graft and meshing of a split-thickness graft
    - Securing of the graft to the recipient site (use a bolster for dressing grafts on a concave surface)
    - Application of a compressive stint dressing
    - Avoidance of dressing removal for at least 3 days if there is a question of wound cleanliness, 5–7 days if the wound is very clean
    - Immobilization of the grafted site
    - Hemostasis at the recipient site

43. **How do you obtain a graft of appropriate size to match the skin defect?**
    Template. These can easily be made from a sterile glove, paper from the sterile glove wrapping, or Esmarch/tourniquet, etc. Press the template material into the wound. Blood/fluid from the wound will stain an imprint of the wound for sizing. When there is more than one wound to cover, make a template for each, place them together for a single larger template, and excise skin graft. The resultant single graft can then be used to fill the first defect, sewing its edge to the largest wound, trimming to fit, and then repeating this process for each wound.

44. **Is there an improved reinnervation density in a digital soft tissue defect with a hypothenar versus thenar skin graft?**
    No. There is no histologic support for this contention. Studies suggest that sensory end-organs may be reinnervated after nerve section and repair, but new sensory end-organs do not form *de novo*.

BIBLIOGRAPHY

1. Buchler U: Traumatic soft-tissue defects of the extremities. Implications and treatment guidelines. Arch Orthop Trauma Surg 109:321–329, 1990.
2. Chao JD, Huang JM, Wiedrich TA: Local hand flaps. J Am Soc Surg Hand 1:25–44, 2001.
3. Eaton CJ, Lister GD: Treatment of skin and soft-tissue loss of the thumb. Hand Clin 8:71–97, 1992.
4. Foucher G, Bishop AT: Island flaps based on the first and second dorsal metacarpal arteries. Atlas Hand Clin 3:93–108, 1998.
5. Foucher G, Khouri RK: Digital reconstruction with island flaps. Clin Plast Surg 24:1–32, 1997.
6. Green DP, Hotchkiss R, Pederson WC (eds): Green's Operative Hand Surgery, 4th ed. New York, Churchill Livingstone, 1999.
7. Lawrence WT: Physiology of the acute wound. Clin Plast Surg 25:321–334, 1998.
8. Levin LS, Condit DP: Combined injuries—Soft tissue management. Clin Orthop 327:172–181, 1996.
9. Lister GD: Local flaps to the hand. Hand Clin 1(4):621–640, 1985.
10. Lister GD: The Hand: Diagnosis and Indications, 3rd ed. New York, Churchill Livingstone, 1993.
11. McCarthy JG, Littler JW, May JW: Plastic Surgery. Philadelphia, W.B. Saunders, 1990.
12. Monaco JL, Lawrence WT: Acute wound healing. An overview. Clin Plast Surg 30:1–12, 2003.

CHAPTER 32

SKIN FLAPS OF THE UPPER EXTREMITY

Donald P. Condit, MD, and Danielle A. Conaway, MD

1. **What is the function of a flap?**
   - A flap provides durable coverage to areas with a loss of tissue.
   - A flap generally consists of skin and subcutaneous tissue with a donor site attachment for circulation. A flap may also include fascia, muscle, bone, and/or nerves as desired.
   - A flap is used to cover exposed bone without periosteum, exposed tendon without peritenon, or exposed nerves with associated hypersensitivity.
   - A flap also permits late reconstructive surgery or replaces poor quality skin in web spaces or across flexion creases.

2. **Why use a flap, instead of a skin graft on the hand?**
   - When there is exposed bare bone, tendon, or cartilage
   - Where less contracture is important, such as over joints and the dorsum of the hand that undergo repeated stretching during daily use
   - Where skin grafts are likely to break down when subject to shear forces created in the hand and may become avascular over convex surfaces when they become taut
   - Where grafts are easily ulcerated with trauma; common in daily use of the hand

   Note: In any instance where a free graft would not provide the best skin, flap coverage is indicated.

3. **Describe the two basic types of flaps by vascular supply.**
   - **Random pattern flaps** have a random pattern of blood supply; nutrition arises from unnamed vessels traveling primarily in the subdermal or subcutaneous plexus.
   - **Axial pattern flaps** receive their blood supply from a single constant vessel (the pedicle). The area of skin supplied by an axial pedicle is known as the vascular territory.

4. **What is an angiotome?**
   An area of skin with a known single arterial supply.

5. **What are the advantages of an axial flap over a random flap?**
   - Axial pattern flaps have a more predictable and dependable circulation than a random pattern flap.
   - Axial pattern flaps also have a relatively small, mobile pedicle that allows greater flexibility for graft mobilization.
   - Axial pattern flaps have a superior blood supply and can be large (greater than 1:1 size).
   - Axial flaps are better at resisting infection.
   - Tissue adjacent to the pedicle does not require attachment to the primary defect (this bridge segment can be sewn to itself instead creating a tube of extra tissue to be used for second stage procedures if needed).
   - Flap can be reduced to its vascular or neurovascular bundle alone, allows for greater flap mobility (island pedicle flap).
   - Pedicle can be divided and anastomosed to vessels at a distant site (free flap).

6. **What is the most common shape for a random flap?**
   Quadrilateral or rhomboid in appearance. This is because the flap is raised on 3 of the 4 sides. The 4th side constitutes the base or pedicle of the flap. As a general rule for a random flap, length and width are in a 1:1 ratio to maintain a reliable blood supply.

7. **Is it safe to extend an axial pattern flap beyond its known vascular territory?**
   Yes. When raising the flap, it is safe to extend the distal portion of the flap by the amount that would be equal to a random pattern flap.

8. **Describe the types of axial flaps based on what they supply.**
   - **Cutaneous:** Skin only
   - **Fasciocutaneous:** Supply fascia and skin; these vessels often lie in the intermuscular septum; vessel can be transferred without skin but not skin without fascia
   - **Musculocutaneous:** Vessel supplies muscle and branches then supply skin; muscle may be transferred without skin, but to reliably supply the skin, muscle must be taken

9. **Describe the two main types of axial pattern flaps.**
   - **Peninsular flap:** With a skin and vessel pedicle
   - **Island flap:** With only a vascular pedicle

10. **What are the most common random pattern flaps used in the upper extremity?**
    - Transposition flap
    - Rotational flap
    - Cross-finger flap
    - Thenar flap
    - Reversed cross-finger flap

11. **Describe the types of flaps based on location.**
    - **Local:** Originates from skin adjacent to the defect
    - **Regional:** Originates from somewhere else on the limb and usually requires two operative procedures—one to raise and inset the flap and one to detach the flap and divide the pedicle
    - **Distant:** Originates from elsewhere in the body and usually also requires two operative procedures unless designed as a free flap

12. **What are the three types of local flaps?**
    - Advancement
    - Rotational
    - Transpositional

13. **What areas on the hand should not be used for flaps?**
    - Skin from the digital web spaces should never be used for flaps.
    - Skin from areas of daily contact should not be used.

14. **True or false: All rotational flaps are a random pattern.**
    True.

15. **What are lines of maximum extensibility?**
    - When deciding on areas to obtain a local flap, one should pinch the skin in various positions of the hand to determine where the flap should be raised.
    - The areas that provide the most available skin (skin pinches up most easily in this area) are along the lines of maximum extensibility.

16. **What is the major difference between a transposition and a rotational flap?**
    - Rotational flaps leave no secondary defect to be closed; the skin is stretched to close the primary defect by differential suturing.
    - Transpositional flaps leave a secondary defect that requires coverage.

17. **What are the key technical features in raising a transpositional flap?**
    - The flap is raised immediately superficial to the paratenon and extensor retinaculum, preserving the opportunity to skin graft the donor site.
    - The flap and its adjacent borders are generously undermined.
    - Hematoma and seroma formation are prevented using thorough hemostasis and carefully applied compression dressings.

18. **What are the two most common pitfalls of the transposition flap?**
    - Underestimating the flap size necessary to fill the involved defect
    - Compromised flap perfusion because of inadequate length-to-width ratio (usually 1:1), excessive tension on the closure, or excessive pressure from the dressing or following hematoma formation

19. **Describe the main advantages of a rotational flap.**
    - Matches nearly identically the characteristics of the tissue that was lost
    - Readily available

20. **Describe the typical indications for using a rotational flap in the hand.**
    - It is suitable for a small- to medium-sized dorsal hand or finger defect that cannot be closed primarily because of the size of the defect or the subsequent functional compromise that would occur with primary wound closure.
    - The tissue surrounding the defect and the proposed flap should be uninjured.
    - Common clinical scenarios include small dorsal hand wounds, mucous cyst excision over the dorsum of the DIP joint in the fingers, and reconstruction of the eponychium.

21. **What are the contraindications for using a rotational flap?**
    - Flap elevation and transfer should never compromise hand function. However, injury of the structures beneath the proposed donor or recipient site does not preclude the use of the rotation flap if its vascular supply is intact.
    - Relative contraindications include peripheral artery disease, smoking, and patients with sympathetic overactivity.

22. **What are the common local flaps used in the hand?**
    - Z-plasty
    - Axial flag flap
    - Dorsal metacarpal artery flap
    - Digital artery island flap
    - V-Y advancement flaps
    - Moberg flap

23. **Describe the two types of V-Y advancement flaps used in digits.**
    - The lateral flap (Kutler) involves two lateral raised flaps, each 6–8 mm in length, on the ulnar and radial borders of the involved digit. These flaps are advanced distally and medially to cover or close a fingertip amputation defect.
    - The volar flap (Atasoy et al, 1970) involves a volar triangular flap with oblique, radial, and ulnar borders 1.5 times the length of the width of the wound bed. It is designed to allow distal advancement of palmar skin to cover an amputation site. The longitudinal limb, after closure, should not cross the distal interphalangeal (DIP) joint flexion crease.

24. **What is the maximal advancement that can be expected of the lateral Kutler or triangular volar V-Y flap?**
    After releasing the deep fibrous septa, it is possible to advance each flap 6–10 mm. Prior to the dressing application, the tourniquet is released, and viability of the flap is confirmed.

25. **If the circulation of the graft is in doubt, what should be done?**
    The sutures should be removed and the flap allowed to return to a relaxed position. Additional bone or fat is then removed, and another attempt is made to advance the flap. If this attempt is unsuccessful, it is preferable to use another technique for coverage. More problems arise from inadequate rather than excessive mobilization.

26. **What are the advantages of using a V-Y advancement flap for fingertip injuries?**
    - It provides wound coverage with tissue similar in type to that destroyed or amputated.
    - Exposed bone, tendon, and/or nerves can be covered with a flap that provides padding, reduced tip tenderness, and fingernail support.
    - The V-Y flap contains Meissner's corpuscles and restores some sensation.

27. **For what size of defect should the V-V flap be used?**
    A V-V flap is ideal for a wound approximately 1 cm^2 with exposed bone. Larger defects are not amenable because of the limited excursion that can be expected of the V-V flap.

28. **Describe the fingertip amputation pattern best suited for each V-Y advancement flap.**
    Both the lateral and the volar triangular V-Y flaps are indicated for a transverse amputation:
    - The volar flap (Atasoy et al, 1970) is ideal for a dorsal oblique fingertip amputation.
    - The lateral flap (Kutler) is best suited for a wound with a greater loss of palmar tissue.

29. **What theoretical increase in length of the long axis is achievable with a 60-degree Z-plasty?**
    Theoretically, a 60-degree Z-plasty results in a 75% increase in length of the long axis. It is critical that all limbs of the Z-plasty be of equal length.

30. **What are the indications for a Z-plasty?**
    - Scar contracture, especially on the volar surface of the fingers
    - Minimal web space contracture

    Note: A central tight scar should be excised.

31. **What are some technical points of a Z-plasty?**
    - The base of the flap should be thicker than the tip.
    - Do not grasp the tips of flaps with forceps.
    - Avoid placing sutures in the midportion of the flap.
    - Only use enough suture to hold flap in place.
    - Limbs of the flap should be in areas of loose skin.

32. **What are the common technical errors to avoid with a Z-plasty?**
    - Tendency to create limbs at 45 degrees instead of 60 (use a protractor if unsure)
    - Poor design
    - Flaps with a scar at the base
    - Suturing flaps with too much tension

33. **What is a Limberg flap?**
    It is a type of transposition flap that is occasionally referred to as a *rhomboid flap*. The soft tissue defect, including adjacent tissue, is resected to create a rhomboid-shaped defect.

An incision is extended from the inferior corner that equals the height of the rhomboid. This incision is extended parallel to one of the sides. The flap is then elevated and transposed into the defect. The wound is closed primarily.

34. **What is the Moberg flap?**
    - A volar advancement flap used for coverage of a thumb tip amputation.
    - The flap is raised through bilateral mid-axial incisions and includes both neurovascular pedicles.
    - The flap restores normal sensation and is particularly useful in the thumb where the preservation of length and normal sensation is important.

35. **What is the maximum amount of advancement for a Moberg flap?**
    Approximately 1.5 cm.

36. **List the advantages that the Moberg volar advancement flap has over local, regional, and distant flaps.**
    - Durable coverage
    - Immediate restoration of essentially normal sensation
    - Preservation of length
    - Low donor site morbidity
    - Single-stage procedure
    - Restoration of pulp contour and character
    - Useful for both primary and secondary reconstructive procedures
    - Straightforward rehabilitation

37. **What are the potential complications of the Moberg volar advancement flap?**
    - Injury to the neurovascular structures with resultant partial or complete flap necrosis (because the dissociation is performed near the neurovascular bundles)
    - Dorsal skin necrosis
    - Interphalangeal (IP) and/or metacarpophalangeal (MCP) joint flexion contracture
    - Dysesthesias or altered sensation

38. **What is a flag flap?**
    - This axial pattern transpositional flap is based on either the dorsal digital (metacarpal) or the proper digital artery at the web space of the donor finger.
    - The donor finger is usually the index or long finger.
    - The flap is mobile and is used to cover soft tissue loss over the volar and dorsal aspects of the proximal phalangeal region of the index finger or the MCP joint of the index or long fingers (Fig. 32-1).

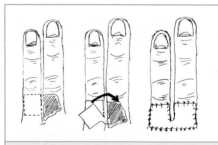

Figure 32-1. An axial flag flap used to cover a dorsal defect of the long finger.

39. **Describe the dorsal metacarpal artery flap.**
    - This axial pattern transpositional flap is an extension of the axial flag flap.
    - The dissection occurs from distal to proximal.

- It contains the 1st or 2nd metacarpal artery, veins, and branches of the radial nerve.
- It may extend from the PIP joint to the base of the 1st interosseous space.
- It is commonly used to cover defects of the thumb and 1st web space.

40. **What is a digital artery island flap?**
    - This is an axial pattern transpositional flap based on either the radial or ulnar digital artery.
    - The most common donor fingers are the ring and long fingers, where these vessels are codominant.
    - It limits reconstruction to the injured digit.

41. **What are the common regional flaps of the hand and upper extremity?**
    - Cross-finger flap
    - Reversed cross-finger flap
    - Innervated cross-finger flap
    - Cross-thumb flap
    - Thenar flap
    - Neurovascular island flap
    - Fillet flap
    - Radial artery forearm flap
    - Ulnar artery forearm flap
    - Lateral arm flap
    - Reversed posterior interosseous artery flap
    - Scapular flap
    - Pectoralis major flap
    - Latissimus dorsi flap

42. **Describe the cross-finger flap.**
    - The cross-finger flap is a random pattern flap usually based on the dorsal surface of the proximal or middle phalanx of an adjacent finger.
    - The flap is designed to cover a volar wound with exposed tendon or bone or an extensive loss of tissue that cannot be covered with an advancement flap.
    - The donor area is covered with a full-thickness skin graft (Fig. 32-2).

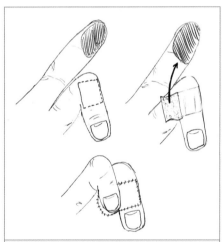

Figure 32-2. A cross-finger flap raised from the dorsum of the middle phalanx of the long finger and used to cover a volar defect in the index finger.

43. **What are some of the advantages of using a cross-finger flap?**
    - There is an excellent color match in lightly pigmented patients. However, in darker pigmented patients, the flaps are less aesthetic.
    - The flap provides a durable surface.
    - The flap has a high rate of success.
    - Most patients regain protective sensation with two-point discrimination of approximately 8 mm, and hyperesthesia rarely occurs.

44. **Describe the reverse cross-finger flap.**
    - This flap is designed to cover defects on the dorsal surface of the finger.
    - It is based on the dorsal surface of the middle phalanx of an adjacent finger.

45. **Describe the innervated cross finger flap.**
    - It is the standard technique for the cross finger-flap.
    - In addition, the dorsal branch of the digital nerve from the donor finger (on the side opposite the injured finger) is divided proximally in its course when raising the free margin of the flap.
    - It is then joined to the proper digital nerve of the injured finger on the opposite side of the donor.

46. **Describe the cross-thumb flap.**
    - It is indicated when the normal donor for a routine cross-finger flap is injured or otherwise unsuitable.
    - The cross-thumb flap is used mainly for defects to the digital pulp or to the radial aspect of the index finger.
    - The flap is taken from the dorsal thumb at the level of the proximal phalanx.

47. **Describe a thenar flap.**
    - Indicated to cover transverse and volar oblique injuries at the fingertip
    - Taken from the proximal and radial aspect of the thumb and raised at the level of the thenar muscles
    - Sutured to the primary defect and the secondary defect is covered with a full-thickness skin graft

48. **What are the advantages of a thenar flap?**
    - The skin is thicker and more durable than a cross-finger flap.
    - The donor site is less obvious to observe.

49. **What are the complications of a thenar flap?**
    - Risk of PIP contracture (although similar flexion has to occur with cross-finger flaps)
    - Tenderness at the donor site

50. **What is a fillet flap?**
    A fillet flap is an axial pattern flap that is harvested from a well-vascularized digit that is otherwise unsalvageable and going to be sacrificed. Surgeons utilize skin and the vascular supply of an otherwise destroyed digit.

51. **Describe the neurovascular island flap.**
    - This axial pattern flap is used primarily to cover a defect involving the volar aspect of the thumb where restoration of sensibility is crucial.
    - The flap is based on the digital neurovascular bundle.
    - Although any finger with adequate blood supply may be used, the ulnar aspect of the long finger is most commonly used (Fig. 32-3).

52. **Which specific physical finding must be assessed in any patient being considered for a neurovascular island flap?**
    The digital Allen's test must confirm intact radial and ulnar digital arterial supply because one of the neurovascular bundles is used in the flap.

53. **Describe the contraindications for using a neurovascular island flap.**
    - Trauma to the vascular supply in the palm or finger from peripheral vascular disease, diabetes, or peripheral vascular disease
    - Absence of normal sensation in the donor skin
    - Lack of useful motion in the hand for gripping or pinching

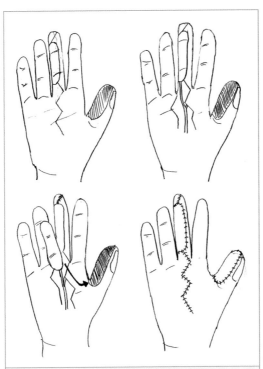

Figure 32-3. The neurovascular island flap is used to cover a defect on the thumb pulp. The flap is based on the digital neurovascular bundle and is transferred to the thumb via a tunnel in the palm.

54. **Describe the radial forearm flap.**
    - The radial forearm flap is an axial pattern fasciocutaneous flap based on the radial artery and its venae comitantes, which supply the fascia and skin of the volar forearm.
    - In addition to the volar forearm skin and portions of the radius, the palmaris longus muscle and tendon, superficial radial nerve, and antebrachial fascia can be used to reconstruct a variety of defects.

55. **What are the types of radial forearm flaps that are commonly constructed?**
    - Fasciocutaneous flap
    - Osteocutaneous flap

56. **What are the indications for using the radial forearm flap?**
    A large dorsal or palmar degloving injury of the hand where deep structures are exposed.

57. **Describe the advantages of each type of radial forearm flap.**
    - **Fasciocutaneous:** The skin is thin, sometimes hairless, and has an excellent texture for skin resurfacing. The flap can be easily contoured. Forearm tendons, nerves, and the radial artery are available for use in more complex reconstructions. Cutaneous nerves of the

forearm can be approximated to suitable recipient nerves to restore sensibility. The entire reconstruction can be completed in the same injured extremity in one step under regional anesthesia. When used as a distally based pedicle flap, microsurgical techniques are not required, and it is a relatively quick operation. Postoperatively, the area may be elevated to avoid dependent position and edema.
- **Osteocutaneous:** The radial volar cortex of the distal radius is ideal for reconstructing a segmental bone loss of the metacarpals and digits.

58. **What is the main contraindication for use of the radial forearm flap?**
    - If the ulnar artery does not perfuse the radial digits, then this flap is contraindicated. This can occur with an incomplete superficial palmar arch or prior ulnar artery injury.
    - Preoperatively, an Allen's test is used to assess the radial side perfusion of the ulnar artery. An equivocal Allen's test requires arteriography or Doppler exam to identify the absence of cross perfusion, which may be found in up to 15% of individuals.
    - If hand ischemia occurs with the use of the flap, reconstruction of the superficial radial artery should be performed with a vein graft.

59. **Why is an ulnar artery forearm flap less commonly used?**
    - Ulnar artery provides major blood supply to the hand in most people
    - Ulnar nerve dysesthesias common
    - Bulkier than the radial forearm flap

60. **Describe the lateral arm flap.**
    An axial-pattern fasciocutaneous flap that is based on the posterior radial collateral artery.

61. **What are the advantages of the lateral arm flap compared to a distant pedicle flap?**
    - A lateral arm flap involves a single operative procedure.
    - This flap brings in a new blood supply to the reconstructed area in contrast to a distant pedicle flap, which becomes parasitic on the recipient side after division of the pedicle.
    - A free lateral arm flap allows elevation of the hand and early motion, which minimizes stiffness and edema. This is not possible with a distant pedicle flap.
    - The flap fulfills an important role because an extensive area of vascularized tissue can be harvested with minimal donor-site morbidity.
    - Cosmesis is excellent for a good color match between the lateral arm and the dorsum of the hand.
    - The flap usually has a thin subcutaneous fat layer.

62. **How reliable is the lateral arm flap?**
    The vascular pedicle is consistently present and can provide up to 8 cm of pedicle length. The artery is usually greater than 1.2 mm in size.

63. **What size of lateral arm flap can be harvested?**
    The skin territory available is very large and can be up to 14 cm by 30 cm in dimension.

64. **What are the disadvantages of the lateral arm flap?**
    - A flap harvested wider than 6 cm requires skin grafting of the donor site.
    - Tenderness of the lateral epicondyle may occur if the flap is not designed so that its bony prominence remains covered with full-thickness skin.
    - Loss of the lateral cutaneous nerve of the forearm results in a small area that with sensory loss on the extensor aspect of the forearm just distal to the elbow.
    - Transferring hair to the palmar surface of the hand is not cosmetically desirable.

65. **Describe the reversed posterior interosseous artery flap.**
    - This axial-pattern fasciocutaneous flap is based on the posterior interosseous artery that originates from the common interosseous artery in 90% of people and the ulnar artery in 10%.
    - The axis is along the lateral epicondyle to the distal radioulnar joint.
    - Because of its short pedicle, the flap is used for coverage of a dorsal hand wound.

66. **Which structure limits elevation of the posterior interosseous flap?**
    - The nerve branch to the extensor carpi ulnaris (ECU).

67. **What are the contraindications to using the posterior interosseous forearm flap?**
    - Significant injury to the dorsum of the wrist and forearm, which may result in thrombosis of the posterior interosseous artery.

68. **What is the scapular flap?**
    - An axial pattern flap of the cutaneous variety
    - Can be raised as an island to cover defects (commonly for burns, axillary scarring) about the shoulder
    - Biggest utility is as a distant free flap

69. **Describe the pectoralis major flap.**
    - An axial pattern flap
    - Main blood supply is the pectoral branch of the thoracoacromial artery, supplemented by the lateral thoracic artery
    - Unlike most regional flaps, remains attached at its insertion and is released at its origin instead
    - Can be split longitudinally to cover two adjacent defects
    - Main indications are to restore elbow flexion lost as a result of arthrogryposis or nerve injury
    - May be used to cover defects about the shoulder and axilla
    - Poor cosmesis, entirely unacceptable to women
    - When available, latissimus dorsi used preferentially over this flap

70. **Describe the latissimus dorsi flap.**
    - This axial flap has a wide arc of rotation and is used for head, neck, shoulder, and breast reconstruction.
    - A large diameter artery (1–2.5 mm) and long pedicle (11 mm) make it the most popular tissue for free transfer for the extremities.
    - The main blood supply is the thoracodorsal artery.
    - It may also split the muscle longitudinally to obtain a narrow flap as it has two main branches.
    - It can also be used for functional reconstruction of elbow flexion or extension.

71. **What are the disadvantages of the latissimus dorsi flap?**
    - Seroma formation (must use suction drains postoperatively)
    - Brachial plexus neurapraxia, often secondary to positioning
    - To avoid neurapraxia, must return the arm to the side for at least 5 minutes every hour

72. **Describe the groin flap.**
    - A distant axial pattern fasciocutaneous flap based on the superficial circumflex iliac artery.
    - The flap is centered 2 cm distal to the inguinal ligament.
    - The soft tissue defect can be closed primarily following elevation of the flap.

73. **Describe the potential indications for the groin flap.**
    - This flap is useful for reconstructing an extensive deep soft tissue deficit on the dorsum of the hand or, on occasion, an entire degloving injury of the thumb.

- It may be a good alternative with minimal morbidity in a patient who is not a good candidate for a free flap microvascular soft tissue reconstructive procedure.

74. **What are the contraindications to the groin flap?**
    - It is relatively contraindicated in the older patient (>65 years) because of the complications of hip joint and upper extremity stiffness.
    - It is also contraindicated in young children who are not capable of participating in postoperative flap care and pedicle monitoring.

75. **What are the important design considerations of the groin flap?**
    - The length and orientation of the pedicle should be designed to allow maximal tension-free insetting of the flap, avoidance of pedicle kinking, freedom of joint motion, and positioning of the patient for comfort.
    - Because the groin flap contains more subcutaneous fat than the hand and often passes over a convex hand surface, it should be larger than the hand defect.

76. **Describe the major pitfalls of the flap.**
    - Failure to divide the deep fascia, which results in kinking of the pedicle vessel and necrosis of the flap or injury to the vascular pedicle during dissection
    - Making the flap too narrow, resulting in a constricting tube
    - Kinking the pedicle either by poorly contouring the flap to the hand or from inadequate postoperative care
    - Undermining the margins of the donor site before closure, thereby predisposing the patient to a groin hematoma.
    - Premature removal of the sutures at the donor site due to erythema, creating a wound similar to a decubitus and almost as difficult to treat
    - Dividing the flap without a preliminary delay when the recipient wound is not soundly healed
    - Abundant subcutaneous thickness in obese patients.

77. **What are the common complications of a groin flap?**
    - Marginal skin necrosis
    - Wound infection
    - Seroma formation
    - Deep vein thrombosis
    - Hand edema
    - Shoulder, elbow, and/or hand stiffness

78. **How do free flaps compare with a pedicle flap?**
    - The operative procedure involved with free flap use is more complex.
    - A free flap can be precisely tailored to cover a soft tissue defect.
    - A free flap has the advantage of a robust predictable blood supply; therefore, it is not parasitic upon the recipient site.
    - A free flap can cover a more extensive soft tissue defect.
    - A second insetting or pedicle division procedure is not required with a free flap.

79. **Describe the role of tissue expansion for soft tissue coverage in the upper extremity.**
    - Tissue expansion is an uncommon but acceptable method for reconstruction of a soft tissue defect that may follow tumor resection or the reconstruction of a burn-scar contracture.
    - The use of tissue expanders requires two stages. The first stage involves insertion of the silicone shell expander and saline. After tissue expansion, the flap is harvested and the defect closed primarily. Soft tissue expansion permits primary closure of the donor site as well as creation of a larger flap.

- The technique has been used to expand a pedicle or free flap prior to transfer or transposition.
- Advantages include less hospital time, lower cost, earlier healing, fewer complications, and good clinical outcomes.

BIBLIOGRAPHY

1. Atasoy E, Ioakimidis E, Kasdan ML, et al: Reconstruction of the amputated finger tip with a triangular volar flap. J Bone Joint Surg 52A:921–926, 1970.
2. Baumeister S, Manke H, Witteman M, Germann G: Functional outcome after the Moberg advancement flap in the thumb. J Hand Surg 27A:105–114, 2002.
3. Birbeck DP, Moy OJ: Anatomy of upper extremity skin flaps. Hand Clin 13:175–187, 1997.
4. Brown RE, Zook EG, Russell RC: Fingertip reconstruction with flaps and nail bed grafts. J Hand Surg 24:345–351, 1999.
5. Buncke HJ: Microsurgery: Transplantation-Replantation. Philadelphia, Lea & Febiger, 1991.
6. Chao JD, Huang JM, Wiederich TA: Local hand flaps. J Am Sac Surg Hand 1:25–44, 2001.
7. Costa H, Garcia ML, Vranchx J, et al: The posterior interosseous flap: A review of 81 clinical cases 100 anatomical dissections: assessment of its indications in reconstruction of hand defects. Br J Plast Surg 54:28–33, 2001.
8. Eaton CJ, Lister GD: Treatment of skin and soft-tissue loss of the thumb. Hand Clin 8:71–97, 1992.
9. Fisher RH: The Kutler method of repair of fingertip amputations. J Bone Joint Surg 49A:317–321, 1967.
10. Foucher G, Braun JB: A new island flap transfer from the dorsum of the index to the thumb. J Plast Reconstr Surg 63:344–349, 1979.
11. Hentz VR, Pearl RM, Grossman JA, et al: The radial forearm flap: A versatile source of composite tissue. Ann Plast Surg 19:405–498, 1987.
12. Lister GD, Pederson WC: Skin flaps. In Green DP, Hotchkiss RN, Pederson WC (eds): Green's Operative Hand Surgery, 5th ed. New York, Churchill Livingstone, 2005.
13. Iselin F: The flag flap. J Plast Reconstr Surg 52:374–377, 1973.
14. Kappel OA, Burech JG: The cross-finger flap: An established reconstructive procedure. Hand Clin 1:677–684, 1985.
15. Katsaros J, Schusterman M, Beppu M, et al: The lateral upper arm flap: Anatomy and clinical applications. Ann Plast Surg 12:489–500, 1984.
16. Levin LS, Aponte RL: Tissue expansion in the upper extremity. Atlas Hand Clin 3:253–260, 1998.
17. McGregor IA, Jackson IT: The groin flap. Br J Plast Surg 25:3–16, 1972.
18. Melone CP, Beasley RW, Carstens JH: Tile thenar flap-An analysis of its use in 150 cases. J Hand Surg 7:291–297, 1982.
19. Moberg E: Aspects of sensation in reconstructive surgery of the upper extremity. J Bone Joint Surg 46A:817–825, 1964.
20. Russell RC, VanBeek AL, Wavak P, Look EG: Alternative hand flaps for amputations digital defects. J Hand Surg 6:399–405, 1981.
21. Schlenker JD: Important considerations in the design and construction of groin flaps. Ann Plast Surg 5:353–357, 1980.
22. Scheker L, Kleinert H, Hanel D: Lateral arm composite tissue transfer to ipsilateral hand defects. J Hand Surg 12A:665, 1987.
23. Soutar DS, Tanner NSB: The radial forearm flap in the management of soft tissue injuries of the hand. J Plast Reconstr Surg 37:18, 1984.
24. Stern PJ, Carey JP: The latissimus dorsi flap for reconstruction of the brachium shoulder. J Bone Joint Surg 70A:526–535, 1988.

CHAPTER 33

ELECTRODIAGNOSTIC TESTING

Hal M. Corwin, MD

1. **What are the components of a motor unit?**
   The motor neuron, its axon, and the muscle fibers on which the motor unit terminates.

2. **What is the difference between type 1 and type 2 muscle fibers?**
   Type 1 fibers are activated with mild effort and utilize oxidative metabolism. They are activated for fine movements. Type 2 fibers are activated with strong effort and use anaerobic metabolism. They are used for stronger movements.

3. **What is the correlation between the diameter of a nerve fiber and its nerve conduction velocity?**
   The smaller the diameter of a nerve fiber, the slower the nerve conduction velocity.

4. **What is the function of the nodes of Ranvier?**
   The myelin sheath covering the axon is interrupted by the nodes of Ranvier. At each node, the axon is not covered by myelin but by Schwann's cell cytoplasm containing gap substance. This arrangement can facilitate a boost in the nerve action potential and is known as *salutatory conduction*.

5. **What is the pathophysiology of conduction block?**
   Conduction block is due to a structural alteration at the nodes of Ranvier without damage to the axons. Conduction block is present when stimulation proximal to the site of the block causes no response or a reduced response when recorded distal to the block. Wallerian degeneration occurs with axonal damage.

6. **How does nerve compression cause slowing of nerve function?**
   Excessive pressure causes successive nodes to be intussuscepted. Therefore, the myelin is detached or demyelinated from the axon. When remyelination occurs, the nodes are closer together. The increased number of nodes is reflected as decreased nerve conduction velocity across that region.

7. **Which of the following can be observed during a clinical examination: Fasciculation, fibrillation, or myotonia?**
   Fasciculation and myotonia may be observed clinically. Fibrillation potentials are observed during the needle electromyography exam and cannot be observed during a clinical examination.

8. **What are denervation potentials on an electromyogram (EMG) recording?**
   When axon loss occurs, each muscle fiber innervated by that axon develops hypersensitivity to acetylcholine along the entire fiber. The muscle fibers may depolarize on their own. An EMG needle inserted near the denervated fibers will record denervation potentials starting at 2 to 3 weeks after the axon loss. Denervation potentials are fibrillations and positive sharp waves.

9. **Denervation potentials during the needle exam can occur with which of the following: wallerian degeneration, demyelination, or conduction block?**
   Denervation potentials only occur with damage to axons, resulting in wallerian degeneration. If the axons are not injured, as with demyelination or conduction block, there is no denervation on the needle examination.

10. **Can upper motor neuron lesions slow the motor nerve conduction studies?**
    Upper motor neuron lesions do not affect the nerve conduction velocity studies that examine the function of the lower motor neurons.

11. **What is the neurotransmitter that is released from the synaptic vesicles of the motor nerve terminal?**
    Acetylcholine.

12. **What is the pathophysiology of myasthenia gravis?**
    Antibodies bind to the acetylcholine receptor sites on the muscle end plate. When acetylcholine is released from the nerve terminal into the synoptic cleft, there is reduced binding to the receptor sites due to these blocking antibodies.

13. **What is the name of the disorder that causes muscle fatigue and weakness due to the presence of antibodies to acetylcholine receptors on the muscle end plate?**
    Myasthenia gravis.

14. **How are motor nerve conduction studies performed?**
    Motor nerve conduction studies are performed by stimulating a peripheral nerve and simultaneously recording from a somatic muscle that is innervated by that nerve.

15. **What is the compound muscle action potential (CMAP)?**
    The CMAP is the electrical response that is recorded during motor nerve conduction studies.

16. **What does the CMAP represent?**
    CMAP is the summation of the electrical activity of the muscles fibers that are in the vicinity of the recording electrode on the surface of the muscle.

17. **True or false: Carpal tunnel syndrome (CTS) may be diagnosed by electrodiagnostic studies.**
    False. Carpal tunnel syndrome is a clinical diagnosis based on the findings on electrodiagnostic testing, the clinical history, and the physical examination. Some patients with an abnormal electrodiagnostic test are free of symptoms and thus do not have a clinical diagnosis of CTS. The electrodiagnostic report may indicate "slowing of median nerve conduction at the carpal tunnel" but cannot diagnose the clinical condition known as CTS.

18. **Define the term *motor latency*.**
    The latency is the time (milliseconds) needed for the electrical stimulus to propagate along the fastest axons to activate the muscle. It is measured by observing when the compound muscle action potential begins.

19. **Will the distal motor latency increase or decrease as the stimulating electrode is moved farther away from the muscle?**
    Increase.

20. **What is meant by the terms *antidromic* and *orthodromic*?**
    - **Antidromic** motor nerve conduction means that the nerve stimulus evokes an action potential that propagates toward the anterior horn cell. Antidromic sensory nerve conduction implies

that the nerve signal propagates from the elbow toward the hand, away from the dorsal root ganglion.
- **Orthodromic** motor conduction means that the nerve stimulus evokes a signal that propagates away from the anterior horn cell (from the elbow to the hand). Orthodromic sensory nerve conduction indicates movement of the nerve signal toward the dorsal root ganglion (from the hand toward the elbow).

21. **How is nerve conduction velocity calculated?**
 The difference between the proximal and distal latencies is divided by the distance between the proximal and distal stimulation sites.

22. **If the distance between the elbow and wrist stimulation sites is measured inaccurately, how will this affect the electrodiagnosis?**
 If the recorded distance is too short, the nerve conduction velocity will appear to be slower. This may lead to a false diagnosis of entrapment neuropathy at the wrist or elbow and possibly unnecessary surgery.

23. **What is meant by the term *anomalous innervation*?**
 Some individuals' nerves do not follow normal anatomic patterns. These less frequent anatomic patterns are termed *anomalous innervation*.

24. **What is the Martin-Gruber connection?**
 The Martin-Gruber connection is an anatomic communication between the median and the ulnar nerves in the forearm. The fibers involved in this connection innervate many intrinsic hand muscles that are usually innervated by the ulnar nerve (e.g., abductor digiti minimi and first dorsal interosseous). The anomaly occurs in 15–31% of the population and may be discovered on nerve conduction studies. An injury to the median nerve near the elbow may result in weakness of the ulnar innervated intrinsic hand muscles in patients with this anomaly.

25. **Why is the Martin-Gruber connection important to recognize on an electrodiagnostic study?**
 The nerve conduction study may falsely indicate the presence of ulnar neuropathy at the elbow in normal patients if a Martin-Gruber connection is not recognized by the electromyographer.

KEY POINTS: ELECTRODIAGNOSTIC TESTING

1. The electrodiagnostic test may differentiate nerve demyelination from axonal injury.
2. Sensory nerve amplitude studies differentiate plexopathy from radiculopathy because they show abnormalities for the former and produce normal results for the latter.
3. The electrodiagnostic study consists of the needle (or EMG) examination and nerve conduction studies, which complement each other and are an extension of the neurologic examination.

26. **A patient has symptoms consistent with carpal tunnel syndrome. The nerve conduction studies reveal slowing of both median distal sensory latencies, both median motor latencies, and both ulnar distal sensory latencies. What are the diagnostic considerations?**
 The patient's study reveals evidence of a polyneuropathy. Additional testing to determine the etiology of the neuropathy is required.

27. **True or false:** Any patient with slowing of median nerve conduction at the wrist should also undergo testing of the opposite hand.
    True. Many patients with carpal tunnel syndrome have bilateral abnormalities on nerve conduction studies even though the symptoms may be unilateral.

28. **Differentiate the electrodiagnostic findings of CTS versus amyotrophic lateral sclerosis (ALS).**
    The sensory latency is always normal in ALS but often abnormal in CTS. The median distal motor latency may be abnormal in both conditions. The needle examination of the patient with ALS is always abnormal in three or more limbs in muscles innervated by different motor nerves. The needle examination in CTS may be normal or show involvement isolated to the median innervated thenar muscles.

29. **Why does nerve root avulsion not affect the sensory nerve conduction studies in spite of clinical evidence of sensory loss?**
    The avulsion injury to the sensory root is proximal to the dorsal root ganglion. The peripheral sensory nerve fibers from the dorsal root ganglion to the hand are intact. Therefore, the sensory nerve conduction studies are normal.

30. **What are the characteristic nerve conduction findings of neurogenic thoracic outlet syndrome?**
    The most common findings are reduced ulnar sensory nerve amplitude, reduced median motor CMAP amplitude, and normal median sensory responses.

31. **True or false:** If cervical radiculopathy affects the sensory root but spares the motor root, the EMG will reveal no abnormality on needle examination.
    True.

32. **True or false:** Patients with polyneuropathy may have a superimposed focal entrapment mononeuropathy due to an increased susceptibility.
    True.

33. **What three findings are indicative of axon loss during electrodiagnostic testing of a peripheral nerve entrapment neuropathy?**
    Axon loss from entrapment indicates severe entrapment. Low amplitude sensory nerve action potentials, low amplitude compound muscle action potentials, and the presence of fibrillation potentials or positive sharp waves on needle examination of affected muscles may be found with axon loss. (Low amplitude sensory and motor potentials may also occur without axon loss.)

34. **What is the importance of monitoring limb temperature during electrodiagnostic testing?**
    A linear relationship between limb temperature and nerve conduction velocity has been confirmed in many studies. Warming of the extremities is necessary in many patients to avoid a false low conduction velocity, slow distal latency, and increased nerve amplitude values, all of which may lead to misdiagnosis. The results of nerve conduction testing should be viewed with skepticism if the limb temperature has not been monitored and recorded.

35. **What are fibrillation potentials?**
    A fibrillation potential is the action potential of a single muscle fiber. It arises from a spontaneously contracting muscle fiber and is recorded through a needle electrode and studied during electromyography. Fibrillation potentials are usually regarded as a sign of denervation, although they may be seen in healthy muscles.

36. **Which conditions result in the presence of fibrillation potentials during EMG?**
    If fibrillation potentials are reproducible during the needle examination of a muscle, then they usually indicate the presence of a lower motor neuron disorder. Fibrillation potentials may be observed in diseases of the anterior horn cell, nerve root, plexus, or peripheral nerve innervating the muscle being studied. Fibrillations may also be found in patients with progressive muscular dystrophy or polymyositis.

37. **How long after an injury to the lower motor neuron will fibrillation potentials first be seen with EMG?**
    Fibrillation potentials are usually not seen until 14–21 days after an injury to the axon because of the delayed onset of denervation hypersensitivity, which follows wallerian degeneration.

38. **Differentiate the electrodiagnostic findings of a C8 motor radiculopathy from carpal tunnel syndrome.**
    With C8 motor radiculopathy, the needle examination may find changes consistent with denervation and/or reinnervation in those muscles innervated by the C8 nerve root (abductor pollicis brevis, first dorsal interosseous, flexor pollicis longus, C8 cervical paraspinous muscles, and others). The distal latencies are normal with a C8 radiculopathy. With carpal tunnel syndrome, the median distal sensory latency and possibly the distal motor latency will often be slow. The thenar muscles may be denervated in some patients with CTS.

39. **Atrophy or weakness of which muscles may differentiate anterior interosseous syndrome from carpal tunnel syndrome?**
    The pronator quadratus, flexor pollicis longus, and flexor digitorum profundus (radial half) may be weak in patients with anterior interosseous neuropathy.

40. **The lateral antebrachial cutaneous nerve is the terminal sensory branch of which nerve?**
    Musculocutaneous nerve.

BIBLIOGRAPHY

1. Brumback RA, Bobele GB, Rayan G: Electrodiagnosis of compressive nerve lesions. Hand Clin 8:241–254, 1992.
2. Campbell WW: Ulnar neuropathy at the elbow: Anatomy, causes, rational management, and surgical strategies. AAEM Course D, pp 7–12, 1991.
3. Campion D: Electrodiagnostic testing in hand surgery. J Hand Surg 21A:947–956, 1996.
4. DeLong RN: The Neurologic Examination, 5th ed Philadelphia, Lippincott-Raven, 1979.
5. DeMyer W: Technique of the Neurologic Examination, 4th ed New York, McGraw-Hill, 1994.
6. Denys EH: The influence of temperature in clinical neurophysiology. Muscle Nerve 14:795–811, 1991.
7. Gilliatt RW, LeQuesne PM, Logue V, et al: Wasting of the hand associated with a cervical rib or band. J Neurol Neurosurg Psychiatry 33:615–624, 1970.
8. Kimora J: Electrodiagnosis in Diseases of Nerve and Muscle, Principles and Practice, 2nd ed. Philadelphia, F.A. Davis, 1989.
9. Warmolts JR: Electrodiagnosis in neuromuscular disorders. Ann Intern Med 95:599–608, 1981.

COMPRESSION NEUROPATHIES OF THE UPPER EXTREMITY

Michael I. Vender, MD, and Prasant Atluri, MD

1. **List the known anatomic sites of median nerve compression.**
   The median nerve is formed by the medial and lateral cords of the brachial plexus and consists of C5, C6, C7, C8, and T1 nerve roots. It runs from the axilla into the arm, coursing first lateral then medial to the brachial artery, before descending toward the cubital fossa. From proximal to distal, the first potential site of median nerve compression is in the distal third of the humerus beneath an anteromedially projecting supracondylar process and the **ligament of Struthers**, located approximately 5 cm above the medial epicondyle while forming an accessory origin for the pronator teres. Farther distally, a second site of compression occurs in the antecubital fossa as the nerve passes deep to the **lacertus fibrosis** or **bicipital aponeurosis**. In the forearm, the median nerve passes between the ulnar and humeral heads of the **pronator teres**, forming a third potential site of compression. A fourth area of compression is encountered as the nerve runs deep to the aponeurotic **arch of the flexor digitorum superficialis (FDS)**, a fibrous band formed by the merger of the ulnar and radial heads of the FDS. It continues distally between this muscle and the flexor digitorum profundus (FDP). Finally, as the median nerve enters the palm through the carpal tunnel, it can often become compressed beneath the **transverse carpal ligament**.

2. **Which structures form the anatomic boundaries of the carpal canal?**
   In a cross section, the carpal tunnel is surrounded on three sides by bone (Fig. 34-1). The roof, the most volar and superficial boundary, is formed by the thick fibrous **transverse carpal ligament (TCL)** (Fig. 34-2). Radially, the TCL attaches to the **scaphoid tuberosity and a portion of the trapezium**. Ulnarly, the TCL attaches proximally to the **pisiform and the hook of the hamate** situated more distally and slightly radially.

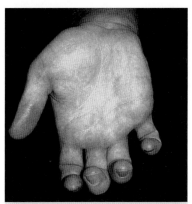

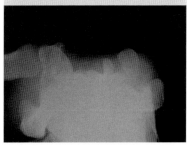

Figure 34-1. Radiographic demonstration of the carpal tunnel. For orientation, the hand in the upper panel is in the same position as the x-ray. The carpal tunnel is formed radially, ulnarly, and dorsally by the carpal bones. (From Concannon MJ: Common Hand Problems in Primary Care. Philadelphia, Hanley & Belfus, 1999.)

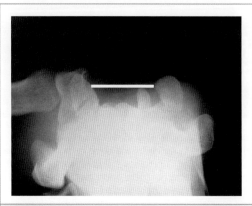

Figure 34-2. The transverse carpal ligament forming the roof the carpal tunnel. The line represents this ligament. This thick, fibrous structure does not yield to expansion, and increased pressure within the carpal tunnel can cause impingement of the median nerve. (From Concannon MJ: Common Hand Problems in Primary Care. Philadelphia, Hanley & Belfus, 1999.)

The floor of the carpal canal is formed by the carpal bones and the deep volar radiocarpal and volar interosseous ligaments.

3. **What are the contents of the carpal tunnel?**
   The median nerve and the nine digital flexor tendons. These tendons include the FDS (index through small), the FDP (index through small), and the flexor pollicis longus (FPL).

4. **What is Kaplan's cardinal line?**
   Kaplan described an anatomic guideline for locating the recurrent motor branch of the median nerve, which can be located by the bisection of a line drawn along the ulnar border of the middle finger, and the line drawn between the radial aspect of the thumb web space with the thumb in abduction and the pisiform.

5. **Describe variations in the motor branch of the median nerve.**
   The median nerve enters the palm through the carpal tunnel, where it gives off the motor branch to the thenar musculature before terminating in sensory branches to the radial digits. The thenar muscles include the opponens pollicis, flexor pollicis brevis, and abductor pollicis brevis. There are three main variations to the motor branch within the palm. In more than half of the cases, there is an extraligamentous branching distal to the TCL with a recurrent pathway to the thenar muscles. Subligamentous branch within the carpal tunnel with a recurrent pathway around the distal extent of the TCL occurs in one third of the cases; in another 20% of cases, subligamentous branching occurs, but a transligamentous route, about 2–6 mm from the distal edge of the TCL, is taken to the thenar muscles. In addition, several other much less common variations have been described.

6. **Describe the pathophysiology of idiopathic nerve compression.**
   There is increasing agreement that the predominant finding in a compressive neuropathy is vascular embarrassment. Mechanical obstruction of venous return from the nerve secondary to positional changes or local anatomy leads to venous congestion, circulatory compromise, and relative anoxia. The secondary effects lead to small vessel and capillary dilatation with resulting endoneurial edema and further compromise of venous return. Persistent interference with the intraneural microcirculation and longstanding endoneurial edema induces fibroblast proliferation.

Fibrosis acts as a barrier to the exchange of vital nutrients and oxygen within the nerve, causing impaired axonal transport, segmental demyelination, and nerve function deterioration.

7. **What electrodiagnostic parameters suggest compressive neuropathy?**
Prolonged sensory and motor latencies and decreased amplitude of nerve conduction potentials are highly suggestive of a compressive or entrapment neuropathy.

8. **What are some of the known risk factors for the development of carpal tunnel syndrome (CTS)?**
   - Female sex
   - Pregnancy
   - Diabetes
   - Rheumatoid arthritis
   - Obesity
   - Alcoholism
   - Hypothyroidism
   - Smoking
   - Aging
   - Mucopolysaccharidosis and mucolipidosis (storage diseases)
   - Lack of fitness or deconditioning

9. **What is pronator syndrome?**
Pronator syndrome is a generalized term describing an entrapment neuropathy of the proximal median nerve at any one of four distinct sites. These sites include the ligament of Struthers, lacertus fibrosis, between the heads of the pronator teres itself, or the FDS aponeurotic arch. Symptoms include pain in the proximal volar surface of the proximal arm and forearm as well as decreased sensation in the thumb, index, and long fingers. Phalen's and Tinel's tests at the wrist are notably negative. Provocative maneuvers may provide some insight into the level of compression. Operative release consists of a detailed exploration of **all** potential sites of compression.

10. **What functional loss is seen in anterior interosseous nerve syndrome?**
The anterior interosseous nerve is a purely motor branch of the median nerve that originates in the proximal forearm. Compression results in a loss of motor function without sensory disturbance. Involved muscles include the FDL, flexor profundus to the index finger, and the pronator quadratus. Clinically, the patient is unable to flex the thumb interphalangeal joint and index finger distal interphalangeal joint as well as experiences mild weakness of pronation, particularly with the elbow maximally flexed.

KEY POINTS: CARPAL TUNNEL SYNDROME

1. The carpal tunnel contains the median nerve and nine flexor tendons.
2. Vibrometry is the most sensitive sensory test (threshold test) for CTS.
3. A digital compression test is most the sensitive provocative test.
4. Electrodiagnostic testing usually reveals prolonged sensory and motor latencies and decreased amplitude.

11. **List the most common anatomic sites of ulnar nerve compression.**
The ulnar nerve is the main continuation of the medial cord of the brachial plexus and consists of the C7, C8, and T1 nerve roots. The nerve runs from the axilla into the arm, medial to the brachial artery, until it pierces the medial intermuscular septum at about the middle of the arm.

From proximal to distal, the first potential site of ulnar nerve compression is the arcade of Struthers, located approximately 8 cm proximal to the medial epicondyle. It is a fascial arcade of the intermuscular septum through which the ulnar nerve passes from the anterior into the posterior compartment of the arm. As its distal course continues, the nerve has a tendency to subluxate anteriorly with the potential for compression by the edge of the thick medial intermuscular septum. In addition, since the ulnar nerve courses along the medial head of the triceps toward the groove between the olecranon and medial epicondyle, a subluxing medial head of the triceps can be a third potential site of compression. A fourth site of compression is the cubital tunnel itself; just distal, a fascial arcade formed by the two heads of the flexor carpi ulnaris (FCU) is a further area of potential compression. The nerve continues into the forearm on the volar surface of the FDP deep to the muscle belly and tendon of the FCU. Finally, as the ulnar nerve enters the palm, it can be compressed in Guyon's canal.

12. **What is the clinical significance of the anconeus epitrochlearis?**
    The anconeus epitrochlearis is an uncommon cause of ulnar nerve compression at the elbow. It is a vestigial muscle that originates from the medial border of the olecranon and inserts into the medial epicondyle. The muscle crosses the nerve over the cubital tunnel, reinforcing the aponeurosis of the two heads of the FCU.

13. **How do the arcade of Struthers and the ligament of Struthers differ?**
    The arcade of Struthers is a fascial arcade of the intermuscular septum through which the ulnar nerve passes as it courses from the anterior to the posterior compartment of the arm. It forms a potential area of compression of the ulnar nerve. The ligament of Struthers, on the other hand, is a ligament extending from the supracondylar process of the distal humerus to the medial epicondyle. It forms a potential site of compression of the median nerve and brachial artery.

14. **Which structures form the boundaries of the cubital tunnel?**
    The cubital tunnel is a fibro-osseous tunnel beginning at the humeral condylar groove and is bordered anteriorly by the medial epicondyle. The lateral border and floor are formed by the ulnohumeral joint and medial collateral ligament. The medial border is formed by the heads of the FCU origin. The roof consists of a fibrous aponeurotic band extending from the medial epicondyle to the olecranon.

15. **Where is Osborne's ligament?**
    The arcuate ligament of Osborne is a thick band of tissue extending transversely between the medial epicondyle and the olecranon, joining the FCU's two heads of origin. The ligament represents the roof of the cubital fibro-osseous tunnel.

16. **What is the most likely cause of postoperative pain along the medial aspect of the elbow following cubital tunnel surgery?**
    The fascia overlying the medial epicondyle and the flexor pronator muscle group contains 100% of the medial antebrachial cutaneous nerve branches and 80% of the medial brachial cutaneous nerve branches. The standard incision for ulnar nerve decompression at the elbow, which involves an extended incision in the area of the humeral groove posterior to the medial epicondyle, potentially jeopardizes these branches. Unless each of the three to seven branches is carefully dissected, they are subject to injury, rendering the skin overlying the medial epicondyle and olecranon anesthetic and potentially painful.

17. **Which structures form the anatomic boundaries of Guyon's canal?**
    A study by Cobb et al (1996) defines the boundaries of Guyon's canal. The roof of the carpal ulnar neurovascular space is formed proximally by the volar carpal ligament and distally by the palmaris brevis. The central roof segment and the radial border of the space are essentially open and formed by adipose tissue. The hook of the hamate does not appear to be a true border of Guyon's canal, and, in many cadaveric specimens, the motor and sensory branches of the ulnar nerve lie

directly palmar or radial to the hook of the hamate. The ulnar wall consists of the pisiform bone proximally and the fascia of the hypothenar musculature more distally. The floor consists of the transverse carpal ligament, the hypothenar muscles, and a portion of the hook of the hamate.

18. **What are the contents of Guyon's canal?**
The ulnar nerve and artery. Within the canal, the ulnar nerve bifurcates into motor and sensory branches that lead to three distinct zones of compression. Zone 1 is proximal to the bifurcation and extends 3 cm from the proximal edge of the volar carpal ligament. Compression at this level produces both motor and sensory disturbances. Zone 2 contains the deep motor branch and zone 3 contains the sensory branch. Compression in zones 1 and 2 is most often due to a ganglion or fracture of the hook of the hamate. Zone 3 compression is often related to ulnar artery aneurysm or thrombosis.

19. **How does the clinical examination help to differentiate the cubital tunnel from Guyon's canal as the site of ulnar nerve compression?**
In the distal third of the forearm, the ulnar nerve gives rise to a large dorsal sensory branch that passes to the dorsum of the wrist between the FCU and the ulna at about the level of the ulnar styloid. The nerve provides sensation to the dorsum of the little finger and dorsal ulnar side of the ring finger. A compressive neuropathy of the ulnar nerve in the cubital tunnel will result in motor dysfunction **and** sensory disturbances of both the volar and dorsal aspects of the ulnar digits, whereas compression within Guyon's canal results in the preservation of dorsal sensation.

20. **List the known anatomic sites of radial nerve compression.**
The radial nerve, the main continuation of the posterior cord, which involves the C5, C6, C7, C8, and T1 nerve roots, enters the arm between the brachial artery and the long head of the triceps. The nerve takes a spiral course behind the humerus, lying in a shallow groove. Here, it is in intimate contact with the humerus and is quite vulnerable to injury from external compression and direct trauma following a humerus fracture. In the distal third of the arm, the radial nerve pierces the lateral intermuscular septum, entering the anterior compartment of the arm. It then passes anterior to the lateral epicondyle between the brachialis and brachioradialis muscles before dividing into superficial sensory and deep posterior interosseous motor branches within the radial tunnel.

The posterior interosseous nerve traverses the radial tunnel, where it encounters four potential sites of compression. The superficial sensory nerve continues distally beneath the brachioradialis before exiting dorsally as it pierces the deep fascia of the distal third of the forearm, supplying sensation to the dorsoradial hand.

KEY POINTS: COMMON SITES OF NERVE COMPRESSION

- **Median nerve:** Supracondylar process, lacertus fibrosis, pronator teres, FDS fibrous bands, carpal tunnel
- **Radial nerve:** Fibrous bands, arcade of Frohse, leash of Henry, medial border ECRB
- **Ulnar nerve:** Arcade of Struthers, medial intermuscular septum, medial head of triceps, cubital tunnel, fibrous bands in FCU, Guyon's canal

21. **What is radial tunnel syndrome?**
Radial tunnel syndrome is an entrapment neuropathy of the radial nerve in an area extending from the radial head to the proximal aspect of the supinator muscle. Four potential sites of compression exist, beginning with a tight fibrous band anterior to the radial head at the entrance

of the tunnel. The leash of Henry consists of radial recurrent vessels that fan out directly over the radial nerve and, if tortuous and engorged, may compress the nerve. In addition, the tendinous margin of the extensor carpi radialis brevis and the proximal supinator can be sites of nerve entrapment. Symptoms include pain in the proximal radial forearm over the extensor-supinator muscle group with radiation into the distal radial forearm and hand. Operative treatment consists of a detailed exploration and decompression of all potential sites of compression.

22. **What is the arcade of Frohse?**
    The arcade of Frohse is the fibrous proximal border of the supinator muscle that acts as a tight band under which the posterior interosseous nerve passes to enter the supinator muscle. The arcade of Frohse is the most common site of posterior interosseus nerve compression in radial tunnel syndrome.

23. **What is cheiralgia paresthetica?**
    Described by Robert Wartenberg in 1932, cheiralgia paresthetica is a superficial radial sensory neuritis. Also known as Wartenberg's syndrome, radial sensory neuritis is most commonly traumatic or iatrogenic in nature although compression has been described in association with muscle anomalies, fascial bands, thrombosis of the radial recurrent vessels or hemorrhage in the proximal forearm, and nerve tumors. Symptoms include dysesthesias in the radial sensory nerve distribution to the dorsum of the hand, positive Tinel's sign, and pain with percussion over the nerve, particularly as it exits dorsally from beneath the brachioradialis in the distal forearm. Treatment includes removing any responsible compressive garments or jewelry, splint immobilization, cortisone injection, and surgical exploration. Results of surgical exploration are unpredictable.

24. **What is the clinical significance of a "double crush" phenomenon?**
    The physiology and function of a nerve are dependent on the unimpeded conveyance of an impulse and the exchange of nutrients, enzymes, neurosecretory granules, polysaccharides, and polypeptides from the area of synthesis (nerve cell body) distally along the axons. Focal compression of the nerve impedes axoplasmic transport, affecting the flow distally. The "double crush" phenomenon refers to a second site of compression along the course of the nerve, usually more proximal, which compounds the distal compressive effect. In terms of the median nerve, a lesser site of compression in the carpal tunnel as manifested by mildly delayed distal palmar sensory latency may have a larger clinical effect if a more proximal cervical compressive lesion is present. This also explains why persistent symptoms can occur following a carpal tunnel release if the proximal site of compression is also not addressed. Therefore, in evaluating the upper extremity for a compressive neuropathy, one needs to consider all potential sites of compression.

25. **Compression of which peripheral nerve causes weakness of the supraspinatus and infraspinatus muscles?**
    Suprascapular nerve entrapment at the suprascapular notch deep to the suprascapular ligament can result in weakness of the supraspinatus and infraspinatus. This is often associated with overhead sports and may be due to local compression by a ganglion cyst.

26. **Symptoms of thoracic outlet syndrome (TOS) can be confused with which major upper extremity nerve entrapment?**
    TOS is an entrapment neuropathy of the lower trunk of the brachial plexus with clinical symptoms similar to cubital tunnel syndrome (numbness in the ring and small fingers and weakness of intrinsic muscles of the hand). In addition, because TOS is a more proximal entrapment neuropathy involving the entire lower trunk, sensory changes in the medial cutaneous nerve distribution of the arm and forearm are also present, unlike in cubital tunnel syndrome.

BIBLIOGRAPHY

1. Campion D: Electrodiagnostic testing in hand surgery. J Hand Surg 21A:947–956, 1996.
2. Cobb TK, Carmichael SW, Cooney WP: Guyon's canal revisited: An anatomic study of the carpal ulnar neurovascular space. J Hand Surg 21A:861–869, 1996.
3. Gelberman RH, Eaton R, Urbaniak JR: Peripheral nerve compression. J Bone Joint Surg 75A:1854–1878, 1993.
4. Green DP, Hotchkiss RN, Pederson W (eds): Green's Operative Hand Surgery, 4th ed. New York, Churchill Livingstone, 1999.
5. Leffert RD: Thoracic outlet syndrome. J Am Acad Orthop Surg 2:317–325, 1994.
6. Lundborg G, Dahlin LB: Anatomy, function, and pathophysiology of peripheral nerves and nerve compression. Hand Clin 12:185–193, 1996.
7. Osborne GV: The surgical treatment of tardy ulnar neuritis. In Proceedings of the British Orthopaedic Association. J Bone Joint Surg 39B:782, 1957.
8. Race CM, Saldanna MJ: Anatomic course of the medial cutaneous nerves of the arm. J Hand Surg 16A:48–52, 1991.
9. Rayan GM: Proximal ulnar nerve compression. Cubital tunnel syndrome. Hand Clin 8:325–336, 1992.
10. Szabo RM, Steinberg DR: Nerve entrapment syndromes in the wrist. J Am Acad Orthop Surg 2:115–123, 1994.

CHAPTER 35

PERIPHERAL NERVE INJURY
Paul S. Cederna, MD

1. **Describe the epineurium, perineurium, and endoneurium of a peripheral nerve.**
   - **Epineurium** consists of loose collagenous connective tissue that either encloses groups of nerve fascicles (external epineurium) or cushions fascicles from external pressure and trauma (internal epineurium).
   - **Perineurium** surrounds individual nerve fascicles and defines the fascicular pattern of a given nerve as mono-, oligo-, or polyfascicular (Fig. 35-1).
   - **Endoneurium** surrounds individual axons.

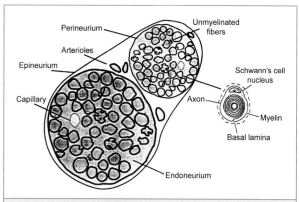

Figure 35-1. Cross-sectional anatomy of a peripheral nerve. (From Kasdan ML: Occupational Hand and Upper Extremity Injuries and Diseases, 2nd ed. Philadelphia, Hanley & Belfus, 1997.)

2. **What is the internal topography of a peripheral nerve?**
   Internal topography refers to the precise location and organization of sensory and motor nerve fascicles within a peripheral nerve. The dramatic plexus formation of fascicles along the course of the nerve creates difficulties during microsurgical repair. Detailed maps have been created to identify the location and contributions of each fascicle within each peripheral nerve of the upper extremity.

3. **Define *wallerian degeneration*.**
   Wallerian degeneration is the process that prepares the injured distal nerve stump for subsequent axonal regeneration. Following division of a nerve, Schwann's cells proliferate within the distal nerve and phagocytose degraded axoplasm and myelin. Macrophages also play a critical role in clearing the endoneurial tubes of debris, allowing an uninhibited path for axonal regeneration.

4. **What happens immediately after a peripheral nerve injury?**
   1. The nerve cell body located in the anterior horn of the spinal cord undergoes metabolic changes to help replenish the structure of the divided axon.
   2. The proximal nerve segment undergoes wallerian degeneration adjacent to the nerve injury.
   3. The distal nerve segment undergoes wallerian degeneration from the site of the nerve injury to the motor end plate.
   4. The muscle begins to atrophy due to the loss of neural input.

5. **What is axonal regeneration?**
   It is the process of nerve regeneration that occurs in an attempt to reinnervate the end organ following a sharp nerve transection. Single nerve fibers begin the process by sprouting growth cones at the most distal node of Ranvier in the proximal nerve segment. These growth cones adhere to the basal lamina of the endoneurial sheath. Axonal regeneration occurs along these endoneurial sheaths in close proximity with Schwann's cells. The nerve fibers become myelinated as they advance along Schwann's cells and mature. Axonal regeneration can be delayed by crush injuries to the nerve and can be compromised by scarring at the nerve coaptation site.

6. **How fast is axonal regeneration?**
   The maximal rate for axonal regeneration in humans is 1–2 mm/day.

7. **How do you predict the time required to achieve maximal recovery after a peripheral nerve injury and repair?**
   The approximate time required for recovery after a peripheral nerve injury is based on the level of nerve injury and the rate of axonal regeneration. For example, if the median nerve is lacerated in the forearm, the number of days after repair required for full recovery of sensibility is ideally the distance in millimeters from the site of repair to the tip of the finger plus 30 days for the clearing of nerve debris and early regeneration. If the nerve is transected 4 cm (40 mm) proximal to the wrist crease, and the distance between the wrist crease and the tip of the long finger is 10 cm (100 mm), the predicted time for full sensory recovery is 30 days (typical delay period) plus 40 days (distance from the nerve repair to the wrist crease). It takes an additional 100 days for sensory recovery in the distance from the wrist crease to the tip of the finger. Thus, the total recovery time is 170 days or approximately 6 months.

8. **How are nerve injuries classified?**
   In 1943, Sir Herbert Seddon classified peripheral nerve injuries into three types based on the anatomic layers of the nerve that have been violated and the potential functional outcomes from these injuries: neurapraxia, axonotmesis, and neurotmesis.
   - **Neurapraxia** is a mild peripheral nerve injury in which the epineurium, perineurium, and endoneurium remain intact. However, there may be disruption of the myelin layer encasing the nerve fascicles. The nerve remains in continuity with no distal wallerian degeneration, but there is a focal area of nerve conduction block. Recovery is usually complete and occurs in days to weeks.
   - **Axonotmesis** is an injury in which the axon is disrupted, and the endoneurium and perineurium may or may not be disrupted. Wallerian degeneration occurs distal to the injury, but the entire nerve remains intact. Recovery is nearly complete because the endoneurial conduits remain intact, and regenerating axons do not need to traverse a nerve coaptation site.
   - **Neurotmesis** is complete transection of the nerve. After microsurgical repair, sensory and motor functional recovery is usually incomplete because of the potential mismatching of fascicles, scarring at the nerve coaptation site, and loss of axons during regeneration.

9. **Define *neuroma in continuity*.**
   Following a nerve injury, it is possible for some nerve fascicles to be intact with normal function, while other nerve fascicles are completely disrupted and devoid of sensory or motor function. The ends of these fascicles may not be in adequate proximity to allow regeneration across the injury site. As axonal sprouting occurs from the proximal nerve stumps, many axonal branches attempt to span the nerve gap unsuccessfully, resulting in a collection of disorganized nerve fibers (neuroma).

10. **How are nerve repairs classified?**
    Nerve repairs may be classified as primary, delayed primary, or secondary.
    - **Primary repair** is the immediate repair of a nerve within hours of injury.
    - **Delayed primary repair** of a nerve occurs within 5–7 days after injury.
    - **Secondary repair** of a nerve is performed more than 7 days after the injury.

11. **Which criteria must be met before a primary nerve repair can be performed?**
    - The wound must have minimal contamination and a healthy, viable, well-perfused tissue bed.
    - The injured extremity also must have adequate circulation, skeletal stability, and soft tissue coverage.
    - The patient should be able to tolerate the anesthesia required to perform the surgical repair.
    - Microsurgical expertise, instruments, and suture are necessary.
    - The nerve must be repairable with minimal tension at the repair site.
    - Primary nerve repair is best performed after a sharp nerve transection. Any injury involving a crushing or avulsion mechanism should not be repaired primarily because it is difficult to determine the extent of nerve injury in such situations.

12. **What is the primary goal of nerve repair?**
    To guide the regenerating axons to the appropriate end organs and to restore normal sensory and/or motor function.

13. **What are the techniques of nerve repair?**
    Several techniques are used for peripheral nerve repair. The most common techniques are the epineurial (Fig. 35-2) or fascicular (Fig. 35-3). Fascicular repair may be performed alone or in combination with an epineurial repair.

    For both techniques, the proximal and distal nerve stumps must be isolated and all damaged neural tissue removed. The nerve stumps are transected perpendicular to the long axis of the nerve. The anatomic vascular landmarks on the epineurium help to align the nerve fascicles during repair. If specific fascicles with

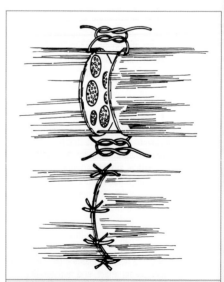

Figure 35-2. Epineurial repair of a transected nerve. (From Green DP, Hotchkiss R, Pederson W [eds]: Green's Operative Hand Surgery, 4th ed. New York, Churchill Livingstone, 1999.)

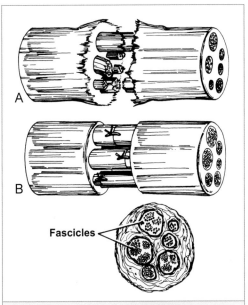

Figure 35-3. Fascicular repair of a transected nerve. (From Green DP, Hotchkiss R, Pederson W [eds]: Green's Operative Hand Surgery, 4th ed. New York, Churchill Livingstone, 1999.)

defined functions can be easily identified in the proximal and distal stumps, a fascicular repair should be performed.

Epineurial suture with 8-0, 9-0, or 10-0 nylon is typically used, depending on the size and location of the injured nerve. The number of sutures used depends on the size of the nerve. Care is taken to avoid placing too many sutures in an effort to prevent fascicular injury and scar formation.

14. **Which techniques are used to match nerve fascicles when an end-to-end neurorrhaphy is performed?**

   Anatomic vascular landmarks within the epineurium can be used to determine the appropriate alignment of proximal and distal nerve stumps during repair.

   Nerve fascicles also may be aligned based on knowledge of nerve topography. Detailed anatomic records are available to delineate the fascicular structure of major peripheral nerves along their course.

   Fascicular stimulation also may be used to map individual nerve fascicles. Selected fascicles are stimulated while the patient is awake, and then the patient identifies the sensation that is elicited. Once mapping has been completed, the nerve fascicles or fascicular groups are repaired. Unfortunately, this technique can be used only to identify proximal sensory and distal motor fascicles. Last, individual motor and sensory nerve fibers can be identified with acetylcholinesterase and carbonic anhydrase histochemical staining techniques.

15. **Which techniques may be used to overcome a nerve gap during peripheral nerve reconstruction?**

   One of the most important principles of peripheral nerve reconstruction is to have a tension-free repair. Tension at the neurorrhaphy site can result in mechanical failure of the repair, increased scarring, ischemia and necrosis, and reduced axonal regeneration across the repair.

A nerve gap can be overcome by one of several techniques, including nerve grafting, mobilization and/or transposition of the nerve, shortening of the involved extremity, or a combination of these techniques. Joint flexion to achieve a tension-free repair is not recommended.

16. **Describe the various nerve grafting techniques.**
    - A **trunk graft** (Fig. 35-4) consists of a whole segment of nerve that is interposed between the proximal and distal ends of the damaged nerve. Trunk grafting has had limited success because of fibrosis of the central portion of the graft during revascularization. The fibrosis inhibits the regenerating axons from reaching the end organs, resulting in reduced sensory and motor function.
    - **Cable grafts** incorporate several strands of nerve graft that are sutured together as a unit to reconstruct a single peripheral nerve gap. Cable grafting also has had limited success because of the lack of fascicular continuity between the injured nerve ends and graft. Cable grafting is best performed in children.
    - **Interfascicular nerve grafting** (Fig. 35-5) involves strands of nerve graft placed between carefully dissected and specifically matched groups of fascicles. This technique has a reasonably high rate of success because it involves numerous nerve grafts placed between specifically matched individual fascicles. If individual fascicles can be identified and matched with their distal counterparts, interfascicular nerve grafting is preferred.
    - **Vascularized nerve grafts** may be used to reconstruct large peripheral nerve gaps, large-diameter nerves, or nerve injuries in poorly vascularized recipient beds.

17. **What are the common donor nerves for nerve grafting?**
    The choice of nerve graft depends on the location and type of nerve injury, the size of the defect to be grafted, surgeon preference, and the associated donor site morbidity.
    - The sural nerve (believed to be the best source of graft material, providing up to 40 cm of nerve graft)
    - The medial and lateral antebrachial cutaneous nerves

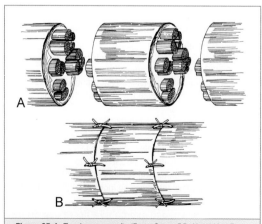

Figure 35-4. Trunk nerve graft. (From Green DP, Hotchkiss R, Pederson W [eds]: Green's Operative Hand Surgery, 4th ed. New York, Churchill Livingstone, 1999.)

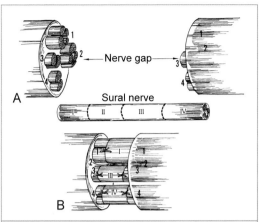

Figure 35-5. Interfascicular nerve graft. **A,** The fascicular groups and the nerve gap are identified. The sural nerve is divided into segments (I–IV) for grafting. **B,** The sural nerve segments are matched with the fascicles and sutured into place. (From Green DP, Hotchkiss R, Pederson W [eds]: Green's Operative Hand Surgery, 4th ed. New York, Churchill Livingstone, 1999.)

- The terminal sensory branch of the posterior interosseous nerve (good for a digital nerve defect)
- The superficial radial nerve

18. **What is Tinel's sign?**
    Tinel's test is used to assess clinically the extent of axonal regeneration after nerve injury and repair. The repaired nerve is percussed along its entire length, starting distally and progressing proximally until paresthesias are elicited. The point at which paresthesias are elicited represents the distal most extent of axonal regeneration. As the axonal growth cones march distally, Tinel's sign progresses along the length of the nerve. The absence of an advancing Tinel's sign after nerve repair is an important clinical finding that may identify an unsuccessful repair.

KEY POINTS: PERIPHERAL NERVE INJURY

1. The three types of nerve injuries, as described by Seddon, are neuropraxia, axonotmesis, and neurotmesis.
2. Primary nerve repair should be performed only if the wound is healthy and viable, the tissue bed is well perfused, circulation is adequate, skeletal stability and soft tissue coverage are present, and a tension-free repair is feasible.
3. The most common donor sites used for nerve grafting include the sural nerve, medial and lateral antebrachial cutaneous nerves, terminal sensory branch of the posterior interosseous nerve, and superficial radial nerve.
4. Treatment of a painful neuroma typically involves transposition or implantation into adjacent muscle or bone.
5. A nerve conduit, such as an autogenous vein or a synthetic polyglycolic acid collagen tube, is used to bridge the gap between nerve ends and guide nerve regeneration.

19. **How can a satisfactory tissue bed be obtained or created after nerve repair or grafting?**
    - The nerve may be transposed to an adjacent healthy tissue bed.
    - The nerve also may be wrapped for protection in healthy tissues, such as a muscle or fascial flap.

20. **What is a neuroma?**
    Following a peripheral nerve injury, the proximal sensory axons begin to regenerate across the nerve gap. If the nerve ends are not approximated adequately, the proximal sensory axons become disorganized as they attempt to bridge the gap. The resulting collection of sensory nerve fibers can become painful (neuroma). Such neuromas can be particularly painful if they occur in an area susceptible to repeated trauma, such as the end of an amputation stump.

21. **What are the various surgical treatment options for a painful neuroma?**
    - Transposition or implantation of the proximal nerve stump into adjacent muscle, bone, or nerve
    - Relocation of the intact neuroma to a scar-free area that is not subject to repeated trauma and irritation
    - Coagulation of the nerve stump
    - Epineurial closure of the nerve stump

22. **Which electrodiagnostic tests are used in the evaluation of a peripheral nerve injury?**
    Electromyography (EMG) and nerve conduction studies (NCS) are used to evaluate the extent of peripheral nerve injuries and to document recovery after repair. Serial studies are useful in monitoring the progression of axonal regeneration and muscle reinnervation.

    If a muscle has been denervated, EMG studies can identify fibrillation and denervation potentials. However, EMG studies are useful only if the muscle has been denervated for at least 3 weeks. As the muscle becomes reinnervated, there is a decrease in the number of fibrillation potentials and insertional activity that can be measured.

    NCS are used to evaluate the development of a nerve action potential across a particular segment of nerve or anatomic region. This study can be quite useful in localizing the injured segment of a nerve.

23. **Define muscle neurotization.**
    Muscle neurotization is a technique used to reinnervate a muscle when no distal nerve segment is available for end-to-end neurorrhaphy. The proximal end of the severed nerve is surgically placed into the belly of the denervated muscle in the region of the motor end plates. Motor axons then sprout from the end of the implanted nerve and are guided to the neuromuscular junctions by various neurotrophic and neurotropic factors.

24. **Can an end-to-side neurorrhaphy be used for peripheral nerve reconstruction when an end-to-end nerve repair is not feasible?**
    - End-to-side neurorrhaphy has been shown to be equally as effective as end-to-end neurorrhaphy for muscle reinnervation in experimental models.
    - Axonal regeneration is more likely to occur when perineurial suturing is performed rather than epineurial repair.
    - If the proximal end of the divided nerve is not available for reconstruction, the distal end of the divided nerve is sutured to the side of an intact donor nerve. Axons from the donor nerve sprout through the epineurial window and regenerate across the end-to-side neurorrhaphy and down the recipient nerve. These motor and sensory axons provide end organ function in the distribution of the injured nerve.

25. **What is a nerve conduit?**
    A nerve conduit is an alternative to a nerve graft. The conduit bridges the gap between nerve ends and permits nerve regeneration across it. Various types of non-neural conduits have been described, but the most popular are autogenous vein and polyglycolic acid collagen tubes (Fig. 35-6).

26. **What is neurotropism?**
    The purposeful nerve regeneration toward the correct distal pathway and end organ.

Figure 35-6. Nerve conduits used to treat a forearm laceration following an assault with a hunting knife. Two polyglycolic acid collagen tubes *(arrow)* have been used as nerve conduits following segmental loss of the ulnar nerve at the wrist.

BIBLIOGRAPHY

1. Brushart TME: The mechanical humoral control of specificity in nerve repair. In Gelberman RH (ed): Operative Nerve Repair and Reconstruction. Philadelphia, J.B. Lippincott, 1991, pp 215–230.
2. Cederna PS, Kallainen LK, Urbanchek MG, et al: Donor muscle structure and function after end-to-side neurorrhaphy. Plast Reconstr Surg 107(3):789–796, 2001.
3. Dellon AL, MacKinnon SE: Treatment of the painful neuroma by neuroma resection and muscle implantation. Plast Reconstr Surg 77:427–436, 1986.
4. Jabaley ME: Electric stimulation in the awake patient. In Gelberman RH (ed): Operative Nerve Repair Reconstruction. Philadelphia, J.B. Lippincott, 1991, pp 241–257.
5. Kalliainen LK, Cederna PS, Kuzon WM: Mechanical function of muscle reinnervated by termino-lateral neurorrhaphy. Plast Reconstr Surg 103:1919–1927, 1999.
6. Lundborg G, Rosen B, Dahlin L, et al: Tubular vs conventional repair of median and ulnar nerves in the human forearm: Early results from a prospective, randomized, clinical study. J Hand Surg 22A:99–106, 1997.
7. MacKinnon SE, Dellon AL, Lundborg G, et al: A study of neurotropism in the primate model. J Hand Surg 11A:888–894, 1986.
8. Nakatsuchi Y, Matsui T, Handa Y: Funicular orientation by electrical stimulation and internal neurolysis in peripheral nerve suture. Hand 12:65–74, 1980.
9. Omer GE, Spinner M: Management of Peripheral Nerve Problems, 2nd ed. Philadelphia, W.B. Saunders, 1997.
10. Rovak JM, Cederna PS, Macionis V, et al: Termino-lateral neurorrhaphy: The functional axonal anatomy. Microsurgery 20:6–14, 2000.
11. Seddon HJ: Three types of nerve injury. Brain 66:237, 1943.
12. Sunderland S: Nerves and Nerve Injuries, 2nd ed. Edinburgh, Churchill Livingstone, 1978.
13. Wood VE, Mudge MK: Treatment of neuromas about a major amputation stump. J Hand Surg 12A:302–306, 1987.

CHAPTER 36

TENDON TRANSFERS
Kevin C. Chung, MD, MS

1. **What is a tendon transfer?**
   A surgical procedure involving the transfer of a musculotendinous unit to restore useful or purposeful motion in an otherwise nonfunctioning limb part.

2. **List the important principles of a tendon transfer.**
   - Sacrificing the tendon must not leave the patient with a substantial deficit.
   - The donor tendon must have adequate strength.
   - The tendon must have adequate amplitude of motion.
   - The tendon must be under conscious voluntary control.
   - The action of the transferred tendon must be synergistic with its new function.
   - The tendon must be able to reach its new insertion point with a straight line of transfer.
   - The patient comprehends the treatment plan and is capable of complying with postoperative rehabilitation.

3. **Which preoperative therapy program should be performed prior to proceeding with a tendon transfer?**
   - Appropriate splinting and joint exercises preoperatively are essential to maintain passive mobility and prevent joint contractures.
   - The soft tissues along the anticipated path of the tendon transfer must be free of scarring to prevent the development of adhesions that restrict gliding of the transferred tendon.

4. **How long should you wait to perform a tendon transfer following a peripheral nerve repair?**
   Approximately 9–12 months.

5. **Which deficit can be anticipated when the wrist flexor tendons are transferred to restore finger extension?**
   Incomplete digital extension because of limited excursion of the wrist flexor tendons compared to the digital extensors.

6. **In a *high* radial nerve palsy, what three functional requirements can be restored with tendon transfers?**
   - Wrist extension
   - Finger extension
   - Thumb abduction and extension

7. **What is the classic combination of tendon transfers used in a *high* radial nerve palsy?**
   - Pronator teres (PT) to the extensor carpi radialis brevis (ECRB)
   - Flexor carpi ulnaris (FCU) or flexor carpi radialis (FCR) to the extensor digitorum communis (EDC)
   - Palmaris longus or FDS ring finger to the extensor pollicis longus (EPL)

8. **Why is the FCU not used as a tendon transfer to restore digital extension in the patient with a posterior interosseous nerve palsy?**
   To avoid radial deviation of the wrist that may be accentuated during wrist extension because the of the intact ECRL.

9. **In patients with an isolated *high* radial nerve palsy, which tendon transfer is appropriate to restore thumb extension if the palmaris longus is absent?**
   The middle or ring finger flexor digitorum superficialis (FDS).

10. **What is the "internal splint" procedure?**
    In patients with a high radial nerve palsy, the subsequent wrist drop and lack of finger extension alter the biomechanical effectiveness of the finger flexors and result in decreased grip strength. In addition, patients may develop an adaptive functional pattern, often consisting of wrist flexion, to assist in finger extension (tenodesis effect). This pattern is difficult to overcome after tendon transfers. The patient may be fitted with a splint that helps extend the wrist and fingers. Although several designs are available, such splints may be cumbersome and uncomfortable for some patients. Therefore, some surgeons perform a PT to ECRB tendon transfer in an end-to-side fashion to assist with wrist extension and eliminate the need to wear a wrist splint during nerve regeneration.

11. **For patients with a *low* median nerve palsy (wrist level), which motor deficits are expected?**
    - The median nerve innervates the thenar musculature, including the abductor pollicis brevis (APB), opponens pollicis, and superficial head of the flexor pollicis brevis (FPB).
    - Denervation of the thenar muscles results in a loss of thumb opposition and abduction.

12. **What are the two most common tendon transfers performed to restore thumb opposition in patients with a median nerve palsy?**
    Extensor indicis proprius (EIP) or ring finger FDS to the APB.

13. **Which tendon transfer may be performed to restore thumb function in the patient with carpal tunnel syndrome and significant thenar atrophy?**
    Palmaris longus to APB to provide thumb abduction, not opposition. The transfer is called a Camitz *abductorplasty* transfer.

14. **What is the Huber transfer?**
    - It is the transfer of the abductor digiti minimi (ADM) muscle to the APB.
    - The transfer is used most often to restore thumb opposition in children with congenital absence or deficiency of the thenar muscles.
    - The transfer restores not only function but improves cosmetic appearance by adding the ADM muscle bulk at the thenar eminence.

15. **Which deficits in hand function occur in an ulnar nerve palsy?**
    - In a low ulnar nerve palsy (wrist level), the interosseous, hypothenar, adductor pollicis, and ring and little finger lumbrical muscles are deficient. Furthermore, the little finger and the ulnar half of the ring finger are insensate.
    - In a high ulnar nerve palsy, the aforementioned muscles and the FCU and flexor digitorum profundus (FDP) of the ring and little fingers are denervated. The dorsal ulnar border of the hand (ulnar sensory branch) is insensate.

16. **Why is clawing not noted in the index and long fingers in a low ulnar nerve palsy?**
    Clawing does not occur in the index and long fingers because the lumbrical muscles to these two fingers are innervated by the median nerve.

17. **What is Froment's sign?**
    Exaggerated thumb interphalangeal (IP) joint flexion during key pinch as a result of denervation of the first dorsal interosseous and adductor pollicis following an ulnar nerve injury. To compensate, patients use the flexor pollicis longus (FPL) to flex and stabilize the IP joint and the EPL to adduct the thumb.

18. **Which tendon transfers are performed to restore key-pinch function in the patient with a *low* ulnar nerve palsy?**
    - The ECRB can be augmented with a tendon graft and inserted into the radial aspect of the thumb metacarpophalangeal (MCP) joint. This transfer not only adducts the thumb but also provides thumb pronation.
    - The long finger FDS tendon may be transferred to the adductor tubercle of the thumb.
    - For index finger abduction, a slip of the abductor pollicis longus (APL) is augmented with a free tendon graft and inserted into the lateral band of the index finger. This transfer stabilizes the index finger during key pinch when combined with an arthrodesis of the MCP joint.

19. **Which procedures are performed to correct "clawing of the fingers" in the patient with an isolated ulnar nerve palsy?**
    In an ulnar nerve palsy, clawing of the fingers is due to paralysis of the interossei and lumbrical muscles. If extrinsic muscle function is intact, the MCP joints hyperextend and the IP joints flex, resulting in clawing.
    Two types of procedures are designed to prevent clawing:
    - **Static block procedures** prevent MCP hyperextension. They include a bone block on the dorsum of the metacarpal head, MCP arthrodesis, or capsulodesis of the MCP joint. Static tenodesis procedures include a tendon transfer (ECRL, ECU) or free tendon graft that is passed volar to the deep transverse metacarpal ligament and inserted at the radial lateral bands of the extensor apparatus.
    - **Dynamic tenodesis procedures** include transfer of an FDS tendon (divided into 2–4 slips) palmar to the deep transverse metacarpal ligament to the lateral band of the dorsal apparatus or A2 pulley; transfer of the FDS through or around the A1 or A2 pulley, after which the FDS is sutured to itself; or a free tendon graft looped through the extensor retinaculum, volar to the deep transverse metacarpal ligament and into the lateral bands.

20. **What is the key technical feature for all tendon transfers to correct the claw deformity associated with a *low* ulnar nerve palsy?**
    The transfer must pass palmar to axis of motion of the MCP joints and the deep transverse metacarpal ligament, which acts as the pulley in the case of the dorsal donor tendon.

21. **Which tendon transfers are performed for an isolated *high* median nerve palsy?**
    - Loss of function of the PT and FCR does not require tendon transfer.
    - To restore index and middle finger flexion, the FDP tendons of the index and middle fingers are sutured to the adjacent ring and small finger FDP tendons.
    - Thumb flexion is restored via a brachioradialis to FPL tendon transfer.
    - Opposition may be restored via transfer of the ring finger FDS or the EIP to the APB.

22. **What is Wartenberg's sign?**
    Paralysis of the hypothenar and the third palmar interosseous muscles following an isolated ulnar nerve palsy. This results in an inability to adduct the extended little finger because the extensor digiti minimi (EDM) is unopposed.

23. **A patient with long-standing rheumatoid arthritis presents with difficulty in fully extending the small finger MCP joints. What is the diagnosis?**
    The most likely diagnosis is rupture of the small finger extensor tendons as a result of direct synovial invasion or attrition via a dorsally subluxed and eroded ulnar head (caput ulna syndrome).

24. **Which tendon transfer is appropriate for the patient described in question 23?**
    - A transfer of the distal remnant of the small finger EDC tendon to the ring finger EDC is possible.
    - Alternatively, the EIP may be transferred to the small finger EDC or EDM.
    - The selected tendon transfer is usually combined with an extensor tenosynovectomy and excision of the distal ulna (Darrach procedure) or fusion of the distal radioulnar joint (Sauve-Kapandji procedure) to prevent additional extensor tendon ruptures.

25. **Which tendon transfer is indicated to improve function in the patient with an anterior interosseous nerve (AIN) palsy and no clinical signs of recovery 1 year after injury?**
    - The AIN is mostly a pure motor branch of the median nerve that innervates the pronator quadratus, FPL, and index finger FDP.
    - Flexion of the index finger DIP joint can be restored by side-to-side suturing of the index FDP to the FDP of the middle finger. Thumb IP joint flexion may be restored by transferring either the brachioradialis or the ring finger FDS to the FPL tendon. Alternatively, if the joint is arthritic, arthrodesis may be performed.
    - The stronger PT usually compensates for loss of function of the pronator quadratus.

KEY POINTS: TENDON TRANSFER

1. The Huber transfer involves transfer of the ADM to the APB to restore opposition. The transfer is most commonly performed in the child with a hypoplastic thumb.
2. Correction of the claw hand deformity secondary to a low ulnar nerve palsy requires a tendon transfer that is passed palmar to the deep transverse metacarpal ligament.
3. Tendon transfers for a high radial nerve palsy must restore wrist extension, finger extension, and thumb abduction and extension.
4. The preferred treatment for the patient with a distal radius fracture and a rupture of the EPL is transfer of the EIP to the EPL.
5. Treatment of the rheumatoid patient with an extensor tendon rupture involves extensor tenosynovectomy and tendon transfer usually combined with surgery to address the dorsally subluxated eroded ulnar head.

26. **A 30-year-old man sustained a closed, nondisplaced fracture of the distal radius during a fall from his bicycle. He underwent cast immobilization. About 6 weeks later, he noticed the sudden inability to extend his thumb. Diagnose this patient and recommend an appropriate treatment.**
    - **Diagnosis:** Attritional rupture of the EPL tendon that occurs in approximately 5% of patients with a distal radius fracture
    - **Treatment:** Transfer of the EIP to the EPL

27. **What is a crossed-intrinsic transfer?**
    Patients with rheumatoid arthritis often present with ulnar deviation ("drift") of the MCP joints. When the articular surfaces remain preserved, soft tissue reconstruction can be used to realign the fingers. The tight ulnar intrinsic tendons are released from the index, middle, and ring fingers. They are transferred to the middle (from the index), ring (from the middle), and little (from the ring) fingers to provide a radial directed force.

    The procedure serves two purposes: removal of the deforming force via release of the tight ulnar intrinsic tendons and transfer of the ulnar intrinsic tendons to the radial side of the fingers to provide a corrective radial pull.

28. **Which muscle-tendon transfer options are available to restore elbow flexion in the patient with biceps paralysis?**
    Available donor muscles include the latissimus dorsi, pectoralis major, and triceps. The surgeon needs to be careful in selecting an appropriate donor muscle because the transfer will result in a loss of one grade of muscle power. The donor muscle needs to have at least M4 strength, which is strength against resistance. After the transfer, these muscles will most likely function at an M3 level, which enables work against gravity.

29. **What is the "key-pinch operation" for the patient with a spinal cord injury?**
    The basic principle of this operation is to use the tenodesis effect that occurs with wrist extension to achieve a forceful pinch between the thumb and side of the index finger. The nonfunctioning FPL tendon is detached proximally and inserted into the volar radius. When the patient actively extends the wrist, the FPL tightens, and the pulp of the thumb will press against the radial side of the index finger. When the wrist is flexed by gravity, the thumb/index web space opens to release objects. Kirschner wire (K-wire) fixation of the IP joint is performed to prevent hyperflexion during the key pinch.

30. **In patients with a spinal cord injury, how is active elbow extension restored to the patient when the triceps muscle is paralyzed?**
    Although active flexion of elbow is useful for personal care, extension of the elbow is important for reaching out to the environment. The ability for a person with a spinal cord injury to extend the elbow actively can be achieved with specially designed devices, such as those designed to help patients use computer keyboards or drive a cart. The most common operation performed for this condition is the deltoid to triceps transfer. When the deltoid muscle is functioning, the posterior deltoid is split from the anterior deltoid and transferred to the triceps muscle using a strip of tensor fascia lata graft. This operation can enhance patients' range of motion and improve their functional capability.

31. **What are the six requirements for the restoration of wrist and hand function in the patient with a combined low median and ulnar nerve palsy?**
    - Improve key pinch.
    - Restore thumb abduction.
    - Restore tip pinch between the thumb and index finger.
    - Improve power flexion with the coordination of MCP and IP joint motion.
    - Restore the metacarpal arch.
    - Restore sensibility for key or tip pinch.

BIBLIOGRAPHY

1. Goldfarb CA, Stern PJ: Low ulnar nerve palsy. J Am Soc Surg Hand 3:14–26, 2003.
2. Grossman JAI, Pomerance J: Staged opposition transfer. J Hand Surg 23A:290–295, 1998.

3. Kalainov DM, Cohen MS: Tendon transfers for intrinsic function in ulnar nerve palsy. Atlas Hand Clin 7:19–39, 2002.
4. Rettig ME, Raskin KB: Tendon transfer for radial nerve palsy. Atlas Hand Clin 7:41–52, 2002.
5. Roach SS, Short WH, Werner FW, Fortino MD: Biomechanical evaluation of thumb opposition transfer insertion sites. J Hand Surg 26A:354–361, 2001.
6. Shin AY, Dao KD: Tendon transfers for thumb opposition. Atlas Hand Clin 7:1–17, 2002.
7. Skoll PJ, Hudson DA, de Jager W, Singer M: Long-term results of tendon transfers for radial nerve palsy in patients with limited rehabilitation. Ann Plast Surg 45:122–126, 2000.

CHAPTER 37

VASCULAR DISORDERS
Brian M. Braithwaite, MD

1. **Which arteries provide blood supply to the hand?**
   The radial and ulnar arteries supply the hand (Fig. 37-1). These arteries anastomose to form the superficial and deep palmar arches. The ulnar artery is usually the main contributor to the superficial palmar arch, and the radial artery is the main contributor to the deep palmar arch. In some patients, a persistent large median artery may also be present.

2. **What does a "complete" superficial arch mean?**
   A complete superficial arch gives off arteries to all five digits (but may be discontinuous) and is present in about 80% of hands. An incomplete arch is always discontinuous, and each artery (ulnar, radial, or, rarely, median) supplies one or more digits.

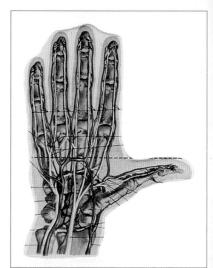

Figure 37-1. Relationship of the radial and ulnar arteries to the palmar arches and digital arteries.

3. **How are vascular disorders evaluated?**
   As with most diseases, evaluation begins with a thorough history and physical exam. Noninvasive and invasive tests are performed as needed to confirm diagnoses and guide treatment.

4. **What is the Allen's test?**
   The Allen's test evaluates the independent circulation to the hand via the radial and ulnar arteries. To perform this test, compress both the radial and ulnar arteries of the same hand at the wrist, and ask the patient to make a tight fist two to three times to exsanguinate the hand. The patient then opens the fingers (about 90%), and pressure over one of the arteries is released. Note the arterial filling time (arterial flush) and the area filled. Repeat the test with release of pressure over the other artery. The test is positive if there is no flush in approximately 5 seconds.

5. **What are the advantages of a Doppler study?**
   Usually available in the emergency department, a Doppler probe will detect blood flow in a vessel that cannot be easily palpated due to pain, swelling, or because it is of small caliber.

6. **What are the advantages of color duplex ultrasonography?**
   The test is an alternative to arteriography and is an inexpensive, noninvasive imaging modality for most forms of vessel pathology, including thrombosis, aneurysm, and stenosis.

7. **How can a bone scan evaluate vascular disorders?**
   A three-phase bone scan can be very helpful. In phase 1, images are taken within 2 minutes after injection and show the outline of the filling arteries. This phase can diagnose large vessel occlusion, hemangiomas, and arteriovenous malformations. In phase 2, soft tissue images provide information about relative perfusion 5–10 minutes after injection. Vasospastic diseases, such as Raynaud's phenomenon, show decreased perfusion in the involved digits. Phase 3 gives information about bones and joints 2–3 hours after injection.

8. **What is hypothenar hammer syndrome?**
   With blunt trauma to the base of the hypothenar eminence (i.e., pounding the palm on a desk, using the hand as a mallet), the distal ulnar artery and the superficial palmar arch are injured. The ulnar artery thrombosis and resulting aneurysm can cause ischemia of the hand and fingers.

9. **What are the signs and symptoms of ulnar artery occlusion and/or aneurysm?**
   Patients may complain of a tender and painful hypothenar mass, cold sensitivity, numbness due to ulnar nerve compression, and ischemic pain with pallor or gangrene. During the clinical evaluation, with compression of the radial artery, the superficial arch Doppler signal will be obliterated.

10. **What is the treatment for an ulnary artery occlusion and/or aneurysm?**
    Usually surgical resection and ligation or reconstruction using a vein graft.

11. **How are true and false aneurysms different?**
    A true aneurysm contains elements of all three layers of the arterial wall, is usually fusiform in shape, and arises from a weakness in the wall due to trauma or disease. False (pseudo) aneurysms usually result from penetrating trauma and are more common in the hand than true aneurysms. They begin as a hematoma, and arterial blood flow produces a cavity in the hematoma. In time, the cavity becomes lined with endothelium and then replaced with scar tissue.

12. **How should an aneurysm be treated?**
    Aneurysms are at risk for rupture and thromboembolic events. Treatment is resection and ligation or reconstruction, depending on the extent of arterial involvement and the severity of symptoms in the hand/fingers.

13. **What is Raynaud's phenomenon?**
    Raynaud's phenomenon is pallor of the digits with or without cyanosis upon exposure to cold. Unlike Raynaud's disease, the phenomenon has an underlying associated disorder: arterial occlusive disease, connective tissue disease (e.g., scleroderma), lupus, cryoglobulinemia, or possibly vibration exposure. Clinical features include absence of peripheral pulses, nail fold changes, raid progression to digital ischemia, and no gender predilection.

14. **What is Raynaud's disease?**
    Raynaud's phenomenon without a causative disorder is known as Raynaud's disease. Attacks are seen mainly in young women and are precipitated by cold or emotional upset. Bilateral, intermittent episodes of digit discoloration, pain, trophic changes, and absence of clinical peripheral artery occlusion upon cold exposure characterize the disease.

15. **What is the treatment for Raynaud's disease?**
    As for all vasospastic conditions, the patient must cease smoking tobacco. Oral calcium channel blockers may provide limited relief. For constant pain or impending tissue loss, a stellate or digital sympathectomy should be considered.

KEY POINTS: FEATURES OF RAYNAUD'S PHENOMENON

1. Digital pallor with cold exposure
2. Associated underlying disorder
3. No gender differences
4. Smoking not recommended

16. **Why does a partial vessel laceration bleed more than a complete transection?**
    A complete laceration of a vessel causes both longitudinal and circumferential spasm and retraction. Such injuries exhibit dramatic initial blood loss with subsequent cessation or slowing of bleeding. In a partial laceration, vessel wall retraction may actually enlarge the opening and/or prevent the vessel from retracting and closing. Such injuries have a much higher mortality following upper extremity trauma and require emergent treatment.

17. **Why is thoracic outlet syndrome important to the hand surgeon?**
    A vascular constriction in the chest and neck region should always be considered when searching for a cause of digital ischemia. Points of obstruction include anterior to a cervical rib, between the clavicle and first rib, or posterior to the anterior scalene muscle.

18. **What is Adson's test (also known as the scalene test)?**
    This test of questionable reliability is widely used to demonstrate potential compromise of the subclavian artery at the thoracic outlet. With the arm at the side, the patient hyperextends the neck, rotates the head to the affected side, and inhales deeply. The test is considered positive if the radial pulse is ablated or significantly diminished.

19. **Which structure is at risk for injury in resection of a volar wrist ganglion?**
    The radial artery is at risk for injury during surgery because it is common for the artery to be adherent to the wall of the cyst.

20. **What is the most common vascular hand tumor?**
    The most common vascular tumor is a hemangioma. Vascular tumors of the hand are relatively uncommon, constituting approximately 5% of all hand tumors.

21. **What is a hemangioma?**
    A hemangioma is a tumor of independently growing blood channels that likely originates from embryonic rudiments of mesodermal tissue. There are cavernous hemangiomas, capillary hemangiomas, or a combination of both.

22. **How is the diagnosis of a hemangioma made?**
    Pain and fullness in the affected area are the most common symptoms. Examination shows a cutaneous or subcutaneous mass that is spongy and compressible or quite firm. Cutaneous lesions are red, and subcutaneous lesions have a darker blue color. A thrill or bruit suggests an arteriovenous fistula.

23. **What is the preferred treatment for a hemangioma?**
    Hemangiomas present since birth should be left alone since most involute. Cryosurgery, argon laser surgery, and long-term compression device use are conservative treatments. Surgical resection is the treatment for a noninvoluting lesion, except the capillary hemangioma (port wine stain) that should be left alone.

24. **What is a glomus tumor?**
    A glomus tumor is a benign vascular tumor arising from the neuromyoarterial apparatus (glomus). A glomus is an arteriovenous anastomosis that regulates temperature and is found within the reticular layer of the dermis.

25. **Where are glomus tumors most commonly located?**
    About 75% of glomus tumors are found in the hand, usually at the distal digits. Most are subungual.

26. **What is the typical clinical presentation of a glomus tumor?**
    Patients present with a triad of pain, tenderness, and cold sensitivity. The tumor may cause a nail deformity, nail bed discoloration, and/or bone erosion. The glomus tumor is usually solitary, but multiple tumors have been described.

27. **How are glomus tumors treated?**
    The treatment of a glomus tumor is surgical excision. Subungual tumors require nail plate removal, tumor excision, and meticulous nail bed repair.

BIBLIOGRAPHY

1. Coleman SS, Anson BJ: Arterial patterns in the hand based upon a study of 650 specimens. Surg Gynecol Obstet 113:409–424, 1961.
2. Green DP, Hotchkiss R, Pederson W (eds): Green's Operative Hand Surgery, 4th ed. New York, Churchill Livingstone, 1999.
3. Lister GD: The Hand: Diagnosis and Indications, 3rd ed. New York, Churchill Livingstone, 1993.

CHAPTER 38

PRINCIPLES OF MICROSURGERY

Robert I. Oliver, Jr., MD

1. **What is the appropriate method of monitoring a free flap postoperatively?**
   There is no consensus about the ideal method of monitoring flap perfusion after microsurgical free tissue transfer. The most time-tested method involves serial clinical examinations to look for changes consistent with arterial (pallor) or venous (congestion) occlusion. The ease of bedside Doppler assessment by the nursing staff makes it a common part of most protocols for monitoring. Alternatively, some think that the routine monitoring of any free flap is futile and that the best method is prevention by attention to technique and the avoidance of inadvertent tension introduced during insetting of the flap, a likely cause of most flap failures.

2. **What should the clinical examination of a free flap include?**
   Four elements can be assessed:
   - Presence of an arterial pulse can be identified by palpation or Doppler (if possible).
   - Excessively pale color should suggest the possibility of arterial occlusion. (Note that some flaps, such as the groin flap, appear somewhat pale even when perfused). Swollen or bluish flaps are telltale signs of venous congestion.
   - Capillary refill should be brisk. However, overly brisk refill can also be associated with venous congestion.
   - Initiating bleeding from the flap is usually reserved for otherwise equivocal exams. Pricking the flap with a scalpel or large needle should demonstrate bright red bleeding. Dark blood or slow oozing suggests venous congestion.

3. **What are the objective ways to monitor free flap or replant perfusion? Are any of these particularly suited for upper extremity reconstruction?**
   Numerous clinical series have been published advocating various methods:
   - Transcutaneous or implantable tissue PO_2 monitoring is one objective method. PO_2 levels greater than 20 mmHg indicate satisfactory perfusion.
   - Tissue pH monitoring is also a common method. A pH <7.3 or a difference of more than 0.35 pH units in adjacent control tissue is consistent with graft occlusion.
   - Photoplethysmography (PPG) uses the difference in infrared light absorption between pulsatile and nonpulsatile flow to establish a waveform, which can be serially monitored for declining trends.
   - Temperature monitoring is effective. Differences of greater than 2.5°C between a replanted digit and a control digit or an absolute temperature of a replanted digit that is below 30°C suggests vessel thrombosis.
   - Surface Doppler probe monitoring is an easy and widely employed method for arterial pulse surveillance. Venous assessment, which produces a characteristic "hum" when adjacent tissue is compressed, has been more difficult to teach to nursing staff and inexperienced observers.
   - Laser Doppler flow meters measure flow values. Values <0.40 units indicate ischemic tissues. With replanted digits, values <0.50 units have been observed with failing grafts.
   - Pulse oximetry allows excellent continuous monitoring of replanted digits. Oxygen saturation >95% is associated with viable tissue. Loss of signal reflects arterial occlusion, while gradual decline in saturation suggests venous occlusion.

4. **Which bedside maneuvers can be attempted to salvage a failing flap?**
   - Correct hypovolemia, which impairs perfusion.
   - Increase ambient room temperature to blunt vasoconstrictive reflexes.
   - Remove any dressing, splint, sutures, or bolster that could be compressing the vessels.
   - Consider the possibility of a hematoma.
   - Elevate the extremity to decrease edema or the limb into a dependent position for better inflow.
   - Administer a regional anesthetic block to potentiate sympathetic nervous system-mediated vasospasm.

   Note: Failure of these maneuvers to rapidly improve the condition of the flap should prompt the patient's immediate return to the operating room for exploration of the vessels.

5. **What is the appropriate role for leech therapy?**
   The medical leech, *Hirudo medicinalis*, has an established role for digital replants with venous failure or the finger with no available vein for anastamosis. Leech therapy has reached salvage rates of 60–70% in some cases. Leech therapy is not a replacement for flap reexploration, but it may be chosen in those patients who are medically unfit for a return to the operating room. Leech therapy is also successful in treating vessel congestion in a pedicled or free flap. The potent anticoagulant secreted by the leech, hirudin, inhibits the conversion of fibrin to fibrinogen, similar to heparin. Other than the direct amount of blood digested by the leech itself, the oozing that persists after removal of the leech is the key for relief of the congestion. Monitoring of a patient's hematocrit should be done routinely while using leech therapy to watch for the development of anemia.

6. **What should all patients receiving leech therapy receive?**
   All patients receiving leech therapy should receive antibiotic prophylaxis for *Aeromonas hydrophilia*, a colonizing organism that can cause a necrotizing soft tissue infection.

7. **Is there a consensus on what constitutes appropriate intraoperative or postoperative free flap/replantation anticoagulation?**
   No. Most authorities agree that *antiplatelet* therapy is essential for thrombotic prevention intraoperatively and postoperatively. Prevention can be achieved with aspirin, which is cost effective, has minimal side effects, and can be delivered per rectum if oral delivery is not available. Ketorolac (Toradol), a nonsteroidal anti-inflammatory drug (NSAID) with potent antiplatelet effects, can be given intravenously and has been used routinely by some microsurgeons. The postoperative infusion of dextran solution has been questioned after an increase in postoperative complications has been reported with head and neck free flaps. Chocolate, caffeine, and nicotine are all strictly prohibited postoperatively because of potential vasoconstrictive effects.

8. **Is there any evidence favoring interrupted versus continuous suture techniques for a microvascular anastamosis?**
   No. Similar patentcy rates have been shown in multiple studies. Attention must be maintained with continuous suturing to avoid "purse-stringing" an anastomosis.

9. **What are the benefits of free flap coverage of the hand following a traumatic wound?**
   Free flap coverage allows elevation with early physical therapy, resulting in less edema and the avoidance of a stiff shoulder and elbow. Often, the use of free flaps can turn what could have been multiple operations into a single reconstructive surgery. Free flaps have some potential advantages in a vascularly compromised limb because they can act as a conduit for vessel reconstruction and because they have fewer perfusion demands than a local flap or skin graft.

10. **What are some of the common free fascial or fasciocutaneous flaps used in upper extremity reconstruction?**
    - **Lateral arm flap** is a fasciocutaneous flap based on the posterior radial collateral artery. The flap can be raised with the inferior lateral brachial cutaneous nerve to provide some sensation.
    - **Proximal forearm flap** is a very pliable and thin fasciocutaneous flap based on the plexus between the radial collateral and posterior interosseus circulations. This flap is advocated for palmar surface defects and can be raised with the posterior cutaneous nerve.
    - **Temporoparietal fascial flap (TPF)** is a large fascial flap that provides excellent coverage of dorsal or palmar defects when combined with a skin graft. An additional benefit is the low donor site morbidity.
    - **Dorsalis pedis flap** is another thin, pliable flap that provides excellent coverage of dorsal hand defects. It can be raised with a sensory nerve (superficial peroneal cutaneous nerve), the underlying tendons, and the second metatarsophalangeal joint as a composite graft, if needed.
    - **Radial forearm flap** provides an excellent vessel conduit for revascularization of the hand, if required. It is also relatively simple to harvest (Fig. 38-1).

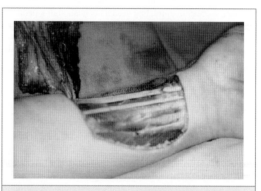

Figure 38-1. Radial forearm flap harvested from the distal forearm.

11. **What are the options for microsurgical reconstruction of the thumb?**
    For the thumb, great and second toe transfers and Morrison's great toe wrap-around flap are the best options. The wrap-around flap contains the skin envelope, pulp, and nail of the great toe and is suited for distal degloving and avulsion type injuries. Each of these flaps is harvested on the first dorsal metatarsal artery or first plantar metatarsal artery, with venous outflow provided by branches of the saphenous vein.

12. **Which digital artery of the thumb is larger?**
    The ulnar digital artery is usually dominant, which is why it is given priority during reconstruction or repair.

13. **What are the contraindications to free flap coverage of a wound?**
    Obviously, a nonfunctional limb precludes the morbidity and expense of a complex surgical reconstruction. A free flap is also not a substitute for poorly performed or inadequate wound debridement with devitalized tissue. With a crush-avulsion injury, planned debridements to assess tissue viability are advisable prior to proceeding with free flap coverage. Traditional teaching has been that the wound should be covered with soft tissue within 72 hours of the injury, with many institutions previously advocating acute reconstruction with the hope of

decreasing the incidence of a postoperative infection. This has been revisited, and some surgeons no longer strictly adhere to this dogma. Advances in local antibiotic delivery with antibiotic-impregnated beads and the vacuum-assisted closure (VAC) device to reduce edema in surrounding tissues has allowed greater discretion in reconstructing what was previously considered an unfavorable wound.

14. **What are the indications for replantation in adults?**
    Amputation of the thumb or multiple digits, of the hand (partial), at the wrist, through the forearm and sharp amputations of the elbow/proximal arm as well as of some individual digits distal to the insertion of the flexor digitorum superficialis tendon (FDS).

15. **What are the potential indications for replantation in children?**
    A distal fingertip amputation can be replaced as a simple composite tissue graft with good results. Overall, digital replants in children have somewhat less favorable results (65–75% survival versus 80–90% in adults) due to the technical challenges presented by the small vessel caliber. Replanted digits that survive have excellent long-term results and should be expected to achieve more than 90% of normal size in adulthood.

KEY POINTS: MICROSURGERY

1. Aspirin is the single most cost-effective antithrombotic regimen used postoperatively following microsurgery.
2. Unrecognized tension during free flap insetting rivals technical errors during the anastamosis as a cause of vessel thrombosis.
3. A free flap or replantation that does not appear normal after bedside maneuvers (e.g., elevation, dressing removal) is a relative indication for urgent reexploration to assess vessel patency.

16. **What should be done with a replanted digit that is unsalvageable?**
    Arguably it should be left in place because there is the potential for only a partial digit loss, allowing some preservation of length with revision procedures.

17. **What are the principles of immediate care and transport of an amputated part?**
    The digit should be moistened with saline or lactated Ringer's solution and placed in a sealed plastic bag that is placed on ice. Direct contact with ice should be avoided to prevent freezing of the tissue, and dry ice should never be used.

18. **How long are amputated limbs and digits viable?**
    A limb can sustain approximately 6 hours of warm ischemia or up to 12 hours of cold ischemia. By virtue of their paucity of skeletal muscle, digits can remain viable somewhat longer.

19. **What should be the goal of a multiple digit replantation?**
    The reestablishment of grasp and pinch functions. Thus, the thumb, long (middle), and ring fingers receive the highest priority with respect to replantation.

20. **What is the sequence for performing the replantation of an amputated limb or digit?**
    1. Bony fixation is first performed.
    2. Repair of flexor/extensor tendons with a digital replant is then performed to avoid disrupting the vessel anastamosis.

3. Arterial repair to reestablish perfusion and distend outflow veins to allow for identification is the third step. This repair also allows the washout of toxic metabolic products produced by ischemic tissue prior to establishing systemic circulation. Both digital arteries are preferentially repaired, but most replanted digits can survive on a patent single vessel.
4. Venous anastomosis is the fourth step.
5. Nerve repair is next to last.
6. Fasciotomies are then performed for a major limb replantation to avoid compartment syndrome during reperfusion.

21. **Which factor is most important with respect to recovering sensation after a digital nerve repair?**
    The age of the patient. Younger patients have the most favorable outcomes.

BIBLIOGRAPHY

1. Chen HC, Buchman MT, Wei FC: Free flaps for soft tissue coverage in the hand and fingers. Hand Clin 15(4):541–554, 1999.
2. Duffy FJ Jr, Concannon MJ, Gan BS, May JW Jr: Late digital replantation failure: Pathophysiology and risk factors. Ann Plast Surg 40(5):538–541, 1998.
3. Goldner RD, Urbaniak JR: Replantation. In Green DP, Hotchkiss RN, Pederson W (eds): Green's Operative Hand Surgery, 4th ed. New York, Churchill Livingstone, 1999, pp 1139–1155.
4. Pederson WC, Sanders WF: Principles of microsurgery. In Green DP, Hotchkiss RN, Pederson W (eds): Green's Operative Hand Surgery, 4th ed. New York, Churchill Livingstone, 1999.

REPLANTATION

Clifford King, MD, PhD, and William M. Kuzon Jr., MD, PhD

1. **Define *replantation*.**
   The surgical attachment, including the revascularization, of a body part that has been completely amputated (i.e., the reattachment of a body part for which no soft tissue connection was maintained between the part and the patient).

2. **Define *revascularization*.**
   The restoration of circulation to a devascularized but not completely amputated part.

3. **Describe the emergency management of a patient with an amputated part.**
   - Emergency management should follow the advanced trauma life support (ATLS) protocol.
   - Major life-threatening injuries take precedence over the replantation of an amputated part.
   - Expeditious transport of the patient and the amputated part to the replantation center should occur to minimize the ischemic interval.
   - Tetanus immunization status should be determined and immunization or booster provided as indicated.
   - Prophylactic antibiotics to cover skin flora should be administered. Broad-spectrum coverage is indicated if significant wound contamination is encountered or if the patient is immunocompromised.

4. **How should an amputated part be transported from the site of injury to the replantation center?**
   - The part should be placed in a bag after being wrapped in a sterile gauze dampened with normal saline.
   - The bag should be placed in a container and submerged in an ice-saline bath to maintain a temperature of 4°C.
   - The amputated part should never be placed in a hypotonic or hypertonic solution.

5. **Which features of the clinical history are relevant when considering replantation of an amputated part?**
   - Mechanism of injury (avulsion or crush)
   - Time of injury (duration of ischemia time)
   - Emergency treatment rendered, including care provided to the amputated part
   - Patient age, hand dominance, and occupation or avocational demands and expectations
   - Previous hand injuries or disability
   - Other major injuries
   - Medical or psychiatric conditions that may preclude replantation

6. **Why is it important to know the hand dominance and occupation of a patient with multiple finger amputations?**
   - The radial-sided fingers are more important in fine pinch and manipulation, which may be relevant with someone such as a musician.
   - Fine manipulation involving the radial side of the hand is more important in the dominant hand.

- The ulnar-sided fingers are important for power grip and would be important for a laborer.
- The nondominant hand typically plays a more important role in power-assist functions.
- In multiple finger amputations, the more viable remnants should be replanted in the most advantageous positions with respect to the patient's needs.

7. Assuming that there are no associated significant injuries, what are the relevant aspects of the physical examination?
   - Location (level) of amputation
   - Single or multiple injury levels in the extremity
   - Single or multiple amputated parts
   - Condition of the amputated part (signs of crush or avulsion)
   - Condition of the amputation stump
   - Amount of soft tissue loss (e.g., skin, vessel, bone, nerve, tendon)
   - Extent of contamination
   - Neurovascular status of the entire upper extremity

8. Which laboratory tests are usually performed in the patient with an amputated part?
   - Plain radiographs of the injured hand or arm and the amputated part
   - Hematocrit and hemoglobin content
   - Blood type and screening
   - Other tests as indicated by the patient's age, general health, and medical history

9. What are the *absolute* indications for attempted replantation?
   - Amputation of thumb
   - Amputation of multiple digits
   - Any amputation in a pediatric patient
   - Amputation at the mid-metacarpal, wrist, or forearm level

10. What are the *relative* indications for attempted replantation?
    - Single finger amputation distal to the flexor digitorum superficialis (FDS) insertion
    - Single digit amputation at other levels, particularly in female patients
    - Proximal limb amputation in a child

11. What are the contraindications for attempted replantation?
    - Other life-threatening injuries that require emergent treatment
    - Major medical conditions with a high perioperative risk
    - Technically challenging procedure due to location of amputation or severe avulsion or crush mechanism of injury
    - Prohibitive duration of ischemia
    - Multiple injury levels
    - Extreme contamination
    - Single digit amputation (relative contraindication)
    - Psychiatric instability (relative contraindication)

12. Since it is theoretically feasible, why is replantation not performed for every amputated part?
    Although replantation has become more feasible with recent technical advancements, the replanted part may actually become an impediment, compromising the overall function of the hand or limb. For example, a single, painful, stiff digit can impair the function of the remaining normal digits. Similarly, a stiff, insensate, minimally functional arm may be less functional than a prosthesis.

13. **What is a heterotopic replantation?**
    The replantation of all or part of an amputated limb or digit to a new position to restore critical function at the expense of an expendable function (e.g., the replantation of a finger in the thumb position).

14. **What is a "spare parts surgery"?**
    The use of components of an amputated part that cannot be replanted /salvaged to repair an injury in the hand or extremity that can be salvaged (for example, using a digital nerve in a nonreplantable finger as a graft in an injured finger with segmental loss of a digital nerve).

15. **What is meant by the term *prohibitive ischemia*?**
    - The ischemic tolerance of an amputated extremity is inversely related to the amount of muscle in the amputated part.
    - Significant muscle necrosis will occur with a warm ischemia time of >4–6 hours.
    - An isolated digit amputation can tolerate up to 12 or more hours of warm ischemia because it contains no muscle.
    - A proximal limb amputation can tolerate only 4–6 hours of ischemia because the limb contains a large amount of muscle mass.
    - Appropriate cooling of an amputated part can significantly extend the ischemic tolerance of tissue.

16. **What is a reperfusion injury?**
    A reperfusion injury occurs when ischemic damage to vascular endothelium and other tissue occurs during ischemia, and the arterial reflow that occurs with revascularization incites a cycle of leukocyte adhesion, migration, and degranulation, resulting in an increase in oxygen-free radical species in the tissues. The radicals are extremely injurious, resulting in cell membrane damage, cellular and interstitial edema, impairment of circulation, and tissue necrosis.

17. **What is reperfusion syndrome?**
    Following a proximal limb amputation and prolonged ischemia, myonecrosis occurs. When the limb is replanted, toxic metabolites are released into the systemic circulation, resulting in acidosis, myoglobinuria, cardiogenic shock, renal failure, multisystem organ failure, and, in extreme cases, death.

18. **What is the recommended sequence for replantation of an amputated finger?**
    1. Wound debridement
    2. Identification of arteries, veins, nerves, and tendons
    3. Bone stabilization
    4. Extensor tendon repair
    5. Flexor tendon repair
    6. Vascular anastomoses
    7. Nerve repair
    8. Skin closure

19. **What is the repair sequence for a major limb replantation?**
    To minimize the risk of reperfusion syndrome, bony stabilization is performed immediately followed by expedient arterial repair to permit arterial inflow and the release of toxic metabolites into the surgical field, effectively "washing out" the toxic metabolites. Venous repair can then be performed followed by nerve and tendon repairs.

20. **When should a fasciotomy be performed following the replantation of a limb?**
    - A fasciotomy of all muscle compartments in the amputated part should be performed prior to revascularization.

- For replantation of a hand, the intrinsic, thenar, and hypothenar compartments should be routinely released.
- For a forearm level replantation, the volar and dorsal forearm compartments should be released.
- A carpal tunnel release should always be performed if the amputation level is proximal to the wrist.

21. **What are the preferred methods for bony stabilization during replantation?**
    - Kirschner wires or interosseous wires are most commonly used because they can be placed rapidly and easily.
    - Other options include lag screw fixation or a miniplate and screws.
    - Plate fixation is recommended for major limb replantation.

22. **Why is skeletal shortening often performed during replantation?**
    - Improves bone-to-bone apposition, thus enhancing healing potential
    - Permits a tension-free vascular anastomosis
    - May obviate the need for a nerve graft
    - Permits better skin and soft tissue coverage

23. **Which technique is used if the available vessels are too short for a tension-free anastomosis during replantation?**
    Interpositional vein grafting.

24. **What is a ring avulsion injury?**
    A specific injury pattern involving the avulsion of soft tissue, including skin, subcutaneous tissue, nerves, and occasionally tendons, without injury to the bony and ligamentous structures. The injury occurs when a ring is forcibly avulsed from a finger, such as catching a wedding ring against a basketball hoop when slam-dunking.

25. **What are the principles of anesthesia management during replantation?**
    - Avoid vasoconstriction and vasospasm, which can result in a thrombosis of the vascular repair.
    - Avoid hypotension with aggressive hydration.
    - Avoid acidosis.
    - Keep the patient warm.

KEY POINTS: REPLANTATION

1. An amputated part should be carefully wrapped in a sterile, saline-soaked gauze and then placed in a sealed plastic bag that is contained in an ice-saline bath to maintain a temperature of 4°C.
2. The absolute indications for attempted replantation include an amputated thumb, multiple digit amputations, any amputation in a child, and an amputation at the mid-metacarpal, wrist, or forearm level.
3. Heterotopic replantation involves the replantation of an amputated limb or digit in a new anatomic location to restore critical and essential functions.
4. The recommended sequence for digital replantation is wound debridement, bone stabilization, extensor tendon repair, flexor tendon repair, vascular repair, nerve repair, and skin closure.
5. If the vessels are not of sufficient length to avoid a tension-free anastomosis in a patient undergoing replantation, an interpositional vein graft is required.
6. Vasoconstriction and hypotension should be avoided following replantation because of the potential for thrombosis at the vascular anastomoses.

26. **What are the important postoperative management concerns in the patient who has undergone replantation?**
    - Administering adequate continuous analgesia
    - Splinting to immobilize and protect the replanted part
    - Frequent and careful monitoring of the vascular status of the replanted part
    - Preventing hypovolemia and hypotension via vigorous hydration and the avoidance of vasodilators and diuretics
    - Preventing vasoconstriction by keeping the patient warm and prohibiting the use of vasoconstrictors, including nicotine, caffeine, and vasoconstrictive medications
    - Careful monitoring of hematocrit and electrolyte status

27. **Which clinical signs are helpful for monitoring the vascular status of the replanted part?**
    The most sensitive indicators of the adequacy of arterial inflow and venous outflow are the following:
    - Color
    - Capillary refill
    - Temperature
    - Tissue turgor

28. **What is the clinical appearance of the replanted part with arterial insufficiency?**
    The replanted part is pale and cool with absent capillary refill and poor tissue turgor.

29. **What is the clinical appearance of the replanted part with venous insufficiency?**
    The replanted part is congested and ruborous with increased tissue turgor and extremely rapid capillary refill.

30. **Which devices are available for monitoring the status of a replanted part?**
    No device exists that unambiguously determines if arterial inflow and venous outflow is adequate. All of the commercially available devices require subjective interpretation of the data and are associated with a high incidence of false-positive and -negative interpretations. Thus, clinical monitoring is preferred. Devices available include the ultrasonic and laser Doppler, plethysmograph, pH monitor, transcutaneous and implantable oxygen electrodes, and temperature monitors.

31. **How is venous insufficiency treated in the replanted digit?**
    - Elevation of the hand
    - Loosening of all dressings and/or removing sutures to relieve constriction
    - Removal of the nail plate, abrasion of the nail bed, and application of a heparin-soaked sponge
    - Application of medicinal leeches (*Hirudis medicinalis*)
    - Systemic anticoagulation
    - Exploration of the venous anastomosis and revision if necessary

32. **What are the management issues when medicinal leeches are used following replantation?**
    - Prophylactic antibiotics must be used because medicinal leeches harbor *Aeromonas* species as endogenous gut flora.
    - The replanted part with the leeches attached should be dressed to confine the leeches and prevent wandering.
    - The leeches should be encouraged to attach to the engorged part and placed by grasping their bodies with forceps.

- When the leeches fall off, they should be killed in a vial of alcohol and discarded.
- Leeches should be applied every 4–6 hours.

33. **How long should medicinal leeches be used in a patient with venous congestion following replantation?**
    The leeches may be used for a minimum of 3 days until collateral circulation has developed.

34. **What are the steps in treating arterial insufficiency following replantation?**
    - Warm the patient.
    - Optimize volume resuscitation.
    - Remove constrictive dressings and sutures.
    - Explore the vessels and perform revision of the anastomoses.

35. **What are the most common complications following a digital replantation?**
    - Vascular compromise
    - Hemorrhagic anemia following medicinal leech use
    - Stiffness
    - Diminished sensation
    - Cold intolerance
    - Chronic pain

36. **Which complications occur following the replantation of a major limb?**
    - Replantation syndrome
    - Infection
    - Soft tissue loss
    - Loss of motor function
    - Diminished sensation
    - Cold intolerance
    - Chronic pain

37. **What are the typical clinical results of digital replantation?**
    - A 50–80% survival rate of the limb
    - Time off from work averaging 7 months
    - Protective sensation in 36% of patients
    - A 50% range of motion
    - About 80% of patients require at least two additional operative procedures

38. **What is hand allotransplantation?**
    The controversial procedure involving the transplantation of a hand from an organ donor to a different recipient. The procedure requires postsurgical immunosuppressive therapy that increases the risk of opportunistic infections and malignancies for what is a nonlethal clinical condition.

39. **Which serum marker has been shown to be prognostic with respect to limb survival following major limb replantation?**
    - A study demonstrated that the serum potassium level was prognostic following a major limb replantation.
    - In the same study, replantation failed in all of the patients with a serum potassium level >6.5 mmol/L immediately after limb perfusion.
    - All of the patients with a successful replantation demonstrated a return to a normal serum potassium level 1 hour after reperfusion.

WEBSITES

1. Emedicine: Hand amputations and replantation
   http://www.emedicine.com/plastic/topic536.htm
2. Emedicine: Replantation
   http://www.emedicine.com/orthoped/topic284.htm
3. The Buncke Clinic
   http://buncke.org/book/contents.html

BIBLIOGRAPHY

1. Atroshi I, Rosberg HE: Epidemiology of amputations and severe injuries of the hand. Hand Clin 17(3):343–350, 2001.
2. Cheng GL, Pan DD, Zhang NP, Fang GR: Digital replantation in children: A long-term follow-up study. J Hand Surg (Am) 23(4):635–646, 1998.
3. Graham B, Adkins P, Tsai TM, et al: Major replantation versus revision amputation and prosthetic fitting in the upper extremity: A late functional outcomes study. J Hand Surg (Am) 23(5):783–791, 1998.
4. Kleinert HE, Juhala CA, Tsai TM, Van Beek A: Digital replantation—Selection, technique, and results. Orthop Clin North Am 8:309, 1997.
5. Lee BI, Chung HY, Kim WK, et al: The effects of the number and ratio of repaired arteries and veins on the survival rate in digital replantation. Ann Plast Surg 44(3):288–294, 2000.
6. Waikakul S, Sakkarnkosol S, Vanadurongwan V, Un-nanuntana A: Results of 1018 digital replantations in 552 patients. Injury 31(1):33–40, 2000.
7. Waikakul S, Vanadurongwan V, Unnanantana A: Prognostic factors for major limb replantation at both immediate and long-term follow-up. J Bone Joint Surg 80B:1024–1030, 1998.
8. Wang H: Secondary surgery after digital replantation: Its incidence and sequence. Microsurgery 22(2):57–61, 2002.

CHAPTER 40

AMPUTATIONS AND PROSTHETICS

Kelly Vanderhave, MD

1. **What are the most frequent causes of an upper limb amputation?**
   - Trauma (approximately 75%)
   - Malignancy
   - Vascular disease

2. **How is a symptomatic neuroma prevented following amputation?**
   - The nerve ending should not be left in a superficial position because it will become entrapped in cutaneous scar.
   - Nerve endings exposed in a wound should be grasped, pulled, cut sharply at the base of the wound, and allowed to retract.
   - The nerve may also be dissected free and transposed into adjacent bone or beneath the cover of local muscle tissue if available.

3. **How does a "lumbrical-plus" deformity occur after an amputation through the distal interphalangeal (DIP) joint?**
   A lumbrical-plus deformity refers to the paradoxical extension of the interphalangeal joints that occurs during grasp because of the shortened lumbrical that originates on the flexor digitorum profundus (FDP) tendon and contributes to the extensor mechanism. When the FDP tendon is cut and allowed to retract, tension on the lumbrical muscle and tendon occurs. Active flexion of the finger further increases tension on the lumbrical and the extensor mechanism and impedes flexion.

4. **At what level of index finger amputation does pinching and fine-motor activity usually transfer to the long finger? What are the implications for treatment?**
   Proximal to the proximal interphalangeal (PIP) joint. Extensive efforts to preserve length via flap coverage are usually not warranted. Shortening of the bone, taking care to preserve the flexor digitorum superficialis (FDS) insertion, and primary closure of the wound are usually sufficient.

5. **Why should the flexor and extensor tendons *not* be sewn to each other to provide coverage after a digital amputation?**
   Because the **quadriga effect** will occur. Sewing the flexor and extensor tendons together limits the motion of both. Because the FDP tendons of the long, ring, and small fingers have a common muscular origin, limitation of motion of one tendon may limit the excursion of adjacent FDP tendons and hence active motion of the adjacent finger(s).

6. **How does a digital amputation through the PIP joint affect function at the metacarpophalangeal (MCP) joint?**
   Because the PIP level amputation is proximal to the FDS insertion, the digit is without extrinsic flexors. The lumbrical alone can be expected to provide approximately 45 degrees of active MCP joint flexion.

7. **Why is a ray (long or ring finger) amputation often preferable to an amputation at the level of the MCP joint?**
   An amputation at the MCP level, particularly centrally, can leave the hand with a gap that results in poor control of small objects carried in the palm. Elective ray amputation and closure of this gap can improve function.

8. **Why is the base of the metacarpal preserved when a ray is amputated?**
   To preserve the insertions of the extensor carpi ulnaris, extensor carpi radialis brevis, extensor carpi radialis longus, flexor carpi radialis, and, indirectly, the flexor carpi ulnaris tendons.

9. **When a central ray is amputated, how is the defect closed?**
   There are two options:
   - The deep transverse intermetacarpal ligaments from the two adjacent rays can be approximated and sutured to close the defect.
   - An adjacent ray can be osteotomized at the base of its metacarpal and the entire ray transposed to the metacarpal base of the excised central ray (i.e., transfer the index ray to the base of the third metacarpal after resection of the third ray). The transposed ray is stabilized with some form of rigid internal fixation.

10. **Which factor has been found to highly influence the outcome of ray amputation?**
    The presence of a worker's compensation claim, which is associated with a poorer outcome overall.

11. **What is a Krukenberg procedure?**
    The conversion of a below-elbow amputee into radial and ulnar rays that act as prehensile forceps. The forearm flexors and extensor muscles are divided, and their action is converted to adduction and abduction of the radial and ulnar rays. The primary motor muscle is the pronator teres. The procedure requires 10 cm of forearm length and should be performed only in patients with good elbow motion. An elbow flexion contracture >70 degrees is considered a contraindication.

12. **What is the ideal indication for the Krukenberg procedure?**
    - A bilateral below-elbow amputee who is blind is a good candidate.
    - Patients with bilateral congenital amputations also may benefit.
    - Performing a Krukenberg procedure does not preclude the subsequent use of a prosthesis.

13. **Discuss the major advantage and disadvantage of a wrist disarticulation compared with a transradial amputation.**
    - The **major advantage** of the wrist disarticulation is the preservation of pronation and supination. An estimated 50% of this motion is transferred through a prosthesis. Pronation and supination are reduced following a transradial amputation.
    - The **major disadvantage** of the wrist disarticulation lies in the bony prominences and accompanying lack of distal padding. This and the additional length (compared with transradial amputation) result in a greater challenge for prosthetic fitting.

14. **Explain the difference between a shoulder disarticulation and forequarter amputation.**
    - A **shoulder disarticulation** consists of an amputation through the glenohumeral joint with preservation of the clavicle, acromion, and scapula.
    - The **forequarter amputation** involves removal of the entire shoulder girdle, including the clavicle and scapula. It is significantly more disfiguring than the disarticulation and, fortunately, is rarely indicated.

15. **Why is a disarticulation preferable to a through-bone amputation in children?**
    Amputations in children with open physes can result in terminal overgrowth, causing the bone to project through the distal skin. The overgrowth phenomenon can be prevented by leaving the bone and distal articular surface intact by performing either a wrist or elbow disarticulation.

16. **What is "phantom limb" sensation?**
    The phenomenon that occurs following an amputation when the amputee feels that the limb is still present.

17. **What are the proposed advantages of immediate prosthetic fitting following an amputation?**
    - Promotes wound healing
    - Assists in edema control
    - Provides early shaping and shrinkage of the soft tissues
    - Promotes early functional return

18. **Name the sources of body power available to produce upper extremity prosthetic motion and function.**
    - Cable-driven prostheses rely on remaining sources of body power to produce the force and excursion length necessary to operate the prosthetic joints and/or terminal devices.
    - Sources include glenohumeral flexion, shoulder elevation and depression, biscapular abduction, and chest expansion.
    - The movement that provides the greatest force and length for prosthetic use is glenohumeral flexion.

19. **What are the two functional categories of body-powered terminal devices? Which is more common?**
    Voluntary opening and closing. The name defines the component of action that is under direct body control via the prosthesis. The other action usually is accomplished with rubber bands or springs. Voluntary opening terminal devices are more common.

20. **How does the patient operate the terminal device for a below-elbow prosthesis?**
    The primary motion to operate the terminal device is glenohumeral flexion and scapular abduction (Fig. 40-1).

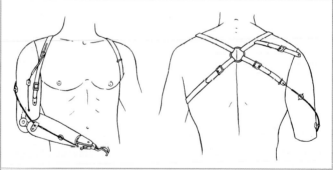

Figure 40-1. Terminal device operation. (Adapted from Bozentka DJ, Beredjiklian PK: Review of Hand Surgery. Philadelphia, W.B. Saunders, 2004, p 255.)

21. **What motion is required to lock and unlock the prosthetic elbow in an above-elbow prosthesis?**
    Shoulder extension and abduction with scapular depression.

22. **How does a myoelectric prosthesis operate?**
    Myoelectric devices use electromyographic (EMG) potentials derived from muscles in the amputation stump and limb to actuate and control the terminal device. The EMG potentials are obtained by skin surface electrodes placed over remaining agonist/antagonist muscle groups (wrist and finger flexors and extensors in the case of the transradial amputee). A battery and motor are contained within the prosthesis.

23. **What are the advantages of a myoelectric terminal device compared with a body-powered device?**
    - Myoelectric devices provide voluntary opening and closing that can be well controlled. The actuation as well as the appearance of the prosthesis is quite natural.
    - The principal advantages of mechanical terminal devices are durability, reliability, and function.

24. **What is the major advantage of a below-elbow myoelectric prosthesis over a standard below-elbow prosthesis?**
    The myoelectric prosthesis is associated with an increased ease of performing activities overhead and close to the body.

25. **Myoelectric prostheses are most successful in patients with what type of amputation?**
    Mid-level below-elbow amputation.

26. **What is the total amount of cable excursion required to operate a transhumeral prosthesis.**
    Approximately 2 inches of excursion are required for full flexion of the elbow and 1.5–2.5 inches are required to open the terminal device fully. Therefore, depending on the style of the prosthesis, up to 4.5 inches of excursion are needed to operate a transhumeral prosthesis.

27. **Why is it important to preserve maximal length with a below-elbow amputation?**
    The longer the forearm segment, the greater the degree of forearm pronation and supination. If a below-elbow amputation is performed in the proximal third of the forearm, consider detaching the biceps and reattaching it to the proximal ulna for improved prosthetic fitting and use.

28. **What is the minimal forearm length necessary for prosthetic fitting in patients with a below-elbow amputation?**
    If elbow flexion is present, 2 cm (approximately 1 inch) of ulna are sufficient.

29. **What is the major advantage of an elbow disarticulation versus an above-elbow amputation?**
    Preservation of humeral length with an elbow disarticulation permits humeral rotation, which improves prosthetic functional use.

30. **What are the most common patterns of congenital upper extremity amputations?**
    - A congenital below-elbow amputation at the level of the proximal forearm is the most common type *(terminal transverse radial)*.
    - Other patterns, in order of incidence, are midcarpal, transmetacarpal, and transhumeral.

KEY POINTS: AMPUTATIONS AND PROSTHETICS

1. The distal radioulnar joint (DRUJ) should be preserved in a wrist disarticulation to permit full forearm pronation and supination.
2. A below-elbow amputation is the most common congenital upper extremity amputation.
3. The Krukenberg procedure is ideally indicated for the blind amputee with bilateral below-elbow amputations.
4. The outcome of a ray amputation is adversely affected by the presence of a worker's compensation claim.
5. A congenital amputation of the upper extremity is frequently associated with other musculoskeletal anomalies in the same limb.

31. **Which upper extremity musculoskeletal anomalies occur in association with a congenital upper extremity amputation?**
    - Congenital radial head dislocation
    - Elbow or forearm synostosis

32. **What is the best time for prosthetic fitting in a child with an upper extremity amputation?**
    - When the infant is 6–9 months of age (sitting upright and the beginning of bimanual activities), a passive device, such as a mitten, is most commonly used.
    - A body-powered prosthesis is considered at 15–24 months of age.
    - A terminal device can be activated at 18–24 months of age.
    - Myoelectric prostheses are used after 2 years of age.

33. **What are the advantages of early prosthetic fitting in patients with traumatic upper extremity amputation?**
    - The earlier the prosthesis is fitted and incorporated into the patient's daily use, the greater the likelihood that the patient will continue to wear and use the prosthesis in the long term.
    - Patients who undergo prosthetic fitting within the first 30 days after amputation are 3 times more likely to use the prosthesis over the long term.

34. **Which level of digital amputation is most amenable to prosthetic use?**
    Between the PIP and DIP joints. The more distal the amputation, the better the prosthetic control (Fig. 40-2).

35. **Which factors are associated with poor digital prosthetic use?**
    - Male gender
    - Amputation stump problems
    - Employment in a manual intensive occupation
    - Distal amputation level
    - Ring finger prosthesis, which is the least likely to be used
    - Small finger prosthesis, which is the most likely to be used

Figure 40-2. The level of digital amputation (between the PIP and DIP joints) most amenable to prosthetic use. The more distal the amputation, the better the prosthetic control. (Adapted from Bozentka DJ, Beredjiklian PK: Review of Hand Surgery. Philadelphia, W.B. Saunders, 2004, p 253.)

36. **What is an osteointegrated prosthesis?**
    The procedure usually involves the thumb and consists of staged implantation of a prosthesis that maintains position via the ingrowth of bone into the implant itself. The prosthetic thumb is matched for skin color, pliability, and overall shape. The first stage involves fixture implantation into the medullary canal of the metacarpal shaft, in combination with autogenous cancellous bone grafting. About 3 weeks later, radiographs are obtained to assess ingrowth of bone into the fixture. The second stage involves soft tissue coverage of the prosthesis.

37. **What are the major surgical considerations for a proximal forearm amputation?**
    - Adequate soft tissue coverage of the stump
    - Prevention of a symptomatic neuroma by burying the nerve end deep within the proximal musculature
    - Biceps tendon attachment to the radius or ulna to preserve elbow flexion

38. **Which nonsurgical intervention should be considered for all patients with an upper extremity amputation?**
    Psychological counseling. Flashbacks, posttraumatic stress syndrome, and a variety of psychologic effects are common following *any* amputation of the upper extremity.

39. **What changes occur in the spinal cord following a distal limb amputation?**
    There is a decrease in the number of anterior horn cells, spinal ganglion cells, and large myelinated fibers of both the anterior and posterior roots.

BIBLIOGRAPHY

1. Barber L: Desensitization of the traumatized hand. In Hunter JM, Schneider LH, Mackin EJ, Callahan AD (eds): Rehabilitation of the Hand, 2nd ed. St. Louis, Mosby, 1984, pp 493–502.
2. Davids JR, Meyer LC, Blackhurst DW: Operative treatment of bone overgrowth in children who have an acquired or congenital amputation. J Bone Joint Surg 77A:1490–1497, 1995.
3. Grunert BK, Smith CJ, Devine CA, et al: Early psychological aspects of severe hand injury. J Hand Surg 8B:177–180, 1998.
4. Hopper RA, Griffiths S, Murray J, Manketlow RT: Factors influencing use of digital prostheses in workman's compensation recipients. J Hand Surg 25A:80–85, 2000.
5. Hunter JH, Mackin EJ, Callahan AD (eds): Rehabilitation of the Hand: Surgery and Therapy, 4th ed. St. Louis, Mosby, 1994.
6. Leonard JA Jr, Meier RH III: Upper and lower extremity prosthetics. In DeLisa JA, Gans BM (eds): Rehabilitation Medicine: Principles and Practice, 3rd ed. Philadelphia, Lippincott-Raven, 1998, pp 669–696.
7. Louis DS, Honter LY, Keating TM: Painful neuromas in long below-elbow amputees. Arch Surg 115:742–744, 1980.
8. Lundborg G, Branemark P, Rosen B: Osteointegrated thumb prosthesis: A concept for fixation of digital prosthetic devices. J Hand Surg 21A:216–221, 1996.
9. Manurangsee P: Osteointegrated finger prostheses: An alternative method for finger reconstruction. J Hand Surg 25A:86–89, 2000.
10. Peimer CA (ed): Surgery of the Hand and Upper Extremity. New York, McGraw-Hill, 1996.
11. Schurr DG, Cook TM: Prosthetics and Orthotics. Norwalk, CT, Appleton & Lange, 1990.
12. Suzuki H, Oyanagi K, Takahashi H, et al: A quantitative pathological investigation of the cervical cord, roots, and ganglia after long term amputation of the unilateral upper arm. Acta Neuropathol 85:666–673, 1993.
13. Swanson AB: The Krukenberg procedure in the juvenile amputee. J Bone Joint Surg 46A:1540–1548, 1964.
14. Wright TW, Hagen AD, Wood MW: Prosthetic usage in major upper extremity amputations. J Hand Surg 20A:619–622, 1995.

BONE TUMORS OF THE HAND AND WRIST

Peter M. Murray, MD, and Theodore W. Parsons, MD, LTC, USAF, MC

1. **How common are primary bone tumors of the hand and wrist?**
   Uncommon, accounting for approximately 5% of all bone tumors.

2. **How common are primary *malignant* bone tumors of the hand and wrist?**
   Rare, accounting for 1% of all bone tumors and 2% of all malignant tumors of bone.

3. **What is the most common primary bone tumor in the hand?**
   Enchondroma, which is a benign cartilaginous growth that arises from within the medullary canal of a single bone (Fig. 41-1).

4. **What is the most common location for an enchondroma in the hand?**
   Proximal phalanx.

5. **Describe the characteristic radiographic appearance of an enchondroma.**
   - Lytic expansile intramedullary lesion with a thin sclerotic border.
   - The endosteal surface is scalloped with expansion or bulging of the cortex.
   - Characteristic punctate, flocculent, or ring-like calcifications also may be noted.

6. **How does an enchondroma of the hand typically present?**
   Most enchondromas are asymptomatic and detected incidentally. Occasionally, patients may present with pain, with or without swelling, or following a pathologic fracture.

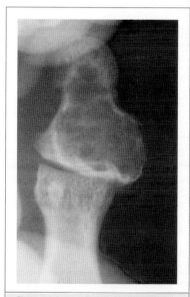

Figure 41-1. An enchondroma of the thumb distal phalanx.

7. **What is Ollier's disease?**
   A *non*hereditary syndrome of multiple enchondromata.

8. **An adolescent boy has a nonhereditary condition typified by multiple enchondromatosis and soft tissue hemagiomas. What is the diagnosis?**
   Maffucci's syndrome.

9. **What percentage of patients with Ollier's disease and Maffucci's syndrome have enchondromas in the hand?**
   Approximately 80%.

10. **What is the suspected diagnosis in a patient with Maffucci's syndrome or Ollier's disease who presents with a painful and enlarging enchondroma of the hand?**
    Malignant degeneration, which occurs in 30–50% of patients with Ollier's disease and Maffucci's syndrome.

11. **What is an osteochondroma?**
    A cartilage-capped bony exostosis or protrusion adjacent to the external surface of a bone near the epiphysial cartilage plate that is usually painless. However, irritation of adjacent tendons or nerve or artery compression may produce symptoms. Osteochondromas may be solitary or multiple. Multiple osteochondromatosis is transmitted in an autosomal dominant fashion. However, up to 20% of patients have a negative family history.

12. **What are the most common sites of involvement for an osteochondroma of the hand?**
    The metacarpals and phalanges (Fig. 41-2).

13. **What complications are seen with an osteochondroma?**
    - Malignant transformation of an osteochondroma of the hand, although extremely rare, may occur.
    - Sarcomatous degeneration is most common in patients with multiple hereditary osteochondromatosis.

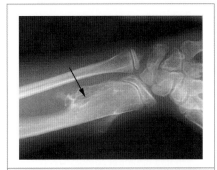

Figure 41-2. Osteochondroma of the distal radius.

14. **What is an osteoid osteoma?**
    A benign osteoblastic lesion consisting of a well-demarcated core (nidus) with a surrounding zone of reactive bone.

15. **What are the classic signs and symptoms of an osteoid osteoma of the hand or wrist?**
    Pain that is characteristically more severe at night and is usually relieved with nonsteroidal anti-inflammatory drugs (NSAIDs). Swelling, limited joint motion, and point tenderness on palpation also may be present, particularly in lesions involving the carpal bones.

16. **Describe the characteristic radiographic features of an osteoid osteoma in the hand.**
    An area of sclerosis with a central radiolucency. Sclerosis may be the only finding in early cases (Fig. 41-3).

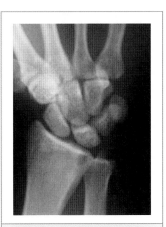

Figure 41-3. Osteoid osteoma involving the distal radius.

17. **What are the most common locations for an osteoid osteoma of the hand and wrist?**
    Osteoid osteomas have been reported in the distal radius and ulna, carpus, metacarpals, and phalanges.

18. **What is the most common location for a *carpal* osteoid osteoma?**
    The scaphoid.

19. **Which test is most helpful for localizing an osteoid osteoma?**
    Computed tomography (CT) scan.

20. **How is an osteoid osteoma of the hand treated?**
    Observation and prolonged NSAIDs until the disease process eventually burns out. Surgical treatment is indicated for failed conservative treatment or unremitting pain. Complete excision of the nidus is usually curative.

21. **What is the most common location for a unicameral (solitary) bone cyst of the hand?**
    The metaphyseal region of the distal radius.

22. **Describe the radiographic features of a unicameral bone cyst.**
    An eccentric, radiolucent cyst of variable size located in the metaphysis of a tubular bone.

23. **What is the treatment for a unicameral bone cyst?**
    Intralesional aspiration and injection of methylprednisolone. Curettage and bone grafting are recommended for large lesions or after failed injection.

24. **What is the most common location in the hand for a giant cell tumor of bone?**
    The distal radius.

25. **What is the most common complication after surgical treatment of a giant cell tumor of bone?**
    Local recurrence.

26. **Describe the characteristic radiographic appearance of a giant cell tumor of the distal radius.**
    An eccentric, expansile, radiolucent lesion involving the epiphysis and subchondral bone of the distal radius. The cortex is usually thinned with a well-developed sclerotic margin. Pathologic fracture also may occur after cortical erosion (Fig. 41-4).

27. **What are the surgical options for a giant cell tumor of the distal radius?**
    - Simple curettage with or without bone grafting (often complicated by local recurrence)

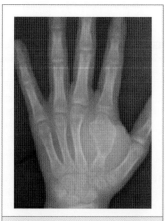

Figure 41-4. Giant cell tumor of bone involving the index metacarpal.

- Wide *en-bloc* excision with skeletal reconstruction using autogenous bone graft
- Allograft replacement
- Joint arthrodesis (which also may be performed after resection of the tumor)

28. **What are the three most common primary malignant bone tumors of the hand and wrist?**
    - Osteogenic sarcoma
    - Chondrosarcoma
    - Ewing's sarcoma

29. **Describe Enneking's system for staging tumors of the musculoskeletal system.**
    Enneking's classification involves three criteria: histologic grade (G), anatomic site (T), and presence or absence of metastasis (M) (Table 41-1). For histologic grading, 0 is benign; 1 is low-grade malignancy, and 2 is high-grade malignancy. The anatomic site may be defined as intracompartmental (T1) or extracompartmental (T2). Metastasis is defined as present (M1) or absent (M2).

TABLE 41-1. STAGING SYSTEM (ENNEKING'S SYSTEM) OF THE MUSCULAR TUMOR SOCIETY

Stage	GTM	Description
1A	G1T1M0	Low grade Intracompartmental No metastases
1B	G1T2M0	Low grade Extracompartmental No metastases
2A	G2T1M0	High grade Intracompartmental No metastases
2B	G2T2M0	High grade Extracompartmental No metastases
3A	G1/2T1M1	Any grade Intracompartmental With metastases
3B	G1/2T2M1	Any grade Extracompartmental With metastases

30. **Name the four types of surgical margins obtainable in tumor surgery (Fig. 41-5).**
    - **Intracapsular (piecemeal):** Leaves gross tumor present
    - **Marginal (shell out the tumor):** Leaves microscopic satellite lesions present
    - **Wide (intracompartmental):** Removes lesion with normal adjacent tissue
    - **Radical (extracompartmental):** Removes the entire compartment of involved and noninvolved tissue

31. **What is an osteogenic sarcoma?**
    A malignant bone tumor consisting of a malignant proliferating spindle cell stroma that produces osteoid or immature bone.

32. **Describe the radiographic features of an osteogenic sarcoma.**
    A sclerotic, expansile, destructive lesion with soft tissue extension that may display transverse or radiating striatas, otherwise known as a sunburst pattern.

33. **How is an osteogenic sarcoma involving the hand treated?**
    Wide *en-bloc* surgical excision, radical removal, or amputation with a wide margin combined with neoadjuvant (preoperative) and adjunctive (postoperative) chemotherapy.

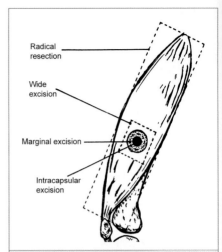

Figure 41-5. The four types of surgical margins available in tumor surgery: intracapsular, marginal, wide, and radical.

34. **Define *chondrosarcoma*.**
    A malignant tumor consisting of a sarcomatous stroma and neoplastic cartilage formation.

35. **Describe the different types of chondrosarcoma.**
    - A primary chondrosarcoma occurs in previously normal bone.
    - A secondary chondrosarcoma arises from a preexisting benign cartilage tumor, such as an enchondroma or osteochondroma.

36. **What is the most common location for a chondrosarcoma in the hand?**
    Proximal phalanx.

37. **How is a chondrosarcoma in the hand treated?**
    - Wide *en-bloc* resection if the lesion is entirely intraosseous
    - Amputation or ray resection if the tumor has extended into the soft tissues

38. **What is Ewing's sarcoma?**
    A primitive malignant tumor of bone consisting of uniform small cells with round nuclei but without a distinct cytoplasmic border.

39. **Describe the typical clinical presentation of Ewing's sarcoma.**
    Pain, swelling, elevated erythrocyte sedimentation rate (ESR), leukocytosis, fever, and the appearance of systemic illness.

40. **How is Ewing's sarcoma of the hand treated?**
    Surgical excision with a wide tissue margin consisting of ray resection or amputation in combination with chemotherapy and radiation therapy.

BONE TUMORS OF THE HAND AND WRIST

KEY POINTS: BONE TUMORS OF THE HAND AND WRIST

1. Under most circumstances, the biopsy of a suspected malignant bone tumor should be performed at the institution providing definitive limb salvage management.
2. Enchondroma is the most common primary bone tumor of the hand and is most frequently located in the proximal phalanx.
3. Malignant degeneration of an enchondroma in the hand occurs in approximately 30–50% of patients with Ollier's disease and Maffucci's syndrome.
4. Simple curettage of a giant cell tumor of the distal radius is associated with a high recurrence rate.
5. The most common metastatic tumor to the hand is lung carcinoma.

---

41. **What are the most common metastatic bone tumors of the hand and wrist (acrometastasis)?**
    - Lung
    - Colon
    - Kidney
    - Breast

42. **What is the significance of an acrometastasis?**
    Metastases to the hand are a poor prognostic sign. Most patients survive less than 6 months.

43. **What is the most common location in the hand for a metastasis?**
    Distal phalanx.

44. **A 52-year-old man presents with a painful, slowly enlarging, fixed 5-cm mass on the volar aspect of the hand. Physical examination is consistent with involvement of both soft tissues and adjacent bone. Which diagnostic tests should be ordered?**
    - Multiplanar radiographs of the anatomic region
    - CT scan of the involved area to assess the osseous characteristics of the lesion and extent of bony involvement
    - Magnetic resonance imaging (MRI) may be useful to assess the extent of soft tissue involvement
    - A chest and abdominal CT scan
    - Bone scan
    - A serum chemistry profile, complete blood count, and platelet count

45. **Describe the characteristic features of a malignant bone tumor of the hand and wrist.**
    - Night or rest pain
    - Rapid growth
    - Destructive radiographic appearance

46. **Where does a giant cell tumor of bone most commonly metastasize?**
    The lungs.

47. **What is the most common primary malignant bone tumor in the upper extremity of children and adolescents?**
    Osteogenic sarcoma.

48. **Describe the typical clinical presentation of an osteogenic sarcoma.**
    A rapidly enlarging, painful mass.

49. **A 14-year-old boy presents with a painful forearm mass, low-grade fever, and elevated white blood cell count. The diagnosis of Ewing's sarcoma is not made until 4 months later. Confusion with which other condition probably contributed to the delay in diagnosis?**
    Infection (osteomyelitis).

50. **What are the primary indications for treating a metastatic bone tumor in the hand?**
    - Disabling pain
    - Preservation of function
    - Impending pathologic fracture

51. **Which supplemental treatment modality has been shown to enhance the cure rate following curettage of an aneurysmal bone cyst?**
    Cryosurgery, which relies on tissue necrosis by rapid freezing.

52. **Describe the Campanacci staging criteria for giant cell tumors of the distal radius.**
    - **Stage 1:** Well-circumscribed lesion with distinct tumor borders and an intact cortex
    - **Stage 2:** Indistinct tumor margins with expansion of the cortex but with the cortex remaining intact
    - **Stage 3:** Cortical perforation with expansion of the tumor into the surrounding soft tissues

53. **Name the different techniques for performing a bone tumor biopsy. Who should perform the biopsy?**
    The biopsy of a bone tumor can be performed in one of three ways: fine needle aspiration, core needle biopsy, or open biopsy. Although fine needle aspiration and core needle techniques are approximately 80% accurate, greater diagnostic accuracy can be obtained using the open biopsy technique. Fine needle aspiration and core needle techniques cause less soft tissue contamination and do not necessarily need to be performed in the operating room. A higher rate of incorrect diagnoses has been documented with the biopsies obtained at a location other than the institution providing the definitive tumor management. Therefore, it is recommended that a biopsy and definitive malignant bone tumor management be performed at an institution with orthopedic surgeons, pathologists, and oncologists who are familiar with musculoskeletal tumor oncology and are able to provide the necessary tumor-related support services. Irrespective of the technique chosen for biopsy, the most direct approach to the bone should be chosen, limiting dissection and resultant surrounding soft tissue contamination.

    Note: The views expressed in this article are those of the authors and do not reflect the official policy of the Department of Defense or the Department of the Air Force.

BIBLIOGRAPHY

1. Akelman E (ed): Tumors of the hand and forearm. Hand Clin 11(2), 1995.
2. Athanasian E: Bone soft tissue tumors. In Green D, Hotchkiss R, Pederson W (eds): Green's Operative Hand Surgery, 4th ed. New York, Churchill Livingstone, 1999, pp 2223–2253.
3. Athanasian E, Wold L, Amadio P: Giant cell tumor of the bones of the hand. J Hand Surg 22A:91–98, 1997.
4. Bedner M, Maywood IL, McCormack RR, et al: Osteoid osteoma of the upper extremity. J Hand Surg 18A:1019–1028, 1993.
5. Bickels J, Wittig JC, Kollender Y, et al: Enchondromas of the hand: Treatment with curettage and cemented internal fixation. J Hand Surg [Am] 27(5):870–875, 2002.
6. Blackley HR, Wunder JS, Davis AM, et al: Treatment of giant-cell tumors of long bones with curettage and bone-grafting. J Bone Joint Surg Am 81(6):811–820, 1999.
7. Bullough P: Orthopaedic Pathology, 3rd ed. Barcelona, Mosby-Wolfe, 1997.
8. Gibbs CP, Weber K, Scarborough MT: Malignant bone tumors. J Bone Joint Surg Am 83A:1728–1745, 2001.
9. Lewis M (ed): Musculoskeletal Oncology: A Multidisciplinary Approach. Philadelphia, W.B. Saunders, 1992, pp 268–273.
10. Machak GN, Tkachev SI, Solovyev YN, et al: Neoadjuvant chemotherapy and local radiotherapy for high-grade osteosarcoma of the extremities. Mayo Clin Proc 78(2):147–155, 2003.
11. Murray P, Berger R, Inwards C: Primary neoplasms of the carpal bones. J Hand Surg 24A:91–98, 1997.
12. Patil S, de Silva MV, Crossan J, Reid R: Chondrosarcoma of small bones of the hand. J Hand Surg 28B:602–608, 2003.
13. Schreuder HW, Veth RP, Pruszczynski M, et al: Aneurysmal bone cysts treated by curettage, cryotherapy, and bone grafting. J Bone Joint Surg 79B:20–25, 1997.
14. Sheth DS, Healey JH, Sobel M, et al: Giant cell tumor of the distal radius. J Hand Surg 20A:432–440, 1995.
15. Vander Griend RA, Funderburk CH: Treatment of giant cell tumors of the distal part of the radius. J Bone Joint Surg 75A:899–908, 1993.
16. Widhe B, Widhe T: Initial symptoms and clinical features in osteosarcoma and Ewing's sarcoma. J Bone Joint Surg 82A:667–674, 2000.
17. Wold L, McLeod R, Sim F, Unni K: Atlas of the Orthopaedic Pathology. Philadelphia, W.B. Saunders, 1990.

CHAPTER 42
SOFT TISSUE TUMORS
Edward A. Athanasian, MD

1. **What are the most common soft tissue tumors or masses in the hand?**
   - Ganglion
   - Giant cell tumor of tendon sheath (localized nodular synovitis)
   - Lipoma
   - Schwannoma
   - Epidermal inclusion cyst
   - Foreign body lesion

   The most common soft tissue mass in the hand is the ganglion (50–70% of all soft tissue masses in the hand). The most common benign soft tissue neoplasm is the giant cell tumor of tendon sheath.

2. **What is a ganglion cyst?**
   A herniation of fluid from a joint or tendon sheath contained in a balloon-shaped pouch. The pouch has a "stalk" that communicates with the joint or tendon sheath. The stalk can function as either a two- or one-way valve, and ganglia may vary in size in a given patient.

3. **What is a pyogenic granuloma?**
   A rapidly growing, benign pedunculated red friable mass that bleeds easily and can arise following a traumatic injury (Fig. 42-1). Although a specific cause has not been identified, a pyogenic granuloma is considered to be a disorder of angiogenesis.

4. **What are the most common locations for a ganglion in the hand/wrist (Fig. 42-2)?**
   - Dorsal wrist (70%)
   - Volar wrist (20%)
   - Distal interphalangeal (DIP) joint (mucous cyst)
   - Flexor tendon sheath (retinacular cyst)
   - Proximal interphalangeal (PIP) joint
   - Carpometacarpal joint
   - Guyon's canal

5. **Where do the majority of volar wrist ganglions originate?**
   Radioscaphoid or scaphotrapezial joint.

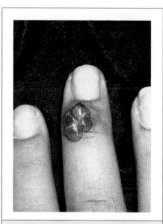

Figure 42-1. Pyogenic granuloma.

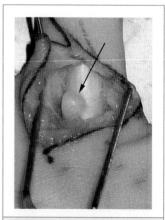

Figure 42-2. A retinacular cyst or ganglion of the flexor tendon sheath.

6. **Where do the majority of dorsal wrist ganglions originate?**
   Scapholunate articulation.

7. **What is the role of transillumination in the evaluation of a suspected wrist ganglion?**
   Transillumination is a simple, cost-effective technique that helps to determine whether a lesion contains fluid. Fluid-containing lesions transmit light more efficiently than soft tissue. Large ganglia, therefore, should transmit light better than surrounding soft tissue. Note that tumors with a high water content (e.g., soft tissue myxoma, myxoid chondrosarcoma) also may transilluminate.

8. **Describe the treatment options for a dorsal wrist ganglion.**
   An asymptomatic ganglion requires only periodic observation. A symptomatic lesion not immediately adjacent to the digital nerves or arteries can be aspirated. Persistent, painful, or enlarging ganglia can be treated with open or arthroscopic excision with a low risk of recurrence (5–10% overall) as long as the associated stalk and a small portion of the capsule are excised.

9. **What is an epidermal inclusion cyst? How is it treated?**
   A common slow-growing, painless, well-circumscribed cystic lesion in the hand that is caused by the implantation of epithelial cells following a penetrating trauma. Treatment consists of marginal excision.

10. **What is a giant cell tumor of tendon sheath (localized nodular synovitis)?**
    - A giant cell tumor of tendon sheath is a benign soft tissue tumor that is the second most common soft tissue tumor affecting the hand.
    - The tumor presents as a slow-growing, nontender, and firm mass.
    - It usually involves the palmar aspect of the finger.
    - Histologically, it is a single or multiple lobulated tumor containing multinucleated giant cells, histiocytes, and hemosiderin deposits (gross yellowish-brown appearance).

11. **How is a giant cell tumor of the tendon sheath treated?**
    - Excisional biopsy (marginal excision) is the usual treatment although entry into the tendon sheath or joint can occasionally be required.
    - The risk of local recurrence ranges between 9% and 50%.

12. **What are the risk factors for a recurrence following excision of a giant cell tumor of tendon sheath?**
    - Prior recurrence
    - Tumor located in the DIP joint
    - Concomitant radiographic evidence of osteoarthritis
    - Associated bone erosion
    - Lesion located in the distal part of the hand

13. **What is the characteristic appearance of a lipoma on magnetic resonance imaging (MRI)?**
    Well-demarcated homogenous mass that is bright on T1-weighted images and dark on T2-weighted images.

14. **What is a lipofibromatous hamartoma?**
    An unusual benign congenital lesion of unknown etiology that infiltrates nerves, leading to progressive neurologic dysfunction. The involved nerve becomes enlarged because of the proliferation of fibrous and adipose tissue within the epineurium.

15. **Which nerve is most commonly affected by a lipofibromatous hamartoma?**
    The median nerve in the wrist and forearm.

16. **Which condition is often associated with a lipofibromatous hamartoma?**
    Macrodactyly.

17. **What is the most common benign nerve tumor of the upper extremity?**
    Neurilemoma or schwannoma.

18. **What is a neurilemoma?**
    A slow-growing benign tumor of Schwann cells also known as a *schwannoma*. Patients typically present with a well-circumscribed mass within a nerve. The volar region of the hand and forearm are more commonly involved.

19. **What is the histologic appearance of a neurilemoma?**
    The tumor is composed of a cellular component (Antoni type A region) and a matrix component (Antoni type B region).

20. **What is the surgical treatment of a neurilemoma?**
    Surgical excision involves "shelling out" the tumor. Neurologic deficit following excision is rare but has been reported in 4% of cases. Recurrence is uncommon.

21. **What is a neurofibroma?**
    A benign, slow-growing tumor that arises within nerve fascicles and contains Schwann cells, fibroblasts, and perineural cells.

22. **Which condition is associated with multiple neurofibromas?**
    Neurofibromatosis type 1 (von Recklinghausen's disease).

23. **What is a hemangioma?**
    - It is the most common vascular tumor of the hand.
    - It is a congenital vascular tumor that has a characteristic growth pattern consisting of early proliferation followed by subsequent involution.
    - Most lesions appear within the first 4 weeks of life.
    - By 5 years of age, 50% of lesions will have involuted; by 7 years of age, 70% will have involuted.

24. **What is Kasabach-Merritt syndrome?**
    A rare complication of infantile hemagiomas that involves a coagulopathy due to entrapment of platelets within the hemangioma.

25. **What is a glomus tumor?**
    A neoplastic proliferation of the smooth muscle cells of the normal glomus, which is an arteriovenous shunt that contributes to thermal regulation.

26. **What is the most common location for a glomus tumor in the hand?**
    Subungual region.

27. **What is the classic presentation of the patient with a glomus tumor?**
    Painful, bluish-discolored lesion that is exquisitely tender and hypersensitive to cold exposure.

28. **Why are persistent or recurrent symptoms common in some patients following excision of a glomus tumor?**
    Multiple tumors occur in up to 25% of patients.

29. **What is the role of MRI in the evaluation of a soft tissue mass?**
    A careful history and physical examination permit an accurate diagnosis for the majority of soft tissue tumors of the hand. Large lesions or lesions that are not readily recognized during physical examination may benefit from evaluation with MRI. MRI is particularly useful if an incisional biopsy of a mass is being considered. The case should be discussed with the radiologist before performing the MRI to be certain that high-quality images of the mass and adjacent tissues can be obtained.

30. **What is a soft tissue sarcoma?**
    A primary malignant cancer of the soft tissues. Although soft tissue sarcomas are uncommon in the hand, the differential diagnosis of a soft tissue mass in this area should always include sarcoma. Lesions may metastasize and are potentially limb and life threatening.

KEY POINTS: SOFT TISSUE TUMORS

1. The most common soft tissue mass in the hand is the ganglion.
2. The majority of dorsal wrist ganglions originate from the scapholunate articulation.
3. The most common benign nerve tumor in the upper extremity is the neurilemoma (schwannoma).
4. A lipofibromatous hamartoma most commonly involves the median nerve.
5. A soft tissue sarcoma most commonly metastasizes to the lungs.
6. The most common soft tissue sarcoma of the forearm and hand is the epithelioid sarcoma.

31. **Name the four most common types of soft tissue sarcoma that occur in the hand.**
    - Synovial sarcoma
    - Epithelioid sarcoma
    - Malignant fibrous histiocytoma
    - Liposarcoma

32. **Which is the most common soft tissue sarcoma of the hand?**
    Epithelioid sarcoma.

33. **How is a soft tissue sarcoma diagnosed?**
    Incisional or excisional biopsy. There is no role for fine-needle aspiration except in centers specializing in this technique. Core needle biopsy may be considered for large lesions in the forearm although the accuracy of the biopsy may be decreased.

34. **What are the most important considerations before performing a biopsy of an unknown lesion in the hand?**
    - Decide between an excisional or incisional biopsy. Lesions >2 cm are probably best assessed with an incisional biopsy before complete removal.
    - Make a longitudinal incision that can be subsequently incorporated into a limb salvage incision or can be completely excised during amputation, if indicated.
    - Minimize potential soft tissue contamination by obtaining good hemostasis and avoiding soft tissue flaps during the approach. *Everything* touched by the surgeon is potentially contaminated with tumor cells.
    - Notify the pathologist in advance. Discuss the differential diagnosis and the handling of the specimen.
    - Culture every specimen for bacteria, tuberculosis, and fungi. Infections may present clinically as a tumor.

35. **How reliable is the diagnosis of a soft tissue sarcoma on frozen section analysis?**
    An accurate diagnosis can be based on the frozen section alone in approximately 80% of cases. Fat-containing lesions are difficult to diagnose accurately on the frozen section. Permanent analysis includes the use of immunohistochemical staining, which considerably improves diagnostic accuracy. An accurate diagnosis is essential in planning appropriate treatment.

36. **Where do soft tissue sarcomas most commonly metastasize?**
    - The lungs
    - The lymph nodes (fewer than 10% in this area)

37. **Which soft tissue sarcoma frequently metastasizes to regional lymph nodes?**
    Epithelioid sarcoma.

38. **Which technique can be used to sample selectively the most distal upper extremity lymph nodes in a patient at risk for the development of metastases?**
    Sentinel lymph node biopsy.

39. **What is the treatment of a soft tissue sarcoma?**
    Wide excision through a plane of normal tissue or amputation. High-grade lesions also may require radiation. Patients with large, high-grade, deep lesions are at high risk for metastases and may benefit from chemotherapy although this approach is controversial.

40. **Which condition should be considered in the patient who presents with a black or dark brown lesion beneath a fingernail?**
    Subungual melanoma.

41. **What is the most important prognostic factor in the patient with a subungual melanoma?**
    Depth of invasion.

42. **What is the most common malignant tumor of the skin of the hand?**
    Squamous cell carcinoma.

43. **Which factors influence the prognosis of a soft tissue sarcoma?**
    - Grade (high versus low grade)
    - Size (intracompartmental or extracompartmental)
    - Location (superficial or deep)
    - Presence or absence of metastasis

BIBLIOGRAPHY

1. Al-Qattan MM: Giant cell tumors of tendon sheath: Classification and recurrence rate. J Hand Surg 26(B):72–75, 2001.
2. Brien CL, Terek RM, Greer RJ, et al: Treatment of soft tissue sarcomas of the hand. J Bone Joint Surg 77A:564–571, 1995.
3. Enzinger FM, Weiss SW: Soft Tissue Tumors, 3rd ed. St. Louis, Mosby, 1995.
4. Green DP, Hotchkiss R, Pederson WC (eds): Green's Operative Hand Surgery, 4th ed. New York, Churchill Livingstone, 1999.
5. Jaoude J, Farah AR, Sargi Z, et al: Glomus tumors: Report on eleven cases and a review of the literature. Chir de la Main 19:243–252, 2000.

6. Kang HJ, Shin SJ, Kang ES: Schwannomas of the upper extremity. J Hand Surg 25(B):604–607, 2000.
7. Lewis JJ, Brennan MF: Soft tissue sarcomas. Curr Probl Surg 33(10):817–872, 1996.
8. McPhee M, McGrath BE, Zhang P, et al: Soft-tissue sarcoma of the hand. J Hand Surg 24A:1001–1007, 1999.
9. Shapiro PS, Seitz WH Jr: Non-neoplastic tumors of the hand and upper extremity. Hand Clin 11(2):133–136, 1995.
10. Steinberg RD, Gelberman RH, Mankin HJ, et al: Epithelioid sarcoma of the upper extremity. J Bone Joint Surg 74A:28–35, 1992.
11. Thornburg LE: Ganglions of the hand and wrist. J Am Acad Orthop Surg 7(4):231–238, 1999.

CONGENITAL HAND DIFFERENCES

Brian D. Adams, MD, and Curtis M. Steyers, MD

1. **What is the sequence of upper limb development?**
   The upper limb bud develops during week 4 of gestation. At 33 days, the hand is recognizable as a paddle, and digital separation occurs at 47–54 days through the breakdown of mesenchymal tissue. At 7 weeks, the complete bony complement of the hand is present. Hand development is complete at birth with the exception of myelination of the nerves, which usually is completed by age 2.

2. **What are the three critical regions that control growth and development of the limb bud?**
   The apical ectodermal ridge (AER), zone of polarizing activity (ZPA), and the wingless-type (Wnt) signaling center.
   - The Wnt pathway, which resides in the dorsal ectoderm, controls dorsoventral limb development (extensor versus flexor development).
   - The ZPA controls differentiation in the anteroposterior (AP) plane (radioulnar limb formation).
   - The AER causes the mesoderm to differentiate, proximal to distal limb development, and causes interdigital necrosis.

3. **Describe the classification of congenital limb anomalies based on embryonic failure during development.**
   There are seven categories:
   I: Failure of formation or arrest of development
   II: Failure of differentiation or separation
   III: Duplication
   IV: Overgrowth
   V: Undergrowth
   VI: Congenital constriction syndrome
   VII: Generalized skeletal abnormalities

4. **What is phocomelia?**
   A form of longitudinal deficiency characterized by intersegmental or intercalary deficiency. Thalidomide taken during the first trimester has been associated with phocomelia.

5. **Describe the classification of longitudinal radial deficiency.**
   The classification is based on the radiographic appearance. Ossification of the radius is delayed in radial deficiency, however, and the difference between total and partial absence cannot be established until approximately 3 years of age.
   - **Type I: Short distal radius.** The distal radial epiphysis is present but delayed in development. The radius has mild shortening, but the carpus is well supported with minor radial deviation of the hand. Thumb hypoplasia is prominent.
   - **Type II: Hypoplastic radius.** Proximal and distal epiphyses are present, but the radius is severely shortened, the carpus is poorly supported, and the ulna is bowed. Moderate radial deviation of the hand is noted.

- **Type III: Partial absence of the radius.** Absence may occur in the proximal, middle, or distal third, but typically the proximal radius is present. The elbow is stable, the ulna is bowed, and the carpus is unsupported. Severe radial deviation of the hand is present.
- **Type IV: Complete absence of the radius.** The ulna is thickened, shortened, and bowed. The hand is unsupported and severely deviated radially.

6. Which type of longitudinal radial deficiency is most common?
   Type IV (Fig. 43-1).

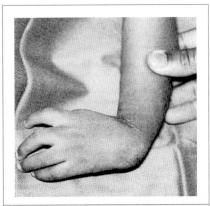

Figure 43-1. Severe radial deviation associated with complete absence of the radius. (From Beredjiklian P, Bozentka DJ: Review of Hand Surgery. Philadelphia, Saunders, 2003, p 227.)

7. What is the preferred surgical procedure for types II, III, and IV radial deficiencies?
   Centralization or realignment of the carpus onto the distal ulna.

8. Which nerve is encountered in the dorsoradial subcutaneous tissue during a centralization procedure?
   The median nerve. The radial nerve usually terminates at the elbow.

9. Which organ systems should be assessed in every child with a radial deficiency?
   - **Cardiac:** Auscultation and echocardiography
   - **Renal:** Ultrasound
   - **Hematologic:** Platelet status via blood count and peripheral blood smear

10. Which four syndromes are commonly associated with absence of the first ray or radial dysplasia?
    - **Fanconi's anemia:** This is an aplastic anemia that is not present at birth but typically develops in early childhood. The condition is fatal unless a successful bone marrow transplantation occurs.
    - **Thrombocytopenia with absent radius (TAR syndrome):** The thrombocytopenia is present at birth but usually improves after 1 year of age.
    - **Holt-Oram syndrome:** This autosomal dominant disorder with variable expression is associated with heart (most commonly cardiac septal defects) and hand anomalies.
    - **VACTERL syndrome:** This syndrome is characterized by **v**ertebral anomalies, **a**nal atresia, **c**ardiac abnormalities, **t**racheoesophageal fistula, **e**sophageal atresia, **r**enal disorders, and/or lower **l**imb anomalies.

11. Compare and contrast the skeletal and hematologic disorders associated with Fanconi's anemia and TAR syndrome.
    Patients with **Fanconi's anemia** most commonly develop aplastic anemia (90%). Less commonly, leukemia (10%) and solid tumors (10%) may be observed. The hematologic disorder usually presents between the ages of 3–12 years (median age of 7 years). The upper extremity musculoskeletal anomalies usually involve hypoplasia or aplasia along the radial border of the limb and are more severe distally than proximally.

Patients with **TAR** typically develop thrombocytopenia within the first 4 months of life. Various upper extremity anomalies have been reported, including disorders of the thumb. However, the thumb is not absent.

12. **Which diagnostic test allows the detection of Fanconi's anemia before the onset of bone marrow failure?**
    A chromosomal challenge test. The child's lymphocytes are subjected to diepoxybutane or mitomycin C, which causes the chromosomes within Fanconi's anemia cells to break and rearrange.

13. **Summarize the essential characteristics of ulnar deficiency.**
    - Failure of parts formation
    - Sporadic and nonhereditary (for most cases)
    - Associated with elbow deficiencies and musculoskeletal abnormalities, such as club foot or syndactyly
    - Not associated with systemic abnormalities

14. **How is ulnar deficiency classified?**
    There are two classification systems for ulnar deficiencies.
    The first system is based on the extent of forearm and elbow involvement:
    - **Type I**: Hypoplastic ulna, minimal shortening, proximal and distal epiphyses present
    - **Type II**: Partial aplasia with absence of the middle or distal third of the ulna
    - **Type III**: Complete aplasia with an unstable elbow, severe deficiencies of the hand and carpus
    - **Type IV**: Synostosis between the radius and humerus

    The second system is based on the anatomy of the first web space:
    - **Type A**: Normal first web space and thumb
    - **Type B**: Mild first web deficiency, mild thumb hypoplasia with intact opposition and extrinsic tendon function
    - **Type C**: Moderate to severe first web deficiency and similar thumb hypoplasia, loss of opposition, dysfunction of extrinsic tendons
    - **Type D**: Absence of the thumb

15. **Compare and contrast the clinical presentations of a "typical" and "atypical" cleft hand (symbrachydactyly).**
    Typical cleft hand:
    - Autosomal dominant inheritance
    - Several limbs often involved
    - Syndactyly (common)
    - V-shaped cleft
    - With increase in severity of deformity, increase in narrowing of the first web space and loss or hypoplasia of the radial rays
    - Hypertrophic skeleton adjacent to the cleft (Fig. 43-2)

    Atypical cleft hand (symbrachydactyly):
    - Sporadic occurrence
    - Usually unilateral
    - Syndactyly (rare)

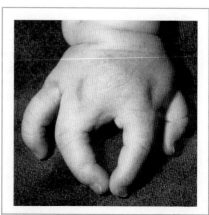

Figure 43-2. Typical cleft hand deficiency with large V-shaped defect.

- U-shaped cleft
- With increase in severity of deformity, increase in severity of hypoplasia or loss of ulnar rays
- Hypoplastic skeleton adjacent to the cleft

16. Define the following terms of hand dysmorphology: *arachnodactyly, brachydactyly, camptodactyly, clinodactyly, hypoplastic, macrodactyly, polydactyly, syndactyly,* and *symphalangism.*
    - **Arachnodactyly:** Unusually long, slender digits
    - **Brachydactyly:** Unusually short digits
    - **Camptodactyly:** Fixed flexion contracture of digit, usually involving the proximal interphalangeal (PIP) joint
    - **Clinodactyly:** Curvature of digit in radial/ulnar plane
    - **Hypoplastic:** Unusually small or underdeveloped
    - **Macrodactyly:** Unusual enlargement of digit
    - **Polydactyly:** Extra digits
    - **Syndactyly:** Two or more partially or completely fused digits, which may involve webbing of skin (simple) or bony fusion (complex)
    - **Symphalangism:** Continuity of middle and proximal phalanx with absence of PIP joint

17. What is the most common congenital hand anomaly in the United States?
    Syndactyly or webbing of digits (Fig. 43-3), which most commonly involves the long and ring fingers and is more common in males. Approximately 20% of cases have an autosomal dominant pattern of inheritance with incomplete penetrance and variable expressivity.

18. Describe how syndactyly is classified.
    - **Complete**: Involves the entire length of the involved digits
    - **Incomplete**: Involves skin bridge not extending the full length of the involved digits
    - **Simple**: Involves only soft tissues
    - **Complex**: Involves soft tissue and bone

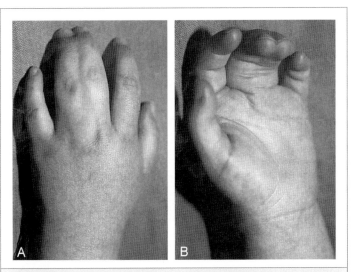

Figure 43-3. Complete simple syndactyly involving the long and ring fingers. *A,* Dorsal view. *B,* Palmar view.

19. **Why is surgical reconstruction staged in the patient with a syndactyly involving three or four adjacent digits?**
   To avoid the risk of ischemia of the finger or skin flaps.

20. **Why is skin grafting needed during the surgical reconstruction of a syndactyly?**
   The circumference of two digits separated is 22% greater than when they are adjoined. Flaps are not sufficient to provide the needed soft tissue coverage.

21. **What is acrosyndactyly?**
   Fusion between the more distal portions of the digits that occurs sporadically and is not hereditary. However, a communication is present between the dorsal and palmar aspects of the conjoined digits proximally. When seen in association with constriction ring syndrome, it is bilateral in as many as 50% of cases but with asymmetric involvement. The condition also may be associated with craniofacial syndactyly syndromes, such as Apert's syndrome, and is symmetrical in such patients. Other terms are terminal fenestrated, exogenous, or amniogeneous syndactyly.

22. **What is Poland's syndrome?**
   Ipsilateral absence of the *sternocostal* portion of the pectoralis major muscle in association with syndactyly (in 10% of patients). The muscle deficiency and breast underdevelopment can also progress to a hollow chest wall deformity with complete absence of the breast. The incidence is 1:20,000 births (Fig. 43-4).

23. **What are the characteristics of Apert's acrocephalosyndactyly?**
   This syndrome includes craniosynostosis and severe complex syndactyly involving the hands and feet. Characteristic deformities in the hand include a shortened, radially deviated thumb with a delta phalanx; complex syndactyly between the middle three rays; simple syndactyly between the small and ring fingers; symphalangism of the central three rays; reduced first web space; and synostosis between the ring and little metacarpals (Fig. 43-5).

24. **Describe the clinical presentation of proximal radioulnar synostosis.**
   - Patients usually present with a fixed forearm pronation deformity.
   - It is usually the result of a sporadic mutation when it is isolated and is an autosomal dominant pattern of inheritance when a positive family history is present.
   - Proximal radioulnar synostosis is the most common upper limb deformity in children with fetal alcohol syndrome. It is frequently bilateral and more common in males.

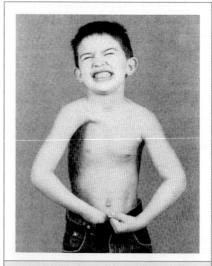

Figure 43-4. Ipsilateral absence of the sternocostal portion of the pectoralis major muscle, which is characteristic of Poland's syndrome. (From Beredjiklian P, Bozentka DJ: Review of Hand Surgery. Philadelphia, Saunders, 2003, p 230.)

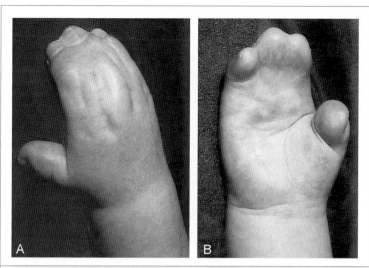

Figure 43-5. Apert's acrocephalosyndactyly. *A,* Dorsal view. *B,* Palmar view.

- The condition can also be seen in patients with trisomy 13 or 21.
- When a severe bilateral pronation or supination deformity is present, surgery (rotational osteotomy) is warranted to provide a more functional position.

25. **What is the most common complication associated with the surgical correction of a severe proximal radioulnar synostosis?**
    Iatrogenic injury of the posterior interosseous nerve (PIN).

26. **Which clinical criteria help to differentiate a *congenital* from an *acquired* radial head dislocation?**
    Sometimes it is difficult to differentiate congenital from acquired forms of radial head dislocation based on radiographic criteria, especially when the patient presents late. Clinical criteria consistent with a congenital dislocation include:
    - Bilateral involvement (50% of cases)
    - Dislocation seen at birth
    - Irreducibility by closed means
    - No history of trauma
    - Positive family history
    - Presence of other congenital anomalies
    - Dome-shaped radial head
    - Hypoplastic capitellum

27. **What is the recommended treatment for a congenital radial head dislocation that becomes symptomatic at 9 years of age?**
    Observation followed by excision of the radial head after skeletal maturity.

28. **What is the classic clinical finding in symphalangism?**
    Loss of both active *and* passive proximal interphalangeal joint motion with an absence of skin creases across the joint.

29. Which syndrome is associated with multiple, nonprogressive joint contractures that are present at birth?
Arthrogryposis multiplex congenita.

30. What is the underlying cause of arthrogryposis?
Lack of motion during fetal life, which is due to various processes such as muscle abnormalities, neurologic anomalies, vascular insufficiency, hydranencephaly, and holoprosencephaly. The neuropathic form is most common and is the result of anterior horn cell degeneration in utero.

KEY POINTS: CONGENITAL HAND DEFORMITIES

1. The most common congenital hand anomaly in the United States is syndactyly.
2. The diagnosis of Fanconi's anemia can be made before the signs of bone marrow failure by a chromosomal challenge test.
3. The most common congenital carpal coalition involves the lunotriquetral articulation.
4. A congenital proximal radioulnar synostosis and a capitohamate coalition have been noted in children with fetal alcohol syndrome.
5. The volar-ulnar portion of the distal radius physis is involved in Madelung's deformity.
6. The four syndromes associated with absence of the first ray or radial dysplasia are Fanconi's anemia, TAR, Holt-Oram, and VACTERL.

31. Describe the clinical features of the upper extremities in the child with the classic form of arthrogryposis (amyoplasia).
The arms are positioned in internal rotation and adduction with the elbows extended, the forearms are pronated, the wrists are flexed, and the hand is in ulnar deviation. The thumb is flexed and clenched in the palm. The fingers are stiff and flexed.

32. Which syndrome was originally reported as a combination of Marfan's syndrome and arthrogryposis?
Contractual arachnodactyly syndrome or Beals' syndrome, which is differentiated from Marfan's syndrome by the presence of contractures and from arthrogryposis by the presence of arachnodactyly.

33. Describe the Blauth classification of thumb hypoplasia. How is each type treated?
Blauth originally grouped hypoplastic thumbs into five types (I–V). Manske later subdivided type III into two subtypes.
- **Type I:** Minor hypoplasia with excellent function
- **Type II:** Smaller thumb with thenar hypoplasia, adduction contracture, and unstable metacarpophalangeal (MCP) joint
- **Type IIIA:** Severe hypoplasia, absent or severely hypoplastic intrinsic muscles, absent and or aberrant extrinsic muscles and tendons, carpometacarpal (CMC) joint present
- **Type IIIB:** Same as type IIIA, but with the base of the metacarpal joint and the entire CMC joint absent
- **Type IV:** Rudimentary thumb with skin attachment only ("pouce floutant" or floating thumb)
- **Type V:** Complete absence of the thumb

34. Which clinical feature determines if a hypoplastic thumb should be reconstructed?
The presence of a carpometacarpal joint.

35. **What is the recommended surgical procedure for types IIIB, IV, and V thumb hypoplasia?**
    Pollicization or neurovascular transposition of the index finger to the thumb position with reconstruction of the intrinsic muscles of the thumb. Although controversial, pollicization is usually performed between 6–12 months of age.

36. **In index finger pollicization for treatment of the congenitally absent thumb, which muscle functions as the new abductor pollicis brevis (APB)?**
    The first dorsal interosseous muscle is reattached to the radial lateral band of the pollicized index finger to assume the function of an APB.

37. **What function do the muscles provide in an index finger pollicization?**
    - The first palmar interosseous muscle becomes the adductor pollicis.
    - The flexor digitorum profundus (FDP) becomes the flexor pollicis longus (FPL).
    - The flexor digitorum superficialis (FDS) becomes the flexor pollicis brevis (FPB).
    - The extensor indicis proprius (EIP) becomes the extensor pollicis longus (EPL).
    - The extensor digitorum communis (EDC) (assuming both are available) becomes the abductor pollicis longus (APL).

38. **What is camptodactyly?**
    A nontraumatic, painless PIP joint flexion contracture that is usually progressive and most commonly involves the small finger (Fig. 43-6). Although the exact etiology is unknown, camptodactyly is believed to be the result of imbalance between the FDS and intrinsic (interossei and lumbricals) mechanisms.

    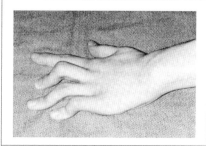

    Figure 43-6. Camptodactyly of the long, ring, and small fingers. (From Beredjiklian P, Bozentka DJ: Review of Hand Surgery. Philadelphia, Saunders, 2003, p 236.)

39. **How is camptodactyly classified?**
    - **Type I:** Unilateral or bilateral small finger PIP joint contracture in an otherwise healthy infant (most common type)
    - **Type II:** Same characteristics as in type I but seen at a later age, usually in adolescence
    - **Type III:** Associated with other congenital anomalies and more severe, usually involving multiple digits.

40. **What is windblown hand?**
    A congenital malformation involving flexion and ulnar deviation of the fingers as well as a fixed flexion deformity of the thumb. All of the digits angulate in an ulnar direction as if they have been blown by the wind. This deformity is associated with craniocarpotarsal dystrophy, digitotalar dysmorphism, and distal arthrogryposis.

41. **Compare and contrast clinodactyly and Kirner's deformity.**
    Clinodactyly is an angular deformity of the digit in the coronal plane (radioulnar) caused by an abnormally shaped tubular bone in the digit. The abnormal shape may range from a trapezoid to a triangle (delta bone). The little finger is most commonly involved, and the middle phalanx usually is affected.

Kirner's deformity is a biplanar angulation of the distal phalanx of the small finger caused by a growth disturbance of the physis of the distal phalanx. It typically presents as a painless swelling at the distal interphalangeal joint of the affected finger, followed by progressive angulation of the distal phalanx in a palmar and radial direction. Radiographs reveal widening of the physis, thinning of the metaphysis, and angulation.

42. **What is a delta phalanx?**
    A triangular bone with an abnormally aligned or oriented epiphysis. The epiphysis covers both ends of the bone and one side, thus forming an epiphyseal "bracket" around the bone. This arrangement prevents normal longitudinal growth and causes an increasing coronal plane deformity during growth. Because this type of pathoanatomy may affect other bones (e.g., metacarpals and metatarsals), it is more properly called a delta bone.

43. **Describe the Wassel classification of thumb duplication.**
    There are six types of thumb duplication (Fig. 43-7). The seventh category of triphalangeal thumb (type VII) is now considered a separate entity.
    - **Type I:** Bifid distal phalanx
    - **Type II:** Duplicated distal phalanx
    - **Type III:** Bifid proximal phalanx
    - **Type IV:** Duplicated proximal phalanx
    - **Type V:** Bifid metacarpal
    - **Type VI:** Duplicated metacarpal

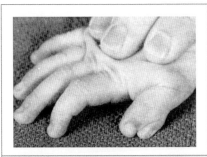

Figure 43-7. Preaxial or thumb duplication with distinct distal parts. (From Beredjiklian P, Bozentka DJ: Review of Hand Surgery. Philadelphia, Saunders, 2003, p 238.)

44. **What is the most common type of thumb duplication?**
    Type IV.

45. **Name the most common form of polydactyly.**
    Ulnar-sided or postaxial polydactyly (Fig. 43-8). Although 10 times more likely in African-Americans than Caucasians or Asians, most radial or preaxial (thumb) duplications are seen in Caucasian and Asian populations.

46. **Describe the radidographic appearance of ulnar dimelia.**
    Ulnar dimelia is also known as *mirror hand*. The radiographs demonstrate two ulnae, no radius, seven or eight fingers, and no thumb. The wrist and elbow appear thickened, and the arm is short.

47. **What is the most common presentation of overgrowth in the upper extremity?**
    Macrodactyly, or congenital enlargement of the digit. An adjacent digit is often affected in 70% of cases.

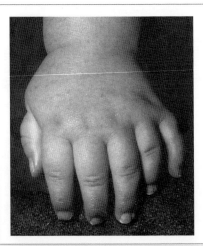

Figure 43-8. Ulnar polydactyly.

Treatment options include amputation, growth plate ablation, realignment osteotomies, and debulking procedures.

48. **Describe the presentation of a hand with fetal hydantoin syndrome (fetal Dilantin syndrome).**
    Fetal hydantoin syndrome is seen in 10% of exposed infants and presents as hypoplasia of the distal digits, nail hypoplasia, and digitalized thumbs.

49. **What is congenital constriction band syndrome?**
    A syndrome caused by bands of amnion that separate from the placenta and wrap around the limbs and trunk of the fetus. Occasionally, they may be swallowed. The strands of amniotic tissue create band ischemia in embryologically defined limbs, resulting in complete transverse amputation as well as digital loss (Fig. 43-9). The most common hand abnormalities in this condition include amputations, constriction rings, and atypical (usually apical) syndactyly. The incidence is approximately 1:15,000 live births.

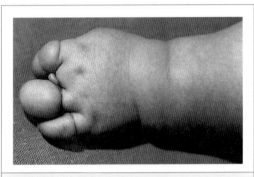

Figure 43-9. Constriction band syndrome.

Additional abnormalities include clubfoot, craniofacial defects including cleft lip and palate, and visceral abnormalities. Other names include Streeter's dysplasia, annular groove syndrome, amniotic banding, and annular constriction rings.

50. **In Madelung's deformity, deficient growth occurs primarily in which region of the distal radial physis?**
    - The deficient growth occurs in the volar-ulnar portion of the distal radial physis.
    - This condition results in volar subluxation of the wrist, radiocarpal deformity, and dorsal subluxation of the distal ulna.

51. **Describe the characteristic clinical features of a patient with Marfan's syndrome.**
    - Arachnodactyly (long, slender digits)
    - Hyperextensibility of the fingers
    - Lens subluxation
    - Aortic dilatation
    - Mitral valve regurgitation or prolapse
    - Severe pes planus (flatfoot)
    - Protrusio acetabuli (arthrokatadysis)

52. **A simian crease is most commonly found in patients with which syndrome?**
    Approximately 50% of patients with Down syndrome (trisomy 21) present with a simian crease.

53. What is the most common carpal coalition (synostosis)?

Lunotriquetral. The "fusion" may be complete or incomplete. The second most frequent coalition is between the capitate and hamate bones, which can be seen in fetal alcohol syndrome. Radial-sided coalitions and coalitions between the proximal and distal carpal rows are the least common.

54. What is brachymetacarpia?

Brachymetacarpia, or short metacarpal, is often associated with a syndrome or systemic disorder, including pseudohypoparathyroidism. The condition may involve any single metacarpal or a combination of metacarpals. The short metacarpal creates a gap or inset when a fist is made (>1-cm shortening may reduce grip). Lengthening is indicated for cosmesis or loss of grip strength.

55. Metacarpal synostosis most commonly involves which metacarpals?

The ring and small metacarpals. The extent of synostosis varies from just basilar involvement to greater than half the length of the metacarpals.

56. What is the most common condition associated with congenital pseudarthrosis of the forearm?

Approximately 50% of cases have clinical findings of neurofibromatosis. The pseudarthrosis or signs of impending pseudarthrosis are present at birth, whereas other signs of neurofibromatosis may not be seen until later in life.

BIBLIOGRAPHY

1. Alter BP: Arm anomalies and bone marrow failure may go hand in hand. J Hand Surg 17A:566–571, 1992.
2. Bayne LG, Klug MS: Long term review of the surgical treatment of radial deficiencies. J Hand Surg 12A:169–170, 1987.
3. Bell DF: Congenital forearm pseudarthrosis: Report of six cases and review of literature. J Pediatr Orthop 9:438–443, 1989.
4. Benson LS, Waters PM, Kamil NI, et al: Camptodactyly: Classification and results of nonoperative treatment. J Pediat Orthop 14:814–819, 1994.
5. Buck-Gramcko D, Wood VE: The treatment of metacarpal synostosis. J Hand Surg 18A:565–581, 1993.
6. Campbell CC, Waters PM, Emans JB: Excision of the radial head for congenital dislocation. J Bone Joint Surg 74A:726–733, 1992.
7. Cohen MS: Thumb duplication. Hand Clin 14:17–27, 1998.
8. Damore E, Kozin SH, Thoder JJ, Porter S: The recurrence of deformity after surgical centralization for radial clubhand. J Hand Surg 25A:745–751, 2000.
9. Ezaki M: Treatment of the upper limb in the child with arthrogryposis. Hand Clin 16:703–711, 2000.
10. Graham TJ, Louis DS: A comprehensive approach to surgical management of the type IIIA hypoplastic thumb. J Hand Surg 23A:3–13, 1998.
11. Gropper PT: Ulnar dimelia. J Hand Surg 8:487–491, 1983.
12. Gupta A, Kay SPJ, Scheker LR: The Growing Hand. St. Louis, Mosby, 2000.
13. Jones KL: Smith's Recognizable Patterns of Human Malformation, 5th ed. Philadelphia, W.B. Saunders, 1997.
14. Kozin SH: Syndactyly. J Am Soc Surg Hand 1:1–13, 2001.
15. Light TR, Ogden JA: Congenital constriction band syndrome pathophysiology and treatment. Yale J Biol Med 66:143–155, 1993.
16. Lourie GM, Lins RE: Radial longitudinal deficiency. A review and update. Hand Clin 14:85–99, 1998.
17. Manske PR, McCarroll HR Jr, James MA: Type IIIA hypoplastic thumb. J Hand Surg 20A:246–253, 1995.
18. McCarroll HR: Congenital anomalies: A 25-year overview. J Hand Surg 25A:1007–1037, 2000.

19. Miura T, Suzuki M: Clinical differences between typical and atypical cleft hand. J Hand Surg (Br) 9:311–315, 1984.
20. Nielsen JB: Madelung's deformity: A follow-up study of 26 cases and a review of literature. Acta Orthop Scand 48:379–384, 1977.
21. Riddle RD, Tabin CJ: How limbs develop. Sci Am 280:74–79, 1999.
22. Sachar K, Mih AD: Congenital radial head dislocations. Hand Clin 14:39–47, 1998.
23. Seidman GD, Wenner SM: Surgical treatment of the duplicated thumb. J Pediatr Orthop 13:660–662, 1993.
24. Smith PJ, Grobbelar AO: Camptodactyly: A unifying theory and approach to surgical treatment. J Hand Surg 23A:14–19, 1998.
25. Van Heest AE: Congenital disorders of the hand and upper extremity. Ped Clin North Am 43:1113–1133, 1996.
26. Wiedrich TA: Congenital constriction band syndrome. Hand Clin 14:29–38, 1998.

CHAPTER 44

UPPER EXTREMITY OCCUPATIONAL INJURIES AND ILLNESS

V. Jane Derebery, MD

1. **Which of the following has the strongest relationship with an increase in claims being filed for work-related upper extremity disorders: highly repetitive work, highly forceful work, work with vibratory hand tools, or job dissatisfaction?**
   Job dissatisfaction. To date, there have been numerous attempts to demonstrate a dose-response relationship between exposure to physical activity and the development of musculoskeletal disorders; however, studies using careful methodology and adjusting for confounding variables have failed to demonstrate a consistent dose-response relationship.

2. **Which factors *do* have an association with an increased number of claims being filed for work-related upper extremity disorders among workers?**
   - Monotonous work
   - Decreased control of work
   - Time pressure
   - Decreased social support

3. **Which of the following is *not* a benefit of ergonomic improvements for employees: increased comfort, decreased fatigue, improved productivity, or less risk in developing musculoskeletal diseases?**
   Ergonomic principles or changes in the workplace have had little impact on reducing the risk of developing musculoskeletal diseases.

4. **Which three types of work lifestyle are most commonly listed by persons diagnosed with carpal tunnel syndrome (CTS)?**
   - Homemaking
   - Unemployment
   - Retirement

   CTS is more common in older and overweight individuals, particularly those with sedentary lifestyles. Although controversial, most cases are not considered to be work related.

5. **Are there any occupations that have been demonstrated to have a higher risk ratio of developing CTS?**
   Meat boners and frozen food workers have a higher risk ratio for developing CTS than the general population. Highly repetitive, forceful work in a cold environment appears to place a person at higher risk for CTS, with the most important factor being prolonged work in the cold. When evaluating a person who works in a cold environment, it is important to determine if he or she has other risk factors for peripheral neuropathy, such as, but not limited to, diabetes or alcoholism.

6. **You have just diagnosed CTS in a 44-year-old woman. She says that her symptoms are worsened by driving a delivery truck all day. Should you restrict her from prolonged driving at work?**
   Driving can be associated with increased pain and numbness from CTS although it does not *cause* CTS. Since the courts have decreed that CTS is not considered disabling, an employer is

not obligated by the Americans with Disabilities Act to provide accommodations; hence, this patient needs to understand that if she is given a driving restriction, the company is not obligated to accommodate her.

7. True or false: Keyboarding is a workplace cause of CTS.
   False. While commonly reported in the lay literature as a cause of CTS, in actuality a causal relationship between keyboarding and CTS has not been demonstrated. In addition, there has not been an increased risk ratio noted among workers who use keyboards regularly.

8. Because of a concern about anthrax contamination in his department, a postal worker has been temporarily reassigned to a different job task that requires prolonged keyboarding and filing of letters. He develops thumb tendinitis. What is the most likely cause of the tendinitis: typing, filing letters, or taking Ciproxin as a prophylaxis for possible anthrax exposure?
   Taking Ciproxin. Highly repetitive and low-force work, such as typing and filing, has not been demonstrated to pose particular risk for developing hand or wrist tendinitis. Tendinitis (and even tendon rupture) is a known complication associated with using Ciproxin and cipro-derivative antibiotics.

9. Which work situation may be a potential risk factor for developing symptoms of CTS: work involving boning and frozen foods, work involving forceful use of hands in a cold environment, work causing upper extremity trauma (e.g., distal radius fracture or forearm hematoma), or all of these?
   All of these situations can increase the risk of developing CTS.

10. Do job candidates who are diagnosed with increased median nerve latencies at the wrist during the preplacement exam later develop symptomatic CTS?
    In an 11-year follow-up of 558 such individuals, it was found that most did not develop CTS. The nerve test abnormality tended to persist (in 82%), while symptoms tended to fluctuate greatly during the 11-year period, with most tending to improve with the passage of time.

KEY POINTS: UPPER EXTREMITY OCCUPATIONAL INJURIES AND ILLNESS

1. Work-related CTS is relatively uncommon. It is associated with forceful wrist work in a cold environment and is not caused by keyboarding.
2. CTS is not considered disabling and hence is not protected by the Americans with Disability Act.
3. The most common associated risk factors for CTS appear to be obesity, gender, and age.
4. Measures of psychological stress and dissatisfaction with work almost always overwhelm ergonomic considerations in evaluating a worker with nontraumatic hand or arm pain.
5. Tendinitis in the upper extremity is classified (medically and legally) as an illness, not an injury. It is likely to be due to a systemic medical condition, such as pregnancy, or linked to the use of a cipro-derivative antibiotic more than it is associated with typing or other low-force and repetitive hand or arm use.
6. Patients who develop arm or hand pain associated with prolonged use of a keyboard generally do so because they do not exercise regularly and are not well-conditioned physically.

11. **Which work-related activity has been implicated as a risk for lateral epicondylitis: highly repetitive and low-force use of arm; prolonged work using the arm in an awkward postural position; or prolonged and forceful work activities with elbow extended, the hand in pronation, and the wrist moving from full flexion to full extension?**
   Prolonged and forceful work activities with the elbow extended, the hand in pronation, and the wrist moving from full flexion to full extension. This type of work was found to cause epicondylitis according to findings from a review of epidemiologic studies of workplace factors by the U.S. National Institute of Occupational Safety and Health (NIOSH). Repetitive and low-force work or work with the arm held in an awkward postural position were *not* found to increase the risk of developing epicondylitis.

12. **True or false: When evaluating a patient with a presumed work-related upper extremity disorder (WRUED), a patient's self-report can be counted on to provide an accurate depiction of risk factors at work.**
   False. Patient self-reports and observer-based measures tend to overestimate exposures according to a number of studies. While a patient self-report regarding ergonomic risk factors is generally sufficient to use in most cases of WRUEDs, a formal ergonomic assessment that includes direct measurement through instrumentation should be obtained in cases for which it is important to have an accurate understanding of the workplace setting.

13. **What is hand-arm vibration syndrome (HAVS)?**
   HAVS is a vascular disorder of the hands caused by prolonged, continuous use of vibratory hand tools. HAVS is uncommon in the United States but well recognized in many parts of the world. Signs and symptoms can be partially or totally reversible depending on the severity of pathological changes. Diagnosis is made by the presence of Raynaud's phenomenon in a patient who has experienced chronic, prolonged vibration exposure and in the absence of other causes, such as Raynaud's disease, collagen vascular disease, or diabetes. There is no satisfactory gold standard diagnostic test. Temporary tingling or paresthesias during or immediately after use of a vibrating tool is not considered vibration syndrome.

14. **What are some of the effective ways to reduce workers' exposure to tool vibration?**
   - Wearing International Standards Association antivibration gloves
   - Using specially designed low-vibration tools
   - Performing rotating tasks to reduce exposure time with vibratory tools
   - Trying to warm the environment

   Note: Wearing antivibration gloves *plus* using low-vibration tools can reduce vibration exposure significantly.

15. **What is a work-related upper extremity disorder (WRUED)?**
   Work-related upper extremity disorder refers to any regional musculoskeletal hand/arm disorder that is believed to be caused by physical activities at work. Previously, WRUEDs were more commonly referred to as *cumulative trauma disorders*, but that term has fallen out of favor since it implies that trauma can be caused by routine, repetitive use. The large number of diagnoses that can be classified as a WRUED can occur with no provoking activity and can be caused by or associated with systemic disease. This can make determination of work-relatedness challenging. The problem is confounded when symptoms are labeled as a disease or injury without objective findings.

16. **When is it appropriate for a patient with a WRUED to wear a splint while working?**
   Joints should not be immobilized unless there is compelling justification. Doing so compromises circulation, rapidly leads to further deconditioning, can encourage illness behavior, and generally prolongs symptoms. A splint should not be prescribed unless the patient has objective evidence

of joint swelling, inflammation, or pathologic changes. Splints lead to compensatory changes in the use of the arm or hand, resulting in symptoms and even specific disorders elsewhere in the limb and body.

17. **A worker who assembles computers is diagnosed with bilateral CTS, right pronator syndrome, and a right cervical root syndrome, confirmed by electrodiagnostic testing. His doctor also provides a diagnosis of multiple crush syndrome. Is it likely that work caused this problem?**
No. Multiple crush syndrome refers to multiple sites of entrapment or compression of a peripheral nerve. The pathophysiological mechanism is not known although it is possible that a significant number of cases are due to hereditary disorders, such as hereditary neuropathy with liability to pressure palsies (HNPP) and hereditary neuralgic amyotrophy (HNA), both of which are autosomal dominant with variable expression. Other disorders associated with the multiple crush phenomenon include alcoholism, diabetes, and collagen disease.

18. **Should patients be restricted from typing if they develop hand or arm pain?**
Usually not. Hand or arm pain is a common complaint among keyboard users. Generally, the exam is normal, and the symptoms are nonspecific. An important factor appears to be poor physical conditioning caused by a sedentary lifestyle—the only regular exercise some people get is keyboarding! Treatment should include education and prescribing of a regular conditioning regimen of low-repetition, high-force exercises of the upper extremities (i.e., weight training) to condition the worker for typing. Periodic stretching at work is helpful, and attention to sitting posture is also important.

19. **Which common upper extremity medical conditions are often mistakenly assumed to be work related?**
Upper extremity disorders are common, regardless of the type of work a patient performs. Although certain types of work exacerbate symptoms after they have developed, there is little sound evidence indicating that physical activities are causative.
   Examples of diagnoses that are mistakenly believed to be caused by repetitive work activities include:
   - Symptomatic osteoarthritis of the hands
   - Heberden's nodes
   - Ganglions
   - Ulnocarpal abutment syndrome
   - Fibromyalgia pain syndrome (defined by a lack of objective findings and the presence of pain)
   - Factitious illnesses

20. **How common are psychogenic hand disorders?**
The workers' compensation system provides an excellent opportunity for somatic expression of distress by patients. Although the true incidence and prevalence rates of upper extremity disorders due to psychosocial factors is unknown, several hand disorders of psychogenic origin have been described.
   - **Secretan's syndrome** (wall-banger's disease) refers to self-induced, persistent edema of the hand dorsum after a minor compensable hand injury. (The patient attempts to prolong symptoms or the disability.)
   - **Sad, hostile, anxious, frustrating, tenacious (SHAFT) syndrome** is a variation of Munchausen's syndrome, in which the patient undergoes multiple, unnecessary surgical procedures of the upper extremity.
   - **Clenched fist syndrome** is a conversion reaction manifested as a persistent clenched fist with normal electrodiagnostic testing of the upper extremity.

21. **Name four circumstances when it might be helpful for a hand specialist to refer a patient to an occupational medicine specialist for consultation or treatment.**
    - When the hand specialist is experiencing difficulty in successfully returning a patient with an upper extremity disorder back to work
    - When there is uncertainty about whether an upper extremity disorder was substantially caused by work
    - When the hand specialist has questions or concerns about the physical demands of a patient's job and whether job modifications are indicated
    - When a patient with an upper extremity disorder is experiencing a delay in recovery or difficulty in returning to work because of the presence of concurrent psychosocial issues.

BIBLIOGRAPHY

1. Andersen J, Thomsen J, Overgaard E: Computer use carpal tunnel syndrome: A 1-year follow-up study. JAMA 289(22):2963–2969, 2003.
2. Bernard B (ed): Musculoskeletal Disorders and Workplace Factors: A Critical Review of Epidemiologic Evidence for Work-Related Musculoskeletal Disorders of the Neck, the Upper-Limb, and Low Back. Washington, DC, NIOSH, 1997.
3. Bongers P, Kremer A, ter Laak J: Are psychosocial factors risk factors for symptoms and signs of the shoulder, elbow, or hand/wrist?: A review of the epidemiological literature. Am J Indust, Med 41:315–342, 2002.
4. Cullip W, Martin C: Hand-arm vibration syndrome: Are we left out in the cold? OEM Report 17(12):85–88, 2003.
5. Demure B, Luippold R, Bigelow C, et al: Video display terminal workstation improvement program: Baseline associations between musculoskeletal discomfort and ergonomic features of workstations. J Occup Environ Med 42:783–791, 2000.
6. Derebery J: Content management: Treatment. In Derebery J, Anderson J (eds): Low Back Pain: An Evidence-Based, Biopsychosocial Model for Clinical Management. Beverly Farm, MA, OEM Press, 2002, pp 65–79.
7. Derebery J: Etiologies prevalence of occupational upper extremity injuries. In Kasdan M (ed): Occupational Hand and Upper Extremity Injuries and Diseases, 2nd ed. Philadelphia, Hanley & Belfus, 1998, pp 49–58.
8. Descatha A, Leclerc A, Chastang F, et al: Medial epicondylitis in occupational settings: Prevalence, incidence, and associated risk factors. J Occup Environ Med 45:993–1001, 2003.
9. Falkiner S, Myers S: When exactly can carpal tunnel syndrome be considered work-related? Austr NZ J Surg 72:204–209, 2002.
10. Jetzer T, Haydon P, Reynolds D: Effective intervention with ergonomics, antivibration gloves, and medical surveillance to minimize hand-vibration hazards in the workplace. J Occup Environ Med 45:1312–1317, 2003.
11. Kasdan ML, Wolens D: Carpal tunnel sydrome not always work related. Kentucky Med 92(8):295–297, 1994.
12. Louis D, Tsai E: Factitious disorders. In Kasdan M (ed): Occupational Hand Upper Extremity Injuries and Diseases, 2nd ed. Philadelphia, Hanley & Belfus, 1998, pp 74–81.
13. MacFarlane G, Hunt I, Silman A: Role of mechanical and psychosocial factors in the onset of forearm pain: Prospective population-based study. BMJ 321:676–679, 2000.
14. Malmivaara A, Viikari-Juntura E, Huuskouen M, et al: Rheumatoid factor and HLA antigen in wrist tenosynovitis and humeral epicondylitis. Scand J Rheum 24:154–156, 1995.
15. Miller RS, Iverson DC, Fried RA, et al: Carpal tunnel syndrome in primary care: A report from ASPN. J Fam Pract 38:337–344, 1994.
16. Nathan P, Meadows KD, Istvan JA, et al: Predictors of carpal tunnel syndrome: Prospective study. J Hand Surg 27A:644–651, 2002.
17. Shaw W, Feuerstein M, Lincoln A, et al: Ergonomic and psychosocial factors affect daily function in workers' compensation claimants with persistent upper extremity disorders. J Occup Environ Med 44:606–615, 2002.

HAND THERAPY

Nan Boyer, OTR, CHT, Cathie Basil, OTR, CHT, Linda Miner, OTR, CHT, Denise Justice, OTR, Trisha Mozdzierz, OTR, Carole Dodge, OTR, Andrea Eisman, MS, OTR, CHT, and Jeanne Riggs, OTR, CHT

1. **When is it appropriate to initiate range-of-motion exercises following split-thickness skin grafting (STSG) and homografting?**
   - Active range-of-motion (AROM) exercises may be initiated as early as 3–5 days after surgery, but the decision is based on the integrity and vascularity of the graft.
   - AROM can begin 24–48 hours after homografting.
   - Exercises should be done with all dressings removed so that the grafted area can be closely monitored. Appropriate splinting should continue at night and during hours of rest.

2. **Which splinting regimen is indicated for a circumferential, deep, partial-thickness hand burn?**
   - An "intrinsic plus" splint with the thumb in palmar abduction and a "paddle" splint with all fingers extended and the thumb radially abducted are indicated.
   - The splints should be alternated at night.
   However, given the accelerated healing process of a burn injury, the rapid formation of hypertrophic scar, and digital tightness, the splinting schedule should be evaluated weekly and modified as necessary to prevent contractures.

3. **When should scar treatment modalities be initiated after grafting of a burn injury?**
   - They can be initiated once the graft has fully healed.
   - Appropriate treatment modalities include pressure garments, silicone gel sheets, scar massage, and inserts for the web spaces or palm. Pressure garments are indicated for all skin grafts. The typical wearing period is 23 h/day for 12–18 months or until the scar matures.

4. **Which therapeutic modalities are used to help preserve the palmar arches and web spaces of the burned hand?**
   - Soft or elastomer inserts can be fabricated and worn beneath a pressure garment.
   - Elastomer or otoform inserts can also be made for palmar burns and used at nighttime to avoid impeding hand function.

5. **Describe the modalities used to minimize edema formation following free flap coverage of an upper extremity wound.**
   - Elevation to heart level is recommended to decrease edema; extensive elevation may impair arterial flow.
   - AROM for adjacent uninvolved joints also aids in decreasing edema.
   - Modalities such as Isotoner gloves, Tubigrip, Coban, and Ace wraps should only be initiated when approved by the surgeon. It is important to begin with light compression to avoid restricting blood flow to the flap.

6. **A 61-year-old man with a history of diabetes mellitus is referred 2 days after a limited palmar and digital fasciectomy for Dupuytren's contracture. When should extension splinting be initiated?**
   - It should be initiated immediately in conjunction with intensive ROM exercises performed 6–8 times per day.
   - A volar or dorsal forearm-based splint can be used with the straps crossing the metacarpophalangeal (MCP), proximal interphalangeal (PIP), and distal interphalangeal (DIP) joints.
   - The extension splint and ROM program should not disrupt the sutures or wound or elicit an inflammatory response.

7. **A patient who was seen postoperatively by another hand therapist has been referred for ongoing care. About 2 weeks ago, the patient underwent a palmar and digital fasciectomy using the open-palm technique. The physician notes mild edema with limited active digital motion. What is the recommended treatment?**
   - Ultrasound is helpful to prevent scar formation.
   - Whirlpool treatments are used to promote wound healing.
   - Contrast baths, elevation, gentle yet intensive ROM, and compression garments (e.g., Coban, digital sleeves, and Isotoner glove) should be used to decrease edema and improve motion.

8. **Which therapeutic modality is used to reduce morning stiffness in the patient with rheumatoid arthritis?**
   Application of heated paraffin wax for 15–20 minutes before ROM exercises.

9. **A 63-year-old woman with long-standing rheumatoid arthritis undergoes MCP implant arthroplasty. How much ROM can be expected?**
   The patient should be expected to achieve approximately 70 degrees of MP joint flexion and extension to neutral.

10. **A 45-year-old kindergarten teacher reports pain in the thumb carpometacarpal (CMC) joint. What type of splint should you recommend?**
    - Chronic basilar thumb pain with key and chuck pinch activities is caused by CMC instability or arthritis.
    - A custom-fabricated orthoplast hand-based thumb spica ("cone") orthosis is recommended.

11. **A 57-year-old carpenter complains that he constantly drops his tools because of an inability to make a complete fist. Which adaptation to his tools can be made to help this problem?**
    Enlarged handles require less ROM and force to manipulate the tool and perform the required task.

12. **A patient undergoes open reduction and internal fixation (ORIF) of a distal radius fracture. When should therapy begin?**
    - Some form of therapy should begin immediately following surgery.
    - The patient is instructed in limb elevation; edema reduction; scar management; and gentle ROM of the digits, thumb, wrist, and forearm (if indicated by the surgeon) as well as of all uninvolved joints.
    - The patient should be encouraged to use the hand for light functional activities as tolerated.

KEY POINTS: HAND THERAPY

1. The three types of mobilizing splints are dynamic, static progressive, and serial static.
2. Early mobilization protocols following a flexor tendon repair should begin on postoperative day 3.
3. Strengthening should not be initiated following a musculoskeletal injury until the pain level is reduced.
4. A partial digit prosthesis can be cosmetic as well as functional if length is needed for functional activities such as keyboard use.
5. Scar management modalities will be required if a burn injury takes more than 14 days to heal.

13. **What common complications should the therapist be aware of during the rehabilitation of the patient with a distal radius fracture?**
    - Tight and/or painful cast, dressing, or splint
    - Cast/splint that blocks MCP and interphalangeal (IP) joint flexion
    - Sympathetic-mediated pain syndrome
    - Tendon rupture
    - Wound infections
    - Contracture of the ipsilateral elbow and shoulder
    - Persistent ulnar-sided pain

14. **Why is the use of a sling inappropriate following a hand injury or surgery?**
    - Muscular inactivity causes venous stasis and edema formation.
    - The arm remains below the heart, resulting in dependent edema.
    - The shoulder is internally rotated and adducted and the elbow flexed, resulting in avoidable loss of motion and contracture formation.
    - The patient is prevented from early use of the injured hand for light functional activities.

15. **Which therapeutic modalities are appropriate during the rehabilitation of a distal radius fracture?**
    - **Mild heat** decreases pain and increases tissue extensibility. Examples include heat packs, warm soaks, paraffin, fluidotherapy, and ultrasound.
    - **Contrast baths** reduce edema and pain.
    - **Electrical stimulation** improves tendon gliding and strengthens select motor groups.
    - **Biofeedback** is appropriate for muscle reeducation in the patient with tendinous cocontraction or muscle substitution patterns.
    - **Icing after treatment** helps reduce swelling and pain.

16. **A 29-year-old man incurs a severe wrist and hand crush injury after getting his left arm caught in a punch press at work. The surgeon performs a mid-shaft below-elbow amputation with primary wound closure. When is it appropriate to initiate prosthetic fitting?**
    - The patient should be referred to a prosthetist and occupational therapist as soon as possible after surgery (ideally within 1 week) so that he can be fitted with a prosthesis and begin prosthetic training.
    - The upper extremity amputee is more likely to accept and integrate the use of a prosthesis into activities of daily living (ADL) if a preparatory (temporary) prosthesis was provided

intraoperatively or within 4 weeks of the amputation. After 4 weeks, many amputees have become well adapted with a one-handed approach and tend to abandon prosthetic use.

17. **A 25-year-old secretary with a primary soft tissue sarcoma underwent amputation of the middle finger through the PIP joint 3 months ago. What kinds of cosmetic digital prostheses are available?**
    - Polyvinyl chloride (PVC) production glove
    - Custom PVC glove—relatively inexpensive but easily stained and irreparable
    - Silicone glove—very expensive, custom-made, quite durable, repels stains
    - Cosmetic partial digital prosthesis—improves function when length is needed

18. **A 36-year-old man is admitted to the hospital after his snowmobile hits a tree. He has no movement in his dominant right arm and no sensation in the C5–C8 or T1 sensory dermatomes. Radiographs of the cervical spine, clavicle, and shoulder regions are normal. A complete right brachial plexopathy is diagnosed. What orthotic should be provided before he leaves the hospital?**
    A figure-of-eight custom-made sling is most effective in approximating the glenohumeral joint and protecting the insensate arm from further injury. In addition, he should be fitted with a custom-made, low-temperature thermoplastic splint with the wrist in neutral position and the digits placed in an intrinsic plus position to prevent contracture formation (Fig. 45-1).

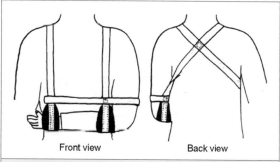

Figure 45-1. A figure-of-eight custom-made sling. This type of sling is most effective in approximating the glenohumeral joint and protecting the insensate arm from further injury.

19. **A 24-year-old man presents with a left brachial plexopathy involving the upper trunk. Diagnostic testing is consistent with avulsion of the C5 and C6 nerve roots and injury but continuity of C7. He has no active motion of the shoulder and significant inferior subluxation of the glenohumeral joint, no active elbow flexion, 4/5 triceps, 5/5 wrist flexors, 4-/5 wrist extensors, and full AROM of the hand. What orthosis should be ordered for the patient?**
    A functional arm orthosis approximates the glenohumeral joint and helps position the hand for function in the absence of elbow flexion. Prosthetic components include laminated humeral and forearm sockets connected with a hinged, externally locking elbow joint, all of which are suspended by a Bowden cable from a shoulder saddle and chest strap. The elbow may be

passively unlocked by pulling the forearm cable with the opposite hand, passively positioned at the desired angle, and then locked again by pulling the cable. A removable palmar strut may be added if wrist support is needed. A qualified orthotist or prosthetist typically provides this type of orthosis (Fig. 45-2).

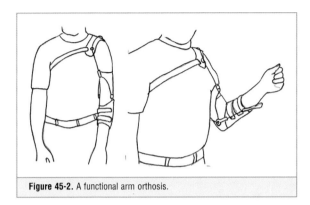

Figure 45-2. A functional arm orthosis.

20. **What is the significance of the juncturae tendinum when the hand is immobilized after extensor tendon repair in zones V and VI?**
    - A **zone V** repair is distal to the juncturae tendinum; therefore, MCP joint flexion in an adjacent digit does not stress the repair site. By splinting the adjacent MCP joints in 30 degrees of flexion, the proximal end of the severed tendon is advanced by the force of the intertendinous connection. This approach helps approximate the repair site.
    - A **zone VI** repair is proximal to the juncturae tendinum; therefore, MCP joint flexion in an adjacent digit stresses the repair site. If the relationship of the repair and juncturae tendinum is unknown, it is recommended that the therapist use Frere's rule and splint adjacent digits in extension.

21. **How much PIP or MCP joint flexion is necessary with an early mobilization program to (a) maintain gliding of the repaired tendon and (b) stimulate cellular activity without rupturing an extensor tendon repair?**
    - The goal is to achieve 3–5 mm of tendon migration with controlled early mobilization.
    - This can be achieved with 30 degrees of PIP joint motion. In the index and middle fingers, 30 degrees of MCP joint motion is permitted. For the ring and small fingers, 40 degrees of MCP flexion is necessary to achieve 5 mm of excursion. In the thumb, 60 degrees of IP joint flexion with the wrist and MCP joint in extension permits 5 mm of tendon excursion at Lister's tubercle.

22. **After a tenolysis procedure, what questions should the therapist ask the surgeon before initiating postoperative therapy?**
    It is important to know about the quality of the involved tendon, the strength of the repair site, and the amount of active and passive ROM in the involved digit(s).

23. **How long postoperatively does an A2 or A4 pulley repair or reconstruction need to be protected?**
    - Immediate protection should include a soft pulley ring, tape, or thermoplastic pulley ring for 6 weeks.

- A metal ring can be substituted once the edema is stable.
- Most surgeons and therapists recommend pulley protection for 6 months after surgery.

24. **When is it appropriate to test grip and pinch strength after a flexor tendon repair?**
Strength testing should not be performed until 12 weeks postoperatively to avoid causing a tendon rupture.

25. **A 45-year-old man underwent radial nerve repair at the mid-humeral level 18 months ago. He has not regained wrist, digital, or thumb extension. He is scheduled to undergo transfer of the pronator teres to the extensor carpi radialis brevis; transfer of the flexor carpi radialis to the extensor digitorum communis of the index, long, ring, and small fingers; and transfer of the palmaris longus to the extensor pollicis longus. What is the recommended position of limb immobilization postoperatively?**
The splint or cast should position the arm with the elbow flexed, the forearm in 15 degrees of pronation, 30 degrees of wrist extension, full extension of the MCP joints, and full extension and abduction of the thumb. The IP joints are left free.

26. **Why do some patients have difficulty performing the desired motion following a tendon transfer? Which therapy techniques are helpful?**
    - The patient may be unable to isolate the donor muscle to perform its new function.
    - Preoperative therapy to help the patient become familiar with the donor may help to initiate the transfer more easily postoperatively. In the mobilization stage, biofeedback to the donor, isolated transfer exercises, and facilitation techniques such as microtaping may help to initiate motion. In addition, having the patient perform the original donor motion concurrently with the new motion, if possible, may help.
    - Later stages of rehabilitation may include light functional activities and functional electrical stimulation to provide motor and sensory feedback and to use the transfer more effectively.
    - It is important to remember that some patients may take 3–6 months to utilize a transferred tendon effectively.

27. **A patient with a low median nerve palsy is scheduled for an opponensplasty procedure. What is the primary reason for failure of an opponensplasty tendon transfer?**
An adduction or adduction-supination contracture of the thumb web space and CMC joint secondary to absent function of the opponens pollicis, abductor pollicis brevis, and the superficial head of the flexor pollicis brevis. Preoperative web space and CMC joint mobilization and stretching are crucial for a successful surgical outcome.

28. **A patient is referred to the therapist from an employee health clinic. The patient describes the onset of pain in the first dorsal compartment over the previous 3 weeks. The patient has a positive Finkelstein's test. Which therapeutic modalities should be initiated at this point?**
    - A forearm-based splint with the wrist in 20 degrees of dorsiflexion, the thumb MCP joint in neutral, and the IP joint free. The splint should be worn continuously with the exception of bathing and cryotherapy should be used 2–3 times per day.
    - An assessment of vocational, leisure, and home-related activities that may have precipitated the symptoms should be undertaken. If activities are identified and they correlate with increasing symptoms, suggestions for activity modification should be provided.

29. **A patient presents with cubital tunnel syndrome. Electrophysiologic testing reveals mild sensory changes with no concurrent motor abnormalities. What are the appropriate therapeutic interventions?**
    - Soft elbow supports are indicated to diminish compression of the ulnar nerve about the elbow, particularly if the patient rests the elbow on a surface for extended periods.
    - Work and activity modification also are indicated if activities involve repeated elbow flexion/extension.
    - Splinting should be performed so that the elbow is in 30 degrees of flexion, the forearm is at neutral rotation, and the wrist is held in neutral. The splint prohibits repetitive flexion of the elbow and is primarily indicated for nighttime use.

30. **When is splinting indicated for a newborn with arthrogryposis multiplex congenita?**
    Usually during early infancy. Static splints or serial casting may be used to gain passive ROM. Wrist or full-hand splints are used to increase wrist/digit extension and correct ulnar deviation. Elbow flexion splints may be used to increase elbow flexion.

31. **A newborn has a birth-related brachial plexus injury. What is the recommended therapy program?**
    - The therapist should educate the family in upper extremity passive ROM techniques to avoid contracture development and keep all joints in the entire involved limb supple.
    - Depending on the extent of motor paralysis and the degree of recovery, periodic static splinting may be indicated.

BIBLIOGRAPHY

1. Arthritis Foundation: The Arthritis Help Book, 4th ed. Atlanta, Addison-Wesley, 1995.
2. Arthritis Foundation: Primer on Rheumatic Diseases, 10th ed. Atlanta, Arthritis Foundation, 1993.
3. Bell-Krotoski JA: Preoperative postoperative management of tendon transfers after ulnar nerve injury. In Hunter JM, Mackin EJ, Callahan AD (eds): Rehabilitation of the Hand: Surgery and Therapy, 4th ed. St. Louis, Mosby, 1995, pp 728–764.
4. Brand PW: Mechanics of tendon transfers. In Hunter JM, Mackin EJ, Callahan AD (eds): Rehabilitation of the Hand: Surgery and Therapy, 4th ed. St. Louis, Mosby, 1995.
5. Clark GL, Wilgis EFS, Aiello B, et al: Hand Rehabilitation, 2nd ed. New York, Churchill Livingstone, 1997.
6. Colditz J: The biomechanics of a thumb carpometacarpal immobilization splint. J Hand Ther 13:228–235, 2000.
7. Evans RB: A clinical report of the effect of mechanical stress on functional results after fasciectomy for Dupuytren's contracture. J Hand Therapy 15(4):331–339, 2002.
8. Gelberman WJ, Pess G: Treatment of de Quervain's tenosynovitis: A prospective study of the results of injection of steroids and immobilization in a splint. J Hand Surgery 73A:219–221, 1991.
9. Hong CZ, Long HA, Kanakamedala RV, et al: Splinting and local steroid injection for the treatment of ulnar neuropathy at the elbow: Clinical and electrophysiological evaluation. Arch Phys Med Rehab 77(6):573–577, 1996.
10. Michlovitz SL: Thermal Agents in Rehabilitation, 2nd ed. Philadelphia, F.A. Davis, 1990.
11. Reynolds CC: Preoperative and postoperative management of tendon transfers after radial nerve injury. In Hunter JM, Mackin EJ, Callahan AD (eds): Rehabilitation of the Hand: Surgery and Therapy, 4th ed. St. Louis, Mosby, 1995, pp 753–764.
12. Stanley BG: Preoperative and postoperative management of tendon transfers after median nerve injury. In Hunter JM, Mackin EJ, Callahan AD (eds): Rehabilitation of the Hand: Surgery and Therapy, 4th ed. St. Louis, Mosby, 1995, pp 765–778.

CHAPTER 46
LEGAL ASPECTS OF HAND SURGERY
Walter E. Harding, JD

1. **What is the single defining purpose of litigation?**
   Conflict resolution. The legal process is concerned not with "the exhaustive search for causative understanding, but for the particularized resolution of legal disputes" (Daubert versus Merrell Dowell, 1993). Law attempts to obtain a finality for conflicts.

2. **Why do physicians have to be cognizant of the legal system, particularly such issues as workers' compensation and medical malpractice?**
   From a practical standpoint, the workers' compensation patient has become a major financial factor in many medical practices. The advent of managed care, insurance adjuster oversight, increased competition between medical practices, and the growing willingness of physicians to critique each other publicly have significantly altered the litigation landscape.
   A surplus of lawyers and shifting definitions of "community standards" have helped to increase medical malpractice actions. As a percentage of actual civil claims, however, medical malpractice cases represent an extremely small number. But the personal anguish and expense involved in defending oneself can be incalculable.

3. **What is the physician's role in workers' compensation?**
   Entitlement is set out arbitrarily by statute. The claimant has to demonstrate an actual measureable "work-related" harmful change in the injured person, that he or she provided due and timely notice to the employer of the workplace where the injury occurred, and that the claim was filed within the requisite statute of limitations.
   Compensable impairment, called *occupational disability* in some jurisdictions, is usually derived from functional impairment ratings from guides provided by some organizations such as the American Medical Association (AMA). The compensable impairment is coupled with an assessment of the claimant's physical restrictions and how those restrictions impact the ability of the worker to earn future wages and compete for employment. Both the plaintiff and defendant will attempt to determine if an amount preexisted and contributed.

4. **What information does the legal system seek in compensation cases?**
   The physician plays a central role in these decisions. He or she will be asked as the treating physician and/or evaluating expert to make a diagnosis and prognosis, to assess residual restrictions, determine causation, and render a functional impairment rating, relate when the disabling condition was communicated to the patient, and to outline "reasonable and necessary" medical care and the expected time of maximum medical improvement.

5. **Are all work-related "harmful" changes compensable under workers' compensation?**
   The requirements for compensability are set out in the workers' compensation statutes, and these make it clear that not all work-related harmful changes are compensable. In general, only physical injuries and their residuals that can be "objectively" measured by standardized methods are compensable. This includes a worsening of a condition due to medical malpractice.

6. **What are "objective medical findings"?**
   The term *objective medical findings* has different meanings to the legal and medical communities. The fact that a particular diagnosis is made in a manner acceptable to physicians for treatment or diagnosis will not necessarily render it legally sufficient for purposes of adjudication. Symptoms may not be evidence of a work-related harmful change in the human organism. Therefore, a diagnosis based on a worker's complaints of symptoms but not supported by objective medical findings is generally insufficient to prove an injury for the purposes of workers' compensation. Thus, *objective medical findings* is defined as that information gained by direct observation and testing, applying objective or standardized methods that can be reproduced by other physicians.

   The legal system recognizes that the term *diagnosis* refers to the process whereby a physician analyzes the nature and cause of the patient's "harmful change" or injury and develops the working differential for a logical basis for treatment and prognosis (Dorland's Illustrated Medical Dictionary, 1965; Taber's Cyclopedic Medical Dictionary, 1985). One arrives at the likely diagnosis from a number of possibilities—noting the symptoms that the patient reports, performing a physical examination, taking a history, observing the patient, and evaluating the significance of the results of objective or standardized testing. However, the ultimate diagnosis must be stated within a degree of medical probability, or it is useless to the adjudicator.

KEY POINTS: LEGAL ASPECTS OF HAND SURGERY

1. Litigation is the legal method of resolving conflicts.
2. The administrative law judge (ALJ) is both the gatekeeper of information (evidence) and the decider of facts.
3. Only physical injuries and their residuals that can be objectively measured are considered compensable.

7. **What is the difference between the terms *certainty*, *probability*, *likelihood*, and *possibilities*?**
   Many physicians in their testimony use these terms interchangeably. Each of these terms has a specific legal effect. This is a critical distinction in that the majority of physicians do not spend a lot of time preparing for a deposition.

   Certainty requires, basically, 100% knowledge. The criminal law system requires a conviction with the standard of certainty—100%, with no "reasonable doubt" allowed. The civil system, however, requires a determination based on probability or likelihood, which is 51% or greater. What is the difference?

   For example, the United States's court system was unable to determine for certain that O.J. Simpson had killed his ex-wife. But the civil court system made a finding that in all "*probability*" or likelihood Simpson had, indeed, killed her. The former standard cannot exist in science. Science has no absolutes and only probabilities or possibilities. "In all medical probability or likelihood" is the standard of proof utilized in medical malpractice and in workers' compensation. In medical malpractice, the plaintiff must demonstrate, through the use of "expert" testimony, that the defendant's conduct "in all probability" constituted a departure from the medical community's recognized standard of care as applicable to the plaintiff. In disproving this claim, however, the defendant must show only that there were other "possibilities."

   Parenthetically, only in those situations where an x-ray reveals that a surgical instrument has been left in the abdomen of a surgical patient can a person state with medical "certainty" that medical malpractice was involved. Of course, the standard for determining liability shifts to probability, or 51% or better, in terms of who had the duty to remove that particular instrument before wound closure.

What are *possibilities*? These are any potential, but not necessarily probable, sources. Physicians will often testify that various types of potential causes are possible. Of course, anything could be construed as possible. To find a cause for what is simply possible should fail within an adjudicative tribunal.

8. **What are the essential elements of a civil case?**
   No civil case can prevail unless three things can be determined within a degree of probability: a breach of a duty owed, causation, and damages. Failure to prove any one of these three defeats the action. For example, if you owe no duty, then you have no liability. If your conduct was not the proximate cause of the alleged injury, then you have no liability. If there are no damages, then there can be no recovery. The defendant does not need to disprove an item, such as causation, but only to present credible evidence to discredit or rebut the plaintiff's evidence. This can be done by submitted evidence of possible causes. There need not be proof within certainty or probability.

9. **Is it difficult to successfully assert a medical malpractice case?**
   In medical malpractice, to establish the proverbial "prima facie case," a plaintiff must introduce expert testimony demonstrating that the defendant's conduct constituted a departure from the medical community's recognized standard of care as applicable to the plaintiff by (1) establish the standard of care in the medical community as it would apply to the plaintiff, (2) show how the plaintiff departed from that standard of care, and (3) show that the defendant's departure was the proximate cause of the plaintiff's injuries. Proximate causation between the negligence and alleged injury must be established by expert testimony. The expert testimony must be within the sense of probability.

10. **What is the standard for the admissibility of scientific evidence in court?**
    The courts have struggled to determine admissibility of scientific evidence but have basically outlined four questions for the decision-making process: (1) Has the theory or technique been tested? (2) Has it been subjected to peer review or publication? (3) What is the rate of error for the technique? (4) Has the technique found acceptance within the relevant scientific community? The trial judge, usually someone who has no scientific background, acts as the "gatekeeper." As in all civil litigation cases, the expert is not required to testify within certainty (100%) but only within the sense of probability and within the sense as to whether the inference or assertion has been derived from the scientific method.

11. **Is there any documented difference with respect to prognosis between those patients with work-related injuries versus those cases where no litigation is involved?**
    A multitude of studies have shown that there is a demonstrable difference, for several reasons, between workers' compensation patients and those for whom litigation is not an issue. Those patients who are not involved in litigation or a workers' compensation claim have a better prognosis overall.

12. **As a physician assessing a patient who is claiming workers' compensation, have you considered dose, temporal relationship, prediction, and prevention?**
    The lawyer is usually trying to determine whether the physician making the diagnosis has actually adequately considered (1) the amount of exposure to the alleged offending activity (dose), (2) whether there is any temporal relationship, (3) whether one could predict that a given number of individuals performing that particular employment would develop that condition, and (4) whether the worker would have developed the condition if he or she had been prevented from performing that alleged offending activity.

13. **What is the purpose of a deposition?**
 To elicit testimony for purposes of later impeaching the witnesses' testimonies or for presentation to the court in lieu of a formal appearance by the physician.

BIBLIOGRAPHY

1. Daubert versus Merrell Dow Pharmaceuticals, Inc., 509 U.S. 579, 113 S. Ct. 2786, 1993, p 485.
2. Dorland's Illustrated Medical Dictionary, 24th ed. W.B. Saunders, Philadelphia, 1965.
3. Thomas CL (ed): Taber's Cyclopedic Medical Dictionary, 15th ed. F.A. Davis, Philadelphia, 1985.

INDEX

Page numbers in **boldface type** indicate complete chapters.

A

Abductorplasty transfer, Camitz, 275
Abductor pollicis brevis tendon, first dorsal interosseous muscle as, 323
Abductor pollicis longus tendon, 11
  insertion and action of, 10
Abscess
  collar button, 83
  subepidermal, 84
Acetylcholine, 255
Acid burns, 100
Acrocephalosyndactyly, Apert's, 320
Acrometastasis, 307
Acro-osteolysis, 40
Acrosyndactyly, 320
Actinomyces israelii infections, 82
Activities of daily living (ADLs), wrist motion required for, 15
Adolescents, osteogenic sarcoma in, 308
Adriamycin, 85
Adson's (scalene) test, 282
Aeromonas hydrophila infections, 82
Aeromonas infections, prophylaxis against, 293
Allen's test, 208, 280
Allergic reactions, to local anesthetics, 33
Allografts, 237
Allotransplantation, of the hand, 294
American Medical Association (AMA), 340
American Society for Surgery of the Hand (ASSH), 20
Americans with Disabilities Act, 328
Amputated digits or limbs. *See also* Replanted digits or limbs
  care and transport of, 142, 289
  viability of, 142
Amputations, **296–300**
  congenital upper extremity, 299
    musculoskeletal anomalies associated with, 300
  emergency management of, 289
  of the fingertip, 228
    complications of, 232
    donor tissue sources for, 232
    with dorsally angulated defect, 228
    with exposed bone, 232
    flap coverage for, 228–233
    revision and closure of, 232

Amputations *(Continued)*
  human bite wound–related, 219
  multiple finger, 289
  as neuroma cause, 233
  nonsurgical interventions for, 301
  prostheses fitting after, 298, 335
    in children, 300
    for traumatic amputations, 300
  ray, 296–297
  as soft tissue sarcoma treatment, 336
  through the ring finger, 233
  of the thumb, 141
  transradial, 297
Amyoplasia, 322
Amyotrophic lateral sclerosis (ALS), 28, 254
Anastomosis, microvascular, suturing techniques for, 140
Anatomic snuffbox, 12
  location of, 25
Anconeus epitrochlearis, 262
Anemia, Fanconi's, 317, 318
Anesthesia, **32–37**
  conscious sedation, 32
  local, 33–34
    allergic reactions to, 33
    low-concentration, 34
    toxicity of, 33
    use in pregnant patients, 36
  regional, 33
    tourniquet use in, 36
  during replantation, 292
Aneurysms
  false, differentiated from true, 281
  treatment of, 281
  of the ulnar artery, 281
Angiotomes, 242
Animal bite wounds
  as *Eikenella corrodens* infection cause, 82
  as *Pasteurella multocida* infection cause, 82
Anterior interosseous nerve, injuries to, 24
Anterior interosseous nerve palsy, tendon transfers in, 277
Anterior interosseous syndrome
  differentiated from median neuropathy, 258
  functional loss in, 261

Anterior oblique carpometacarpal ligament, in Bennett's fracture, 13
Antibiotic therapy, for frostbite, 95
Anticoagulation therapy, 140
Antidotes, for extravasation injuries, 87
Antidromic, definition of, 255
Apert's acrocephalosyndactyly, 320
Arachnodactyly, 319
Arcade of Frohse, 264
Arcade of Struthers, 262
Arc injuries, greater, lesser, and inferior, 165
Arterial insufficiency, in replanted body parts, 293, 294
Arteries, of the hand 280. *See also names of specific arteries*
  of the thumb, 141
Arthritides, **70–74**
Arthritis. *See also* Osteoarthritis; Rheumatoid arthritis
  of the basilar joint, 25
  carpometacarpal, 25
  gouty, 82
  inflammatory bowel disease-associated, 74
  psoriatic, 71–72
    differentiated from rheumatoid arthritis, 40
    of the distal interphalangeal joints, 73
  septic, 70, 73, 82
    gonorrhea-related, 83
    human bite wounds-related, 71
    penetrating trauma related, 83
    radiographic features of, 44
Arthritis mutilans, 73
  bone loss prevention in, 74
Arthrodesis (fusion)
  of the fingers, 195,
    of the distal interphalangeal joints, 197
    headless cannulated screw use in, 197
    of the metacarpophalangeal joints, 195
    with posterior interosseous nerve (PIN) resection, 56
    of the proximal interphalangeal joints, 53, 195
  of the thumb, 195, 198
  of the trapeziometacarpal joint, 56, 57
  of the wrist
    capitate-lunate-hamate-triquetral, 194
    capitolunate, 194
    contraindications to, 193
    indications for, 193
    intercarpal (limited), 194, 195, 197
    lunotriquetral, 194
    nonunion rate of, 197
    preferred position for, 193
    scaphocapitate, 194, 197
    in the SLAC wrist, 198
    surgical goals for, 193
    triscaphe, 194
    unilateral, 194
Arthrography
  definition of, 45

Arthrography *(Continued)*
  of the wrist
    for carpal instability assessment, 160
    magnetic resonance, 48
    triple-injection (three-compartment), 45
    for wrist pain evaluation, 46
Arthrogryposis, 322
  classic form of, 322
  with Marfan's syndrome, 322
Arthrogryposis multiplex congenita, 322
  splinting treatment of, 339
Arthroplasty
  interpositional
    hemiresection, 198
    of osteoarthritic first carpometacarpal joints, 197
  of the metacarpophalangeal joints, 334
  of the proximal interphalangeal joints, 53
  in rheumatoid arthritis, 63
  silicone, 130, 197
  total wrist
    complications of, 193
    contraindications to, 193
    indications for, 193
Arthroscopic-assisted reduction and fixation (AARF)
  of distal radius fractures, 135, 143, 146, 148
    intra-articular ligament injuries associated with, 148
    Ringer's solution use with, 148
  fluid extravasation during, 135
  of the scapholunate ligament, 135
Arthroscopy
  diagnostic wrist, 160
  of the trapeziometacarpal joint, 57
Autografts, 237
Avulsion injuries, 257
  to the distal phalanx, 40
  to the nail bed, 227
  ring, 292
  to the ulnar collateral ligament complex, 180
Axillary blocks, 35
  contraindication to, 36
Axonal loss, in peripheral nerve entrapment neuropathy, 257
Axonal regeneration, 267
Axonotmesis, 267

B
Baclofen, 114
Bacterial colonization, of extravasation injuries, 85
Bacterial count
  determination of, 238
  effect on wound healing, 238
Barton's fractures, 140
Basilar joint, osteoarthritis of, 25, 51, 53
  arthrodesis treatment of, 56
Beals' syndrome, 322
Beau's lines, 228
Bennett's fracture, 180–182
  anterior oblique carpometacarpal ligament in, 13

Bennett's fracture *(Continued)*
   definition of, 180
   first metacarpal subluxation in, 181
   reverse, 182
   treatment of, 182
Benzhexol, 114
Biceps paralysis, elbow flexion restoration in, 278
Bier's block, 33
   intravenous catheter use in, 36
Biofeedback, 335
Biomechanics, of the hand and wrist, **15–19**
Biopsy
   of bone tumors, 308
   of distal upper extremity lymph nodes, 314
   of hand lesions, 313
Bite wounds
   human, 219
      as *Eikenella corrodens* infection cause, 82
      as septic arthritis cause, 71
      with zone V extensor tendon lacerations, 220
      as *Pasteurella multocida* infection cause, 82
Blauth classification, of thumb hypoplasia, 322
Bone grafts
   as Kienböck's disease treatment, 130
   for scaphoid nonunion, 170
Bone scans, 47
   for carpal instability assessment, 160
   definition of, 46
   for differentiation of gouty and septic arthritis, 82
   of electrical burns, 100
   for evaluation of vascular disorders, 281
   focal increased carpal bone uptake on, 47
   of the frostbitten hand, 95
   of osteomyelitis, 47
   of scaphoid fractures, 25, 169
   three-phase, 47
      of osteomyelitis, 47
Bone tumors, of the hand and wrist, **302–308**
   aggressive or malignant, 44, 302
   biopsy of, 308
   metastatic, 307, 308
Bony prominences, removal of, 66
Botulinum toxin, as hand dystonia treatment, 28
Botulinum toxin nerve block, 117
Bouchard's nodes, 51, 59
Boutonnière deformity, 24, 221
   burn injury-related, 100
   classification system for, 64
   clinical stages of, 222
   delayed development of, 221
   reconstruction of, 221, 222
   rheumatoid arthritis-related, 63, 64
Bowstringing, of the tendons
   of the flexor tendons, 215
   role of pulleys in prevention of, 211
Boxer's fractures, 24, 178
Brachial plexopathy, 29
   birth-related, 339

Brachial plexopathy *(Continued)*
   orthoses for, 336
Brachial plexus blocks, 33, 34, 35
   axillary technique, 35
   in complex regional pain syndromes, 36
   failed, supplementation of, 35
   interscalene technique, 35
   supraclavicular technique, 35
Brachioradialis muscle
   innervation of, 11
   spasticity of, 118
Brachydactyly, 319
Brachymetacarpia, 326
Brain injury. *See also* Traumatic brain injury
   as cubital tunnel syndrome cause, 118
Bunnell technique, modified, 218
Bupivacaine, 36
Burn patients, transfer to burn centers, 98
Burns, **98–100**
   acid, 100
   arches and web spaces preservation in, 333
   chemical, 100
   degrees of, 98
   electrical, 100
   healing of, 98
   hypertrophic scars associated with, 100
   palmar, 99
   skin grafts for, 99, 333
   splinting of, 333
   tangential excision of, 99
   topical agent treatment for, 98
   zones of, 98

C

Cable nerve grafts, 270
Calcification
   flexor carpi ulnaris tendinitis-related, 204
   of the triangular fibrocartilage complex, 41, 72
Calcium gluconate infusions, intra-arterial, 88
Calcium pyrophosphate dehydrate deposition (CPPD) disease, 72, 200
Camitz abductorplasty transfer, 275
Campanacci staging criteria, for giant cell bone tumors, 308
Camptodactyly, 319, 323
Capitate bone
   fractures of, 171
   ossification of, 12
   shortening of, 129
Capsulitis, adhesive, as axillary block contraindication, 36
Caput ulnae syndrome, 60
Carpal alignment, radiographic assessment of, 38
Carpal bones. *See also specific carpal bones*
   axial instability of, 166
   effect of distal radius malunion on, 147
   fractures of, 167, 171–172
   ossification of, 12, 202

Carpal bones *(Continued)*
  ulnar translocation of, 167, 185, 191
Carpal boss, 56, 200
Carpal coalition, 326
Carpal force transmission, in perilunate injuries, 164
Carpal height, 9
  measurement of, 41
Carpal instability, **158–173**
  classification of, 162
  evaluation of, 159–162
    imaging modalities in, 160–162, 164
    intercarpal angles in, 42
  most common form of, 165
Carpal instability complex (CIC), 165
Carpal instability dissociative (CID), 165
Carpal instability nondissociative (CIND), 165
Carpal row, proximal
  effect of scapholunate ligament sectioning on, 16
  during wrist deviation, 16
  during wrist flexion/extension, 16
Carpal tunnel
  anatomic boundaries of, 9, 259
  contents of, 9, 260
Carpal tunnel release, 104
Carpal tunnel syndrome
  bilateral, 257
  definition of, 27
  diagnosis of, 255, 256, 257
  differentiated from C8 motor radiculopathy, 258
  diseases that mimic, 29
  electrodiagnostic evaluation of, 256, 257
  first-time treatment for, 200
  Linburg's syndrome-associated, 79
  occupational factors in, 328–329, 331
  as paresthesia cause, 23
  during pregnancy, 200
  risk factors for, 261
  thumb function restoration in, 275
  work lifestyles associated with, 328
Carpectomy, proximal row, 194
  contraindication to, 194
  with radial styloidectomy, 194
Carpometacarpal (CMC) boss, 56, 200
Carpometacarpal (CMC) joints
  arthritis of, 25
  dislocations of
    dorsal, 186
    multiple, 186
    of the thumb, 187
    treatment of, 186
  first, interpositional arthroplasty of, 197
  injuries to, 180, 186
    fracture-dislocations, 186
    injuries associated with, 180
  motion of, 17
  osteoarthritis of, 197, 198

Carpometacarpal (CMC) joints *(Continued)*
  of the thumb
    arthrodesis-related complications of, 198
    dislocations of, 187
    osteoarthritis of, 55, 198
    pain in, 334
Carpus. *See* Carpal bones
Cascade posture, of the hand, 210
Casting
  as distal radius fracture treatment, 142
  as occult fracture treatment, 205
  as pediatric Kienböck's disease treatment, 129
  postoperative position of, 338
Causalgia, 125
Centralization procedure, 317
Central nervous system diseases, as carpal tunnel syndrome mimics, 29
Central slip, laceration or rupture of, 221
Cerebral palsy, 114–116
  classification of, 114
  definition of, 114
  treatment of, 114
  upper extremity physical examination in, 116
Certainty, 341–342
"Chauffeur's" fractures, 141
Cheiralgia paresthetica, 264
Chemical burns, 100
Chemotherapy agents, vesicant, extravasation of, 85
Child abuse, as multiple fracture cause, 206
Children
  carpal bone ossification in, 202
  disarticulations in, 297
  distal radius fractures in, 138
  epiphyseal plate location in, 14
  fingertip injuries in, 233
  flexor tendon injury treatment in, 216
  Kienböck's disease in, 129
  most common thumb fracture in, 182
  occult fractures in, 208
  osteogenic sarcoma in, 308
  palmar burns in, 99
  prosthetic fitting in, 300
  replantation in, 142
Chloroprocaine, 36
Chondrocalcinosis, 72
  of the triangular fibrocartilage complex, 41
Chondrosarcoma, 305
  definition of, 306
  most common location of, 306
  treatment of, 306
  types of, 306
Chromosomal challenge test, 318
CIC (carpal instability complex), 165
CID (carpal instability dissociative), 165
CIND (carpal instability nondissociative), 165
Ciproxin, 329
Cisplatin, 85
Civil lawsuits, 342

INDEX

Claw hand deformity
  isolated ulnar palsy-related, 276
  low ulnar palsy-related, 276
Cleft hand (symbrachydactyly), 318
Clenched fist syndrome, 331
Cline, Henry, 108
Clinodactyly, 319, 323
Closed head injuries, evaluation of the hand in, 22
Clostridium botulinum infections, 117
Clostridium perfringens infections, as gas gangrene cause, 84
Clostridium tetani infections, risk factors for, 20
CMC joints. *See* Carpometacarpal joints
Coagulation, burn-related, 98
Cold therapy, 335
Cold tolerance, postfrostbite, 97
Collagen, in normal skin and in scar tissue, 235
Collar button abscesses, 83
Collateral ligaments, in fracture open reduction and internal fixation, 183
Color duplex ultrasonography, advantages of, 280
Compartments, of the hand, 104
  first dorsal
    extensor, anatomic variations in, 80
    pain in, 338
    release of, 75, 76
    retinaculum excision of, 79
  release of, 105
  second dorsal, tenosynovitis of, 76
Compartment syndrome, **102–106**
  acute, 102
    treatment of, 104
  carpal tunnel release in, 104
  classification of, 102
  definition of, 102
  diagnosis of, 103, 104
  "five Ps of," 103
  of the forearm, 104
  impending, 103
  physiology of, 102
  pressure levels in, 102, 104
    clinical usefulness of, 104
  untreated, 103
Complex regional pain syndromes, **122–125**
  brachial plexus blocks in, 36
  diagnostic confusion regarding, 122
  misdiagnoses of, 124
  nondisplaced distal radius fractures-related, 208
  psychological risk factors for, 125
  signs and symptoms of, 122
  treatment of, 123–124
  type 1, differentiated from type 2, 125
Compound muscle action potentials (CMAPs), 255
Computed tomography (CT)
  of proximal scaphoid pole sclerosis, 44
  of the scaphoid bone, 45
  of scaphoid fractures, 169
  of the wrist, 44

Condylar fractures, 176
Congenital constriction band syndrome, 325
Congenital hand differences, **316–326**
Congenital limb anomalies, classification of, 316
Conscious sedation, 32
Constriction band syndrome, congenital, 325
Contractual arachnodactyly syndrome, 322
Contractures
  arthrogryposis multiplex congenita-related, 322
  burn injury-related, 100
  cerebral palsy-related, 114, 116
  Dupuytren's. *See* Dupuytren's disease
  flexor, 110
  Volkmann's ischemic, 103
Contrast baths, 335
Corticosteroid injection therapy
  for de Quervain's disease, 80
  for tendinitis/tenosynovitis, 79
  for trigger finger, 80
Corticosteroids, contraindication as high-pressure injury treatment, 92
Cotton-Loder position, 140
Crank test, 54
Crohn's disease, 74
Cross-intrinsic tendon transfers, 278
Crush injuries, 37
Cubital tunnel
  anatomic boundaries of, 262
  as ulnar nerve compression site, 23, 263
Cubital tunnel surgery, postoperative pain associated with, 262
Cubital tunnel syndrome
  definition of, 27
  differentiated from thoracic outlet syndrome, 264
  in stroke and brain-injured patients, 118
  treatment of, 339
Cumulative trauma disorders, 330
Cysts
  bone
    aneurysmal, 308
    unicameral, 304
  epidermal inclusion, 311
  of the fingernails, 228
  ganglion. *See* Ganglion
  mucous, 51–52
  silicone implants-related, 196

D

Dantrolene, 114
Darrach procedure, 61, 156
  failed, 197
Debridement, 237
  of the frostbitten hand, 96
Degenerative joint disease. *See* Osteoarthritis
Delta phalanx, 324
Denervation potentials, 254
  wallerian degeneration-related, 255
Depositions, legal, 343

de Quervain's disease, 75–76
　clinical presentation of, 75
　differential diagnosis of, 75
　treatment of, 75
　　complications of, 80
　　success rates of, 80
Diabetes mellitus
　neuropathy associated with, 36, 124
　subepidermal abscesses associated with, 84
　trigger finger associated with, 79
Diabetic patients, anesthesia use in, 36
Diagnosis, legal definition of, 341
Diazepam (Valium), 114
"Die-punch" fractures, 141
Digital artery island flaps, 247
Digital cords, in Dupuytren's disease, 108, 110, 111
Digital flexor tendons, cerebral palsy-related tightness of, 115
Digital flexor tendon sheath, 211
　healing of, 212
Digital nerve blocks, 33
DIP joints. *See* Distal interphalangeal joints
Disability, occupational, 340
DISI (dorsal intercalated segmental instability), 162
　radiographic features of, 42
Dislocations, **185–192**
　of the carpometacarpal joints
　　dorsal, 186
　　multiple, 186
　　of the thumb, 187
　　treatment of, 186
　of the distal interphalangeal joints
　　complex, 191
　　joint reduction prevention in, 191
　　with proximal interphalangeal joint dislocations, 190
　of the distal radioulnar joint, 155
　　dorsal, 185
　　effect of radial malunions on, 155
　　imaging of, 45
　　treatment of, 155
　Essex-Lopresti, 156, 201
Distal interphalangeal joints
　amputations through, 296
　arthrodesis of, 197
　dislocations of
　　complex, 191
　　joint reduction prevention in, 191
　　with proximal interphalangeal joint dislocations, 190
　frostbite-related radial deviation of, 96
　osteoarthritis of, 50, 52–73
　passive range of motion of, 221
　psoriatic arthritis of, 73
　ring finger amputation through, 233
　septic, 73
Distal phalanges, fractures of, 175
　avulsion fractures, 40

Distal phalanges, fractures of *(Continued)*
　complications of, 175
　displaced transverse, 175
　mallet fractures, 175
Distal phalanx tuft
　fractures of, 175
　radiographic loss of, 73
Distal radial physis, in Madelung's deformity, 325
Distal radioulnar joint, **151–157**
　definition of, 150
　dislocations of, 155
　　dorsal, 185
　　effect of radial malunions on, 155
　　imaging of, 45
　　treatment of, 155
　in distal radius fractures, 147
　dorsal subluxation of, 149
　effect of distal radius deformity on, 147
　fracture-dislocation of, 185
　in Galeazzi's fracture-dislocation, 192
　instability of
　　subtle, 155
　　treatment of, 61
　stability of, 15, 16
Distal radius bone
　anatomic relationship to the lunate and scaphoid bones, 18
　deformity of, effect on distal radioulnar joint, 147
　dorsal angulation of, 18
　epiphyseal injuries to, 142
　force transmission through, 16, 17
　fractures of, 10, **138–149**, 277
　　arthroscopic-assisted reduction and fixation of, 135, 143, 146, 148
　　articular step-off of, 143
　　Barton's fractures, 140
　　"chauffeur's" fractures, 141
　　classification of, 138
　　complications of, 335
　　"crossed pinning" of, 143
　　"die-punch" fractures, 141
　　with distal radioulnar joint subluxation, 149
　　distraction of, 143
　　epiphyseal, 142
　　external fixation of, 145
　　fragment-specific fixation of, 146
　　hand therapy for, 334, 335
　　intracarpal injuries associated with, 142
　　ligamentotaxis of, 142, 143
　　limited open reduction of, 144
　　malunion of, 147
　　median nerve injuries associated with, 146
　　minimally displaced, 206
　　nondisplaced, 208
　　nonunion of, 147, 148
　　occult, immobilization of, 205
　　open reduction and internal fixation of, 144, 334
　　as osteoarthritis risk factor, 56

Distal radius bone *(Continued)*
   fractures of *(Continued)*
      osteochondral injuries associated with, 135
      pinning of, 143, 144
      plating of, 145
      reduction of, radiographic assessment of, 42
      Smith's fractures, 140
      soft tissue injuries associated with, 135
      superficial radial nerve in, 146
      treatment of, 138–149
      unstable, 143
      upper extremity examination in, 139
   giant cell tumors of, 304, 308
   injuries to, **138–149**
   loading pattern on, 143
   premature epiphyseal closure in, 142
Doppler studies, advantages of, 280
Dorsal intercalated segmental instability (DISI), 162
   radiographic features of, 42
Dorsalis pedis flaps, 140
Dorsal metacarpal artery flaps, 246
Dorsal trans-scaphoid perilunate fracture-dislocation, 172
"Double crush" phenomenon, 264
Down's syndrome (trisomy 21), 325
Doxorubicin, 85
Driving, relationship to carpal tunnel syndrome, 328
DRUJ. *See* Distal radioulnar joint
Dupuytren, Guillaume, 108
Dupuytren's diathesis, 109
Dupuytren's disease, 26, **108–113**
   anatomic distribution of, 109
   biochemical abnormalities associated with, 110
   cellular fascial components in, 110
   clinical conditions associated with, 109
   clinical findings in, 108
   definition of, 108
   differential diagnosis of, 110
   ectopic, outside the palmar fascia, 109
   epidemiology of, 109
   etiology of, 109
   family history of, 109
   myofibroblasts in, 110
   natural history of, 108
   occupational factors in, 110
   palmar and digital fasciae in, 108, 110, 111
   palmar pulley triggering associated with, 79
   stages of, 110
   surgical treatment of, 112–113
Dynamic tenodesis procedures, 276
Dysplasia, radial, 317
Dystonia, of the hand, 27
   treatment of, 28, 114

E
Edema
   fasciectomy-related, 334
   free flap coverage-related, 333
Eikenella corrodens infections, 71, 82

Elbow disarticulation, 299
Elbow supports, 339
Electrical burns, 100
Electrical stimulation, 335
Electrodiagnostic testing, **254–258**
   in peripheral nerve injury evaluation, 272
Electromyography, 254–258
   denervation potentials in, 254
   limb temperature monitoring with, 25
Encephalopathy, lead poisoning-related, 208
Enchondroma, 302
   clinical presentation of, 302
   location of, 302
   Maffucci's disease-related, 302, 303
   multiple, 302
   Ollier's disease-related, 302, 303
   with a pathologic fracture, 49
   radiographic appearance of, 302
Endoneurium, 266
Enneking's classification, of musculoskeletal system tumors, 305
Epicondylitis, lateral, 330
Epinephrine, in combination with local anesthetics, 34
   contraindication to, 34
Epineural nerve repair, 268–269
Epineurium, 266
Epiphyseal fractures, Salter-Harris classification of, 41, 142, 182, 222
Epiphyseal injuries, to the distal radius, 142
Epiphyseal plates
   location in children, 14
   premature closure of, 96
      in the distal radius, 142
   previous injuries to, evaluation of, 41
Epithelialization
   of skin graft donor sites, 240
   of skin grafts, 237
Eponychium, 225
Ergonomic improvements, in the workplace, 328
Escharotomy, 99
Essex-Lopresti fracture/dislocation, 156, 201
Expert testimony, 341
   depositions for, 343
Extension
   sudden loss of
      differential diagnosis of, 66
      in rheumatoid arthritis, 65
   of the wrist, 15, 16
      for maximal power grip, 17
Extensor carpi radialis brevis tendon, 11
   innervation of, 11
   insertion of, 11
Extensor carpi radialis longus tendon, 11
   insertion of, 11
Extensor carpi ulnaris tendon, 11
   dislocation of, 222, 223
   snapping, 204

Extensor digiti minimi tendon, 11, 78
  anatomic relationships of, 10
Extensor digitorum communis tendon, 11
Extensor digitorum tendon, anatomic relationship with extensor indicis tendon, 10
Extensor indicis proprius tendon, 11
Extensor indicis tendon, anatomic relationship with extensor digitorum tendon, 10
Extensor pollicis brevis tendon, 11
  injury to, 191
  insertion and action of, 10
Extensor pollicis longus tendon, 11
  insertion and action of, 10
  lacerations to, in zone IV, 219
  location of, 25
  rupture of
    attritional, 277
    distal radius fracture-related, 10, 147
    rheumatoid arthritis-related, 61
    thumb in, 206
    treatment of, 66
  tenosynovitis of, 78
Extensor tendon grafts, 224
Extensor tendons
  extrinsic tightness of, 223
  injuries to, 218–224
  injury repair in
    as flexion loss cause, 219
    mobilization programs following, 337
    poor outcome in, 219
  lacerations to, 219
    as flexion loss cause, 219
    human bite wound-related, 220
  rupture of
    rheumatoid arthritis-related, 277
    systemic lupus erythematosus-related, 70
  sagittal band tears of, 222
  staged reconstruction of, 223
  subluxation or dislocation of, 222
Extravasation injuries, 85–88
  cellular destruction mechanisms in, 85
  irritants-related, 85
  nontoxic agents-related, 85
  prevention of, 88
  risk factors for, 87
  treatment of, 86–87
  vesicants-related, 85
Extrinsic extensor tendon tightness, 223
Extrinsic muscles, of the hand, 23

F

Fascial flaps, 140
Fascicular nerve repair, 268
Fasciculation, 254
Fasciectomy
  as Dupuytren's disease treatment, 112
    complications of, 113
  as edema cause, 334

Fasciitis, necrotizing, 82, 84
  treatment of, 84
Fasciocutaneous flaps, 140
Fasciotomy
  as Dupuytren's disease treatment, 112
  following replantation of limbs, 291
  "prophylactic," 104
  surgical approach in, 106
  wound management after, 104
FCR tendon. See Flexor carpi radialis tendon
FCU tendon. See Flexor carpi ulnaris tendon
FDP tendon. See Flexor digitorum profundus tendon
FDS tendon. See Flexor digitorum superficialis tendon
Felons, 82
Fetal alcohol syndrome, 326
Fetal hydantoin (dilantin) syndrome, 325
Fibrillation, 254
Fibrillation potentials, 257, 258
  in lower motor neuron injuries, 258
Fibroblasts, in wound contraction, 236
Fibromatoses, 108
"Fight bite," 24, 219
Finger bones. See Phalanges
Finger joints. See also specific joints
  joint contact area of, 17
Fingernail bed, injuries to, **225–233**
  epidemiology of, 226
  repair of, 227
Fingernails
  anatomy of, 225
  avulsion of, 23, 227
  infection resistance in, 11
  new, growth rate of, 226
  plate ridging of, 228
  production of, 11
  in psoriatic arthritis, 72
Fingers, 20. See also Index finger; Long finger; Ring finger; Thumb
  arthrodesis of, 195
    complications of, 195
  flexor pulleys of, 9
  joints of, 21
  maximum flexion arc of, 13
Fingertips
  amputations of
    complications of, 232
    donor tissue sources for, 232
    with dorsally angulated defect, 228
    with exposed bone, 232
    flap coverage for, 228–233
    revision and closure of, 232
  anatomy of, 225
  injuries to, **225–233**
    in children, 233
    evaluation of, 226
Finkelstein's test, 25, 75, 338

INDEX

Fishhooks, removal of, 89–90
Fist, "functional" spastic clenched, 120
Flaps, **242–252**
  axial
    extension of, 243
    versus random flaps, 242
    types of, 243
  blood circulation in, 245
  contraindications to, 243
  cross-finger, 229, 247
    innervated, 248
    reverse, 247
  cross-thumb, 248
  digital artery island, 247
  dorsal metacarpal artery, 246
  fillet, 248
  flag, 246
  free
    anticoagulation use with, 140
    benefits of, 140
    clinical examination of, 138
    comparison with pedicle flaps, 252
    contraindications to, 141
    dorsalis pedis, 140
    as edema cause, 333
    fascial, 140
    fasciocutaneous, 140
    lateral arm, 140
    postoperative monitoring of, 138
    proximal forearm, 140
    radial forearm, 140
    temporioparietal fascial, 140
    for traumatic wounds, 140
  functions of, 242
  groin, 251–252
  lateral arm, 250
  latissimus dorsal, 251
  Limberg, 245
  lines of maximum extensibility in, 243
  local, 244
    versus split-thickness skin grafts, 232
    types of, 243
  local pedicle, morbidity associated
    with, 231
  Moberg volar advancement, 228, 229, 246
    advantages of, 246
    complications of, 229, 246
    maximal advancement for, 229, 246
  neurovascular island, 248
    contraindications to, 248
  pectoralis major, 251
  pedicle
    comparison with free flaps, 252
    local, morbidity associated with, 231
    morbidity associated with, 231
  posterior interosseous artery, 251
  radial forearm, 21, 140, 249–250
    harvesting of, 12

Flaps *(Continued)*
  random
    most common types of, 243
    rotational flaps as, 243
  regional, 247
  reversed posterior interosseous artery, 251
  rotational
    contraindications to, 244
    differentiated from transpositional flaps, 244
    indications for, 244
    as random flaps, 243
  salvage of, 138
  scapular, 251
  versus skin grafts, 242
  thenar, 248
    advantages of, 248
    complications of, 248
  transpositional, 244
    advantages of, 244
    differentiated from rotational flaps, 244
    pitfalls of, 244
  types of, based on location, 243
  ulnar artery forearm, 250
  vascular supply to, 242
  V-Y advancement
    amputation pattern for, 245
    for fingertip injuries, 245
    Kutler, 228, 244, 245
    size of defect and, 245
    volar, 228, 244, 245
  Z-plasty
    60-degree, 245
    indications for, 245
    technical errors associated with, 245
    technical points of, 245
Flexion, in the wrist, 16
Flexor carpi radialis tendon
  in ligament reconstruction and tendon interposition (LRTI) procedures, 57
  tendinitis of, 204
    as rupture cause, 204
Flexor carpi radialis tunnel syndrome, 78
  clinical presentation of, 78
  treatment of, 78
Flexor carpi ulnaris tendon
  calcific tendinitis of, 204
  contraindication as tendon transfer, 275
  tenosynovitis in, 78
Flexor digitorum longus tendon graft, gliding resistance of, 216
Flexor digitorum profundus tendon
  lacerations of, 210
  "overtightening" of, 216
  in ring finger amputation, 233
  ruptures of, 212
    attritional, 173
  transection of, 233
  weakness in, 24

Flexor digitorum superficialis tendon
  absence of, 9
  advancement during laceration repair, 217
  evaluation of, 210
  excursion at the metacarpophalangeal joint, 17
  lacerations of, 216
Flexor pollicis longus tendon
  location of, 9
  rupture of
    attritional, 148
    rheumatoid arthritis-related, 61
  weakness in, 24
Flexor tendons
  anatomic zones of, 210, 211
  avulsed, retracted into the palm, 213
  bowstringing of, 215
  cerebral palsy-related tightness of, 115
  healing of, 212
  injuries to, **210–217**
    anatomic zones in, 210
    partial, 210
    sheath closure in, 213
    suturing techniques for, 213
  injury repair in, 213, 214
    analysis of results of, 215
    in children, 216
    of lacerations, 216
    strength testing after, 338
  lacerations of
    in the fibro-osseous sheath, 213
    partial, 216
    retracted into the palm, 213
    treatment of, 216
    in zone VIII, 220
  nutrition of, 212
  during passive range of motion, 18
  tenosynovitis of, 24
  transfers of, 274
Flexor tendon sheath, functions of, 9, 19
Flexor tenosynovitis, 24
Fluoric acid burns, 88
Fluoroscopy, 48
  real-time, 160
Forearm
  cerebral palsy-related pronation deformity of, 115
  compartments of, 104
  compartment syndrome of, 104
  congenital pseudoarthrosis of, 326
  innervation of, 30
  proximal amputations of, 301
  replantation of, 141
  rotation of, 150
    force distribution during, 16
    longitudinal axis of, 18
    radioulnar joints and, 17
    radius/ulna relationship during, 150
    trauma-related loss of, 157
Foreign bodies, in the hand, 88–90

Foreign-body inflammatory response, to silicone
    implants, 196
Forequarter amputation, 297
Fractures
  boxer's, 24, 178
  of the carpal bones, 167, 171–172
  child abuse-related, 206
  of the distal phalanges, 175
    avulsion fractures, 40
    complications of, 175
    displaced transverse, 175
    mallet fractures, 175
  of the distal radius, **138–149**
    arthroscopic-assisted reduction and fixation
        of, 135, 143, 146, 148
    articular step-off of, 143
    Barton's fractures, 140
    "chauffeur's" fractures, 141
    classification of, 138
    complications of, 335
    "crossed pinning" of, 143
    "die-punch" fractures, 141
    distraction of, 143
    epiphyseal, 142
    external fixation of, 145
    fragment-specific fixation of, 146
    hand therapy for, 334, 335
    intracarpal injuries associated with, 142
    ligamentotaxis of, 142, 143
    limited open reduction of, 144
    median nerve injuries associated with, 146
    minimally displaced, 206
    occult, immobilization of, 205
    open reduction and internal fixation
        of, 144, 334
    as osteoarthritis risk factor, 56
    osteochondral injuries associated with, 135
    pinning of, 143, 144
    plating of, 145
    reduction of, radiographic assessment of, 42
    Smith's fractures, 140
    soft tissue injuries associated with, 135
    superficial radial nerve in, 146
    treatment of, 138–149
    unstable, 143
    upper extremity examination in, 139
  epiphyseal, Salter-Harris classification of, 41, 142,
      182, 221, 222
  Essex-Lopresti, 156, 201
  Galeazzi's, 156, 185, 192
  of the hook of the hamate, 172, 205
    complications of, 172
    imaging of, 45
    nonunion of, 173
  metacarpal and phalangeal, **175–183**
  occult, in children, 208
  radial, effect on the triangular fibrocartilage
      complex, 154

Fractures *(Continued)*
  scaphoid, 24, 25, 167, 200
    arthroscopic-assisted percutaneous fixation of, 136
    with avascular necrosis, 40
    classification of, 168
    displaced, 169
    healing rates of, 168
    malunion of, 169
    mechanism of, 168
    nonunion of, 170
    occult, immobilization of, 205
    treatment of, 168
    vascularity assessment in, 45
  ulnar collateral ligament avulsion, 180
  of the ulnar styloid process, 154
Frohse, arcade of, 264
Froment's sign, 23, 30, 276
Frostbite, **94–97**
  classification of, 94
  definition of, 94
  first aid for, 94

G
Galeazzi's fracture, 156, 185, 192
"Gamekeeper's thumb," 188, 189
Ganglion, 310
  definition of, 310
  of the dorsal wrist, 310, 311
  most common locations for, 310
  transillumination evaluation of, 311
  of the volar wrist, 310
    resection of, 282
Gangrene, gas, 84
Garrod's knuckle pads, 109
Gas gangrene, 84
Geissler arthroscopic grading system, for intercarpal ligament injuries, 136
Germinal matrix, of the fingernails, 225
  incomplete removal of, 228
Giant cell tumors
  of bone, 202, 304, 308
    metastatic, 307
  of the tendon sheath, 310, 311
Glass, embedded in hand, diagnostic imaging of, 206
Glomus tumors, 283, 312
Gonorrhea, 83
Gouty arthritis, 82
Granuloma, pyogenic, 310
Gravel, as foreign body, 89
Grease, high-pressure injection of, 90
"Great masqueraders", 82
Grind test, 25, 54
Grip strength test, after flexor tendon repair, 338
Guyon's canal
  anatomic boundaries of, 262
  contents of, 10, 263
  as ulnar nerve compression site, 23, 263

H
Hamartoma, lipofibromatous, 311–312
Hamate bone
  fractures of, 171
    types of, 172
  hook of, fractures of, 172, 205
    complications of, 172
    imaging of, 45
    nonunion of, 173
  ossification of, 12
Hand
  anatomic zones of, 210, 211
  anatomy of, **9–14**
  arterial blood supply to, 280
  biomechanics of, **15–19**
  fixed unit of, 12
  innervation of, 23
  maximal sensory innervation in, 12
  physical examination of, **20–26**
  primary functions of, 15
  topographic areas of, 21
Hand-arm vibration syndrome (HAVS), 330
Hand dominance, in multiple finger amputation patients, 289
Hand pain, 20
Hand therapy, **333–339**
Hansen's disease (leprosy), 83
HAVES (hand-arm vibration syndrome), 330
Heat therapy, 335
Heberden's nodes, 51, 59
Hemangioma, 282, 312
  multiple enchondromatosis associated with, 302
Hematoma
  splinter, 226
  subungual, 175, 226
Hemiarthroplasty, resection, 61
Hemiplegia, spastic, 114, 116
Hemochromatosis, 55
Hemorrhage, from blood vessel lacerations, 282
Hereditary neuralgic amyotrophy (HNA), 331
Hereditary neuropathy with liability to pressure palsies (HNPP), 331
Herpes simplex virus type 1, 82
High-pressure injection injuries, 90–92
Hirudis medicinalis. See Leech therapy
Holt-Oram syndrome, 317
Homografts, 333
Hook nails, 228
Hook of the hamate, fractures of, 172, 205
  complications of, 172
  imaging of, 45
  nonunion of, 173
Horner's syndrome, 35
Huber transfer, 275
Hueston's tabletop test, 112
Human bite wounds, 219
  as *Eikenella corrodens* infection cause, 82
  as septic arthritis cause, 71

Human bite wounds *(Continued)*
  with zone V extensor tendon lacerations, 220
Humerus, distal, occult fractures of, 208
Humpback deformity, of the scaphoid bone, 203
Hydrofluoric acid burns, 100
Hyperemia, burn-related, 98
Hyperparathyroidism, radiographic findings in, 40
Hyponychium, 225, 226
  infection resistance in, 11
Hypoplasia, 319
Hypothenar hammer syndrome, 200, 281
Hypuricemia, 73

I

Imaging, diagnostic, **38–49**. *See also* Computed tomography (CT); Magnetic resonance imaging (MRI); Radiographic evaluation
  of carpal instability, 160–162, 164
  of cervical radiculopathy, 28
  of foreign bodies, 88–89
  of Kienböck's disease, 127, 128
  of scaphoid fractures, 169
Immobilization. *See also* Casting; Slings; Splinting
  intrinsic plus position of, 175
  juncturae tendinum in, 337
  of occult fractures, 205
  postoperative position of, 338
  in skin graft patients, 99
Index finger
  amputation of, 296
  fractures of, 177
  metacarpophalangeal dorsal dislocations in, 187
  pollicization of, 323
Infections, **81–84**
Inflammation, physical signs of, 235
Inflammatory bowel disease, 74
Infraspinatus muscle, in suprascapular nerve entrapment, 264
Innervation, of the hand, 23. *See also specific nerves*
  anomalous, 256
Intercarpal fusion, effect on wrist flexion/extension, 16
Intercarpal ligaments, strength of, 18
Intercompartmental supraretinacular artery (ICSRA) bone grafts, 130
Interfascicular nerve grafts, 270
"Internal splint" procedure, 275
International Classification of the Hand in Tetraplegia, 119
Interosseous muscles
  actions of, 10
  distal attachment of, 10
Interphalangeal joints. *See also* Distal interphalangeal joints; Proximal interphalangeal joints
  checkrein ligament of, 13
  degrees of freedom of, 17
  extended, in cerebral palsy, 116
  open dislocation of, 191

Interscalene nerve blocks, 35
Intersection syndrome, 79, 204
  definition of, 76
  differentiated from de Quervain's disease, 76
  treatment of, 76
Intramedullary nailing, contraindication to, 183
Intravenous drug use, as necrotizing fasciitis cause, 84
Intrinsic muscles, of the hand, 23
Intrinsic tightness, test for, 25
Ischemia
  extravasation injury-related, 85
  high-pressure injury-related, 92
  prohibitive, 291

J

"Jersey finger," 212
Job dissatisfaction, 328
Joint capsule, injury to, 191
Joint contact area, of the finger joints, 17
Joint-leveling procedures, in Kienböck's disease, 129
Joint pain
  in osteoarthritis, 26
  in rheumatoid arthritis, 26
Joints, 21. *See also specific joints*
  contact area of, 17
  dislocations of, **185–192**
Juncturae tendinum, in immobilized hand, 337

K

Kanavel, Allen Buchner, 81
Kanavel's cardinal signs, 24, 81
Kapandji technique, of intrafocal pinning, 144
Kaplan's cardinal line, 260
Kasabach-Merritt syndrome, 312
Kessler technique, modified, 218
Keyboarding
  as carpal tunnel syndrome cause, 329
  as hand or arm pain cause, 331
  relationship to carpal tunnel syndrome, 329
  as tendinitis cause, 329
Key-pinch function, restoration in low ulnar nerve palsy, 276
"Key-pinch operation," in spinal cord injury patients, 278
Kienböck's disease, **126–130**, 203
  in children, 129
  diagnostic imaging of, 127, 128
  joint-leveling procedures in, 129
  radiographic staging of, 128
  silicone replacement arthroplasty in, 130
Kirner's deformity, 324
Kirschner wires, flexible intramedullary, 182
Kleinman shear test, 159
Krukenberg procedure, 297

L

Lacerations
  of blood vessels, as hemorrhage cause, 282

Lacerations *(Continued)*
   of the fingernail bed, 226
   of the flexor digitorum profundus, 210
Lateral antebrachial cutaneous nerve, 258
Lateral arm flaps, 140
Lead poisoning, 208
Ledderhose's disease, 109
Leech therapy, 82, 138, 139, 293–294
Legal aspects, of hand surgery, **340–343**
Leprosy (Hansen's disease), 83
Leukocyte scans, 48
Levodopa, 114
Lhermitte's sign, 29
Lichtman radiographic classification, of Kienböck's disease, 128
Ligament of Struthers, 262
Ligament of Testut. *See* Radioscapholunate ligament
Ligament reconstruction and tendon interposition (LRTI) procedure, 57
Ligaments. *See also specific ligaments*
   injuries to, **185–192**
   arthroscopic grading system for, 136
Limb bud, growth and development of, 316
Limbs
   amputated
      care and transport of, 142, 289
      viability of, 142
   congenital anomalies of, classification of, 316
   replantation of
      complications of, 294
      fasciotomy following, 291
      prognostic serum marker for, 294
      recommended sequence for, 291
Linburg's syndrome, 78–79
Linscheid ulnar snuffbox compression test, 159
Lipoma, 310
   magnetic resonance imaging of, 311
Lister's tubercle, 12, 61
LOAF mnemonic, 11
Local anesthetics
   allergic reactions to, 33
   low-concentration, 34
   toxicity of, 33
   use in pregnant patients, 36
Long finger
   extensor injuries to, 218
   in sagittal band tears, 222
   as trigger finger, 77
Lowenstein (Sauve-Kapandji) procedure, 61, 198, 277
LRTI (ligament reconstruction and tendon interposition) procedure, 57
Lumbrical muscles, 12
   anatomic site of origin of, 210
   location of, 210
Lumbrical muscles, separation from the interossei muscles, 10
Lumbrical-plus deformity, 25, 216, 233, 296
Lunate at risk, 127

Lunate bone
   anatomic relationships of, 16, 18
   avascular necrosis of. *See* Kienböck's disease
   dorsal intercalated segmental instability (DISI) in, 162
   extraosseous blood supply to, 126
   fractures of, 171
   ossification of, 12
   relationship to the scaphoid bone, 16
   synostosis of, 326
   volar dislocation of, 203
Lunatomalacia. *See* Kienböck's disease
Lunocapitate joint, in midcarpal joint contact, 18
Lunotriquetral ligament, 158
   arthroscopic visualization of, 132
   competency assessment of, 159
   instability of
      signs and symptoms of, 166
      treatment of, 166
   rupture of, 162
   strength of, 15
Lunotriquetrum, synostosis of, 326
Lunula, 225

M

Macrodactyly, 312, 319, 324
Madelung's deformity, 325
   radiographic features of, 40
Mafenide acetate cream (Sulfamylon), 98
Maffucci's disease, 302, 303
Magnetic resonance arthrography, 48
   of the wrist, 48
Magnetic resonance imaging (MRI), 48
   for carpal instability assessment, 160
   of foreign bodies, 89
   of Kienböck's disease, 128
   in multiple sclerosis patients, 31
   of scaphoid fractures, 169
   of soft tissue masses/tumors, 48, 313
Mallet finger deformity, 25, 220
   classification of, 220
   with swan-neck deformity, 223
Mallet fractures
   definition of, 175
   treatment of, 175
Malpractice litigation, 340
   prima facie cases in, 342
Malrotation, relationship to digital overlap, 179
Mannerfelt syndrome, 61
Marfan's syndrome, 325
   with arthrogryposis, 322
Martin-Gruber connection, 256
MCP joints. *See* Metacarpophalangeal joints
Medial antebrachial cutaneous (MABC) nerve, in hand numbness, 29
Median nerve, 23
   anatomic variations in, 260
   bowstringing of, 215

Median nerve *(Continued)*
  crossover in, 12
  distal radius fracture-related injuries to, 146
  muscles innervated by, 11
Median nerve blocks, at the wrist, 36
Median nerve compression, anatomic sites of, 259
Median nerve palsy
  as carpal tunnel syndrome risk factor, 329
  high, tendon transfer in, 276
  low, 275
    motor deficits associated with, 275
    opponensplasty for, 338
    with ulnar nerve palsy, 278
    wrist and hand function restoration in, 278
  thumb opposition restoration in, 275
Melanoma, subungual, 314
Meshing, of skin grafts, 240
Metacarpal base fractures
  Bennett's fractures, 180–182
    anterior oblique carpometacarpal ligament in, 13
    definition of, 180
    first metacarpal subluxation in, 181
    reverse, 182
    treatment of, 182
  metacarpal length restoration in, 180
  Rolando's fractures, 182
Metacarpal bones
  fifth, fractures of, 24
  shortening of, 182
  synostosis of, 326
Metacarpal fractures, 177–182
  plate and screw fixation of, 179
  rotational malalignment of, 178, 179
Metacarpal head, fractures of, 177
  comminuted fractures, 177
Metacarpal neck, fractures of, 178
  metacarpal length restoration in, 180
Metacarpal shaft, fractures of, 178
  angulation in, 179
  displacement patterns of, 179
  dorsal plating of, 183
  malunion of, 183
  metacarpal length restoration in, 180
  open reduction and internal fixation of, 179
Metacarpophalangeal implant arthroplasty, 334
Metacarpophalangeal joints
  amputations at level of, 296
  arthrodesis (fusion) of, 195
  dislocations of
    complex, 190, 192
    complex dorsal, 187
    dorsal, 187
    in the index finger, 187
    open reduction of, 190
    reduction maneuvers for, 187
    simple, 187
    simple differentiated from complex, 187

Metacarpophalangeal joints *(Continued)*
  effect of proximal interphalangeal-level amputations on, 296
  flexed, in cerebral palsy, 116
  flexor digitorum superficialis excursion at, 17
  human bite wounds to, 219
  hyperextension instability of, 57
  intrinsic tightness assessment in, 25
  joint contact area of, 17
  in mobilization programs, 337
  osteoarthritis of, 55
  rheumatoid arthritis-related deformity of, 62, 63, 278
  in systemic lupus erythematosus, 70
  ulnar deviation of, rheumatoid arthritis-related, 278
Metastases, of bone tumors, 307, 308
Microsurgery, **284–288**
Microwave exposure, 205
Midcarpal joints
  contribution to wrist flexion/extension, 17
  force distribution through, 18
  instability of
    assessment of, 159, 162
    causes and treatment of, 166
    triangulation method for determination of, 162
  stabilizing ligaments of, 158
Middle phalanges, fractures of
  displacement of, 183
  open reduction and internal fixation of, 183
Mithramycin, 85
Mitomycin, 85
"Mobile wad," 14
Mobilization programs, following tendon repair, 337
Monitoring
  during conscious sedation, 32
  during regional anesthesia, 33
Morning stiffness, rheumatoid arthritis-related, 334
Motor latency
  definition of, 255
  relationship to stimulating electrodes, 255
Motor nerve conduction studies, 255
Motor neuron injuries
  lower, fibrillation potentials in, 258
  upper, differentiated from lower, 30
Motor units, components of, 254
Multiple crush syndrome, 331
Multiple sclerosis, magnetic resonance imaging studies in, 31
Mummification, of frostbitten digits, 96
Muscle fibers, type 1 and type 2, 254
Muscle neurotization, 272
Muscles, of the hand
  extrinsic, 23
  intrinsic, 23
Muscle-tendon transfers, in biceps paralysis, 278
Muscle wasting, cervical rib-related, 28
Muscle weakness, myasthenia gravis-related, 255
Musculocutaneous nerve, 258
Musculoskeletal tumors, classification of, 305

Myasthenia gravis, 255
  pathophysiology of, 255
Mycobacterial infections, as the
  "great masqueraders," 82
Mycobacterium leprae infections, 83
Mycobacterium marinum infections, 81, 207
Myofibroblasts
  in Dupuytren's disease, 110
  in wound contraction, 236
Myonecrosis, clostridial, 84
Myotonia, 254

N

Nail fold, 225
Nails
  of the fingers. See Fingernails
  from nail guns, removal of, 90
Nail spikes, 228
Nalebuff classification system, of thumb deformities, 65
Necrosis
  avascular
    radiographic diagnosis of, 44
    of the scaphoid bone, 14, 40
  extravasation-related, 85, 87
  high-pressure injury-related, 92
Necrosis interval, 87
Neisseria gonorrhoeae, 83
Neonates, brachial plexus injuries in, 339
Neovascularization, of skin grafts, 237
Nerve blocks
  effect of local anesthetics on, 34
  at the elbow, 36
Nerve compression. See Neuropathies, compression
Nerve conduction
  in motor neuron lesions, 255
  in neurogenic thoracic outlet syndrome, 257
Nerve conduction block, pathophysiology of, 254
Nerve conduction velocity
  calculation of, 256
    inaccurate, 256
  correlation with nerve fiber diameter, 254
Nerve conduits, 273
Nerve fiber diameter, correlation with conduction velocity, 254
Nerve grafts
  donor nerve to, 270
  tissue bed creation after, 272
  types of, 270
Nerve injuries
  diagnostic tests for, 21
  repair of
    classification of, 268
    return of sensation following, 142
Nerve tumors. See also Neuroma
  of the upper extremity, 312
Neurilemoma, 312
Neurofibroma, 312
  multiple, 312
Neurofibromatosis, 326
  type 1 (von Recklinghausen's disease), 312
Neurologic deficits, tetraplegia-related, 118
Neurologic examination, of the upper extremity, **27–31**
Neuroma
  amputation-related, 233, 296
  in continuity, 268
  definition of, 272
  surgical treatment of, 272
Neuropathies
  anterior interosseous, 258
  compression, 254, **259–264**
    electrodiagnostic parameters of, 261
    idiopathic, 260
  diabetic
    anesthesia use in, 36
    misdiagnosed as complex regional pain syndromes, 124
  peripheral
    axon loss in, 257
    common causes of, 30
Neuropraxia, 267
Neurorrhaphy, end-to-end, 272
  fascicle matching in, 269
Neurotmesis, 267
Neurotropism, 273
Nicotine, effect on wound healing, 233
Nodes, Bouchard's, 51, 59
Nodes of Ranvier, 254
Nodules
  palmar, 108, 110
  rheumatoid, 60
"No man's land," 211
Nuclear medicine tests, for osteomyelitis evaluation, 48
Numbness, of the hand, 29

O

OA. See Osteoarthritis
Objective medical findings, 341
Occupational injuries and illnesses, of the upper extremity, **328–332**
Occupational medicine specialists, referrals to, 332
"Okay sign," 24
Ollier's disease, 302, 303
Open palm technique, 112
Open reduction and internal fixation (ORIF)
  as collateral ligament detachment cause, 183
  of distal radius fractures, 334
  of metacarpal shaft fractures, 179
Opponensplasty, 338
Orthodromic, definition of, 255
Orthotics, for brachial plexopathy, 336
Osborne's ligament, 262
Ossification, heterotopic, 117
Osteoarthritis, **50–57**
  of the basilar joint, 25, 51, 53
    arthrodesis treatment of, 56

INDEX

Osteoarthritis *(Continued)*
  of the carpometacarpal joints, 197, 198
  clinical conditions associated with, 54
  clinical features of, 50
  comparison with rheumatoid arthritis, 59
  differential diagnosis of, 55
  erosive, 50
  frostbite-related, 96
  functional limitations in, 50
  genetic factors in, 51
  joint pain associated with, 26
  nonoperative treatment for, 52
  prevalence of, 50
  primary generalized, 50
  radiographic features of, 41
  secondary, 54
  treatment goals in, 52
  triscaphe, 56
  in the wrist, 205
    patterns of, 55
    SLAC pattern of, 55, 198, 202
Osteoarthrosis, 50
Osteochrondroma, 303
  complications of, 303
  definition of, 303
Osteoid osteoma, 303–304
  carpal, 304
Osteolysis, psoriatic arthritis-related, 72
Osteomyelitis, of the hand or wrist, 308
  bone scans of, 47
  leukocyte scans of, 48
  prevalence of, 84
  radiographic features of, 44
  treatment of, 84
Osteotomy, ulnar shortening, 18

P

Pain, sensory pathways for, 29
Palmar cutaneous nerve, lacerations of, 209
Palmar (extrinsic) ligaments, 150
Palmaris longus tendon, presence of, 9
Palmar vascular arch, incomplete, 208
Paresthesia, carpal tunnel syndrome-related, 23
Parkinson's disease
  onset age of, 29
  upper extremity symptoms in, 28
Parona's space, 83, 204
Paronychia, 81, 225
Pasteurella multocida infections, 82
Penetrating trauma, as septic arthritis cause, 83
Perilunate injury, carpal force transmission in, 164
Perineurium, 266
Perionychium, 225
Periosteum, 227
Peripheral nerve(s)
  anatomy of, 266
  correspondence to upper extremity reflexes, 30
  entrapment neuropathy of, 257

Peripheral nerve blocks, 33
Peripheral nerve grafts, types of, 270
Peripheral nerve injuries, **266–273**
  classification of, 267
  electrodiagnostic evaluation of, 272
  maximal recovery time in, 267
  repair of
    classification of, 268
    goal of, 268
    techniques, 268–273
    with tendon transfers, 274
    tissue bed creation in, 272
Peyronie's disease, 109
Phalangeal fractures
  of the distal phalanx
    avulsion fractures, 40
    complications of, 175
    displaced transverse, 175
    mallet fractures, 175
  with persistent fracture line, 177
  plate and screw fixation of, 179
  of the proximal phalangeal neck, 176
  with residual deformity, 182
  rotational malalignment of, 179
  stable transverse, 177
  unstable, 177
Phalangeal head, condylar fractures of, 176
Phalangeal neck fracture, 176
Phalanges
  delta, 324
  distal. *See* Distal phalanges
  fractures of. *See* Phalangeal fractures
  middle. *See* Middle phalanges
  proximal. *See* Proximal phalanges
Phalanx. *See* Phalanges
Phalen's test, 23, 261
Phantom pain, 37, 297
Pharyngeal defect, radial forearm free flaps for, 21
Phenol block, 117
Phocomelia, 316
Physical examination, of the hand, **20–26**
Piano key sign, 60
Pilon fractures, 176
Pinch strength test, after flexor tendon repair, 338
PIP joints. *See* Proximal interphalangeal joints
Pisiform bone
  fractures of, 171
    mechanism of injury in, 171
    undetected, 171
  ossification of, 12
  tendon insertion of, 13
Pisotriquetral joint, radiographic views of, 42
Poirier, space of, 132
Poland's syndrome, 320
Pollicization, of the index finger, 323
Polyarthralgias, migratory, 83
Polydactyly, 319, 324

Polyneuropathy, 256
  with superimposed focal entrapment mononeuropathy, 257
Possibilities, 341
Posterior interosseous nerve (PIN), 11
  palsy of
    finger extension loss in, 65
    tendon transfers in, 275
  resection of, 56
Posterior interosseous syndrome, 11
Power grip, maximal, 17
Pregnant patients
  anesthesia in, 36
  carpal tunnel syndrome during, 200
Prehension, 15
Preiser's disease, 170, 200
Probability, 341
Pronator syndrome, 23, 261
Proprioception, 29
Prostheses, 298–299, 300
  above-elbow, 298
    comparison with elbow disarticulation, 299
  below-elbow, 298, 299
    minimum forearm length in, 299
  body-powered, 298
  fitting of, 298, 335
    in children, 300
    in traumatic amputations, 300
  myoelectric, 298–299
  osteointegrated, 300
  types of, 336
Proximal forearm flaps, 140
Proximal interphalangeal joints
  amputations through, 296
  arthrodesis (fusion) of, 195
  dislocations of, 190
    dorsal, complications of, 192
  fracture-dislocations of, volar articular surface in, 192
  injuries to
    classification of, 189
    treatment of, 189
  lateral instability in, 191
  in mobilization programs, 337
  osteoarthritis of, 52–53
  pilon fractures of, 176
  stability of, 189
Proximal phalangeal neck, fractures of, 176
Proximal phalanges
  fractures of
    displacement of, 183
    open reduction and internal fixation of, 183
    Salter-Harris type II epiphyseal separation of, 182
Proximal scaphoid pole, sclerosis of, 44
Prune test, 22, 23
Pseudoarthrosis, congenital, of the forearm, 326
Pseudogout, 72, 200
Pseudo-Terry Thomas sign, 13

Psoriatic arthritis, 71–72
  differentiated from rheumatoid arthritis, 40
  of the distal interphalangeal joints, 73
Psychogenic hand disorders, 331
Psychological disorders
  as complex regional pain syndromes risk factor, 124
  misdiagnosed as complex regional pain syndromes, 124
Pulleys, 211
  annular, 211
  cruciform, 211
  flexor, of the fingers, 9
  oblique, 212
  palmar triggering of, 79
  postoperative care for, 337
  in trigger finger, 77
  in zone 2 flexor tendon repairs, 9
Puncture wounds, 73

Q

Quadriga effect, 25, 215, 217, 233, 296

R

Radial artery, 280
  arterial line insertion into, 208
  distal, in radial forearm flap harvesting, 12
  dorsal branch of, 12
  volar wrist ganglion resection-related injury to, 282
Radial collateral ligament, injury to, 191
Radial head
  congenital
    differentiated from acquired, 321
    treatment of, 321
  excision of, 19
  occult fractures of, 208
Radial nerve, 23
  crossover in, 12
  dorsal sensory branches of, 209
Radial nerve compression, anatomic sites of, 263
Radial nerve palsy, tendon transfer in, 274, 275
Radial sagittal band, tears of, 222
Radial tunnel syndrome, definition of, 263
Radiculopathy, cervical, 28
  C7, 29
  C8 motor, differentiated from carpal tunnel syndrome, 258
  diagnostic studies of, 28
  electromyographic studies of, 257
  sensory nerve conduction studies of, 30
  signs and symptoms of, 28
Radiocarpal joints
  contribution to wrist flexion/extension, 17
  dislocations of, 185
  fusion of, 15
  stabilizing ligaments of, 158

Radiocarpal ligaments, extrinsic, 132
Radiocarpal/ulnocarpal load transfer ratio, 133
Radiographic evaluation, of the wrist and hand, 38–44
　of distal radius fractures, 138
　of fractured digits, 175
　of high-pressure injection injuries, 91
　of osteoarthritis, 59
　posterioanterior (PA) views in, 38
　of rheumatoid arthritis, 59
Radioscapholunate ligament, 159
　function of, 19
Radius bone
　absence of, 316, 317
　distal. *See* Distal radius bone
　dysplasia of, 317
　fractures of
　　effect on the triangular fibrocartilage complex, 154
　　malunion of, 155
　longitudinal deficiency of, 316–317
　shortening of, 129
Range of motion, passive, digital flexor tendon force generation during, 18
Range-of-motion exercises, 333
Ray amputations, 296–297
Raynaud's disease
　definition of, 281
　treatment of, 281
Raynaud's phenomenon, 281
Reagan ballottement test, 159
Reconstruction, upper extremity
　free flap or reperfusion monitoring in, 138
　free flap use in, 140
Reflexes, upper extremity, 30
Reflex sympathetic dystrophy, 112, 122. *See also* Complex regional pain syndromes
　bone scan findings in, 47
Regional anesthesia, 33
　tourniquet use in, 36
Reperfusion injury, 291
Reperfusion syndrome, 291
Repetitive activities, as intersection syndrome cause, 79
Replantation, **289–294**
　anesthesia management during, 292
　bony stabilization during, 292
　complications of, 294
　contraindications to, 290
　definition of, 289
　heterotopic, 291
　indications for, 141
　　absolute, 290
　　in children, 142
　　relative, 290
　interpositional vein grafting in, 292
　obstacles to, 290
　polydigit, 142
　postoperative management in, 293

Replantation *(Continued)*
　sequence of events in, 142
　　for amputated fingers, 291
　　for amputated limbs, 291
　　skeletal shortening during, 292
Replanted digits or limbs
　arterial insufficiency in, 293, 294
　unsalvageable, 142
　vascular status monitoring of, 293
Replant reperfusion, monitoring of, 138
"Retinacular windows," 213
Revascularization, definition of, 289
Rewarming, rapid, of frostbitten hand, 95
Rheumatoid arthritis, 26, **59–68**
　comparison with osteoarthritis, 59
　cross-intrinsic tendon transfers in, 278
　diagnosis of, 59
　differentiated from psoriatic arthritis, 40
　finger extension loss in, 65
　joint pathology in, 59
　juvenile, 66–68, 202
　　diagnostic criteria for, 66
　　differentiated from adult rheumatoid arthritis, 68
　　pauciarticular onset, 68
　　polyarticular onset, 68
　metacarpophalangeal implant arthroplasty in, 334
　metacarpophalangeal ulnar deviation in, 278
　morning stiffness treatment in, 334
　radiographic evaluation of, 43
　　preoperative, 43
　surgical treatment of, 59
　　arthroscopic synovectomy, 135
　tendon ruptures associated with, 61, 72, 277
　tenosynovitis associated with, 72
　wrist changes associated with, 62
Rheumatoid factor, 71
Ribs, cervical, 28
Ring avulsion injury, 292
Ringer's solution, lactated, 148
Ring finger
　amputation though, 233
　in "jersey finger," 212
　as trigger finger, 77
Rolando's fracture
　definition of, 182
　treatment of, 182
Rotational deformities, assessment of, 24

S

Sad, hostile, anxious, frustrating, tenacious (SHAFT) syndrome, 331
Sagittal band tears, 222
　with extensor tendon subluxation or dislocation, 222
Salter-Harris classification, of epiphyseal fractures, 41, 142, 182, 222
Salter-Harris type II epiphyseal separation, 182
Sarcoma
　epithelioid, 313, 314

Sarcoma *(Continued)*
  Ewing's, 305
    clinical presentation of, 306
    definition of, 306
    treatment of, 306
  osteogenic, 305, 306
    in children and adolescents, 308
    clinical presentation of, 308
    radiographic features of, 306
    treatment of, 306
  soft tissue, 313
    diagnosis of, 314
    metastatic, 314
    prognosis for, 314
    treatment of, 314, 336
Sauve-Kapandji (Lowenstein) procedure, 61, 198, 277
Scalene (Adson's) test, 282
Scaphocapitate joint, in midcarpal joint contact, 18
Scaphocapitate syndrome, 170
Scaphoid bone
  anatomic relationships of, 18
  avascular necrosis of, 14, 40
  blood supply to, 14, 168
  computed tomography of, 45
  fractures of, 25, 167, 171, 200
    arthroscopic-assisted percutaneous fixation of, 136
    with avascular necrosis, 40
    classification of, 168
    displaced, 169
    healing rates of, 168
    malunion of, 169
    mechanism of, 168
    nonunion of, 170
    occult, immobilization of, 205
    percutaneous screw fixation of, 172
    treatment of, 168
    vascularity assessment following, 45
  humpback deformity of, 203
  location of, 25
  neglected nonunion of, 56
  ossification of, 12
  osteoid osteomas of, 304
  radiographic views of, 42
  relationship to the lunate bone, 16
Scaphoid nonunion advanced collapse (SNAC) pattern, of wrist arthritis, 56, 170
  radiographic features of, 41
Scaphoid shift test, 159
Scaphoid view radiograph, 42
Scaphoid waist, fractures of
  nondisplaced, 172
  as proximal scaphoid pole sclerosis cause, 44
Scapholunate advanced collapse (SLAC) pattern, of wrist arthritis, 55–202
  four-corner arthrodesis for, 198
Scapholunate ligament, 158
  function of, 16

Scapholunate ligament *(Continued)*
  instability of, 159
    radiographic findings in, 162
    treatment of, 166
  rupture of, 162
    arthroscopic-assisted reduction and fixation of, 135
  sectioning of, 16
Scaphotrapeziotrapezoid joint, in midcarpal joint contact, 18
Scar tissue, 235, 236
Schwannoma, 312
Scientific evidence, 342
Scleroderma, 73
Sclerosis, of the proximal scaphoid pole, 44
Secondary intention, 237
Secretan's syndrome, 331
Secrets, Top 100, **1–8**
Sedation, conscious, 32
Sensory nerve conduction studies
  of avulsion injuries, 257
  of cervical radiculopathy, 30
Sensory pathways, 29
Septic arthritis, 70, 73, 82
  gonorrhea-related, 83
  human bite wounds-related, 71
  penetrating trauma-related, 83
  radiographic features of, 44
Septic joints, 73
  as surgical emergency, 73
SHAFT (sad, hostile, anxious, frustrating, tenacious) syndrome, 331
Shoulder, "frozen," 208
Shoulder disarticulation, differentiated from forequarter amputation, 297
Sigmoid notch
  in distal radioulnar joint stability, 15, 16
  incongruous, 198
Silicone implants
  foreign-body inflammatory response to, 196
  as synovitis cause, 196, 197
  radiographic features of, 41
Silicone replacement arthroplasty, 130, 197
Silver nitrate solution, 98
Silver sulfadiazine cream (Silvadene), 98
Simian crease, 325
Simpson, O. J., 341
Sinuses, yellow sulfur granule drainage from, 82
"Skier's thumb," 188
Skin deficiency, fasciectomy-related, 112
Skin flaps. *See* Flaps
Skin grafts
  allografts, 237
  autografts, 237
  for burns, 99, 333
  donor site epithelialization of, 240
  for the dorsum of the hand, 237
  following fasciectomy, 112

Skin grafts *(Continued)*
full-thickness
advantages of, 239
contraindications to, 238
disadvantages of, 239
donor sites for, 240
errors in, 240
technical features of, 238
harvesting of, 239
hypothenar, reinnervation density in, 241
inosculation of, 237
neovascularization of, 237
size of, 241
versus skin flaps, 242
split-thickness, 333
adherence of, 239
advantages of, 238
contraindications to, 239
depth of, 239
disadvantages of, 238
meshing of, 240
survival of, 237
thickness of, 238
wound maturation and contraction following healing, 238
survival of, 237
technical features of, 241
thenar, reinnervation density in, 241
timing of, 238
types of, 237
xenografts, 237
Skin tumors, malignant, 314
SLAC (scapholunate advanced collapse) pattern, of wrist osteoarthritis, 55, 202
four-corner arthrodesis for, 198
Slings
figure-of-eight, 336
inappropriate use of, 335
Smith's fractures, 140
Smoking
as complex regional pain syndromes risk factor, 124
as Dupuytren's disease risk factor, 109
Smoking cessation, in fingertip injury patients, 233
Soft tissue coverage, **235–241**. *See also* Flaps; Skin grafts
principles of, 236
"reconstructive ladder" for, 236
in systemic lupus erythematosus, 70
tissue expansion in, 252
Soft tissue tumors, **310–314**
magnetic resonance imaging of, 48
Space of Parona, 204
Space of Poirier, 13, 132
"Spare parts surgery," 291
Spasticity
cerebral palsy-related, 114, 115
stroke-related, 117
traumatic brain injury-related, 117
treatment of, 117

"Spilled tea cup" sign, 43, 164, 203
Spinal cord, effect of distal limb amputations on, 301
Spinal cord injury/disease
biceps muscle paralysis associated with, 278
elbow extension restoration in, 278
"key-pinch operation" in, 278
misdiagnosed as complex regional pain syndromes, 124
as tetraplegia cause, 118
triceps muscle paralysis associated with, 278
Spinal roots, correspondence to upper extremity reflexes, 30
Spiral cords, 12
Splinting
as arthrogryposis multiplex congenita treatment, 339
of boutonnière deformity, 221
as carpal tunnel syndrome treatment, 200
carpometacarpal, 334
as cubital tunnel syndrome treatment, 339
as de Quervain's disease treatment, 80
as dorsal compartment pain treatment, 338
of Dupuytren's contractures, 334
of mallet finger injuries, 220
of occult fractures, 205
of partial-thickness hand burns, 333
postoperative position of, 338
as work-related upper extremity disorder treatment, 330
Squamous cell carcinoma, 314
Staphylococcus aureus infections, 70, 71
as osteomyelitis cause, 84
as suppurative flexor tenosynovitis cause, 81
Static block procedures, 276
Stener's lesion, 188
Sterile matrix, of the fingernails, 225, 227
loss of, 228
Still's disease, 68
Strength testing, after flexor tendon repair, 338
Streptococcus pyogenes group A infections, 82
Stroke
as cubital tunnel syndrome cause, 118
as spasticity cause, 117
surgical reconstruction following, 116
Stroke patients
functional surgical procedures in, 120
nonfunctional surgical procedures in, 120
Struthers, arcade and ligament of, 262
Styloidectomy, radial, 194
Suave-Kapandji procedure, 157
Subfascial spaces, infections of, 83
Sulfur granule drainage, 82
Superficial arches, complete, 280
Supraclavicular blocks, 35
Suprascapular nerve entrapment, 264
Supraspinatus muscle, in suprascapular nerve entrapment, 264

Suturing techniques
  epitendinous, 66, 213
  in flexor tendon repair, 213
Swan-neck deformity, 24, 54
  causes of, 223
  classification system for, 63
  definition of, 223
  rheumatoid arthritis-related, 63
  secondary, 223
Symbrachydactyly (cleft hand), 318
Symphalangism, 319, 321
Syndactyly, 319–320
Synostosis, 326
  metacarpal, 326
  proximal radioulnar, 320, 321
Synovectomy
  arthroscopic, in rheumatoid arthritis, 135
  with extensor tendon realignment, 63
Synovial fluid analysis, in osteoarthritis, 54
Synovitis, silicone implants-related, 196, 197
  radiographic features of, 41
Systemic lupus erythematosus, 70

T
TAR syndrome, 317, 318
"Telescoping," of the digits, 73
Temperature, sensory pathways for, 29
Temporioparietal fascial flaps, 140
Tendinitis
  calcific, 80
  Ciproxin-related, 329
  corticosteroid injection therapy for, 79
  of the flexor carpi radialis tendon, 204
  of the flexor carpi ulnaris tendon, 204
Tendon grafts, 62
  tight, 216
Tendons. *See also specific tendons*
  injuries to, anatomic zones in, 218
  rupture of
    psoriatic arthritis-related, 72
    rheumatoid arthritis-related, 61, 66, 72
    treatment of, 62
Tendon sheath, giant cell tumors of, 310, 311
Tendon transfers, 62, **274–278**
  cross-intrinsic, 278
  definition of, 274
  hand therapy following, 338
  in high radial palsy, 274, 275
  preoperative program for, 274
  principles of, 274
  in rupture-related finger extension loss, 66
  tendon motion following, 338
Tenodesis
  dynamic, 276
  passive, 22, 23
Tenolysis, postoperative therapy
  following, 337
Tenosynovectomy, in rheumatoid arthritis, 66

Tenosynovitis, 60
  corticosteroid injection therapy for, 79
  of the extensor pollicis longus, 78
  flexor, 24
  of the flexor carpi ulnaris, 78
  magnetic resonance imaging findings in, 47
  psoriatic arthritis-related, 72
  pyogenic flexor, 81
  rheumatoid arthritis-related, 72
  of the second dorsal compartment, 76
  stenosing
    de Quervain's, 75, 76
    trigger finger, 76–78
  suppurative, 81
  systemic lupus erythematosus-related, 70
"Terry Thomas sign," 13, 162, 202
Tetraplegia
  neurologic deficits associated with, 118
  surgical reconstruction following, 117
  upper-limb functional surgery in, 119
Thenar space infections, 83
Thomas, Terry, 13, 162, 202
Thoracic outlet syndrome, 282
  differentiated from cubital tunnel syndrome, 264
  neurogenic, 257
Thrombocytopenia, with absent radius
  (TAR syndrome), 317, 318
Thumb
  amputations of, 141
  anesthetization of, 37
  arthrodesis of, 195,
    complications of, 195
  blood supply to, 12
  carpometacarpal joints of
    arthrodesis-related complications of, 198
    dislocations of, 187
    osteoarthritis of, 55, 198
    pain in, 334
  congenitally absent, 323
  digital arteries of, 141
  Dupuytren's disease of, 113
  extensor injuries to, anatomic zones in, 218
  in extensor pollicis longus tendon rupture, 206
  flexor tendon injuries to, 210
  "gamekeeper's," 188, 189
  hypoplasia of, 322
    reconstruction of, 322, 323
  metacarpophalangeal joint of
    hyperextension instability of, 57
    osteoarthritis of, 57
  microsurgical reconstruction of, 140
  phalangeal fracture-related residual
    deformity in, 182
  proximal phalanx ulnar base fractures of, 180
  pulley system of, 9, 211
  replantation of, 141
  rheumatoid arthritis-related deformities of, 65
  "skier's," 188

Thumb *(Continued)*
  tendinitis of, 329
  trapeziometacarpal osteoarthritis of, 51, 53, 54, 57
  as trigger finger, 77
Thumb duplication
  most common type of, 324
  Wassel's classification of, 324
Thumb-in-palm deformity, 117
  cerebral palsy-related, 115
  traumatic brain injury-related, 120
Thumb opposition, in median nerve palsy, 275
Thumb-to-index fingertip pinch, force generation during, 18
Tinel's test, 261, 271
Tissue expansion, for soft tissue coverage, 262
Tizanidine, 114
TMC joint. *See* Trapeziometacarpal joint
Tools, ergonomic modifications to, 334
Tool vibration, workers' reduced exposure to, 330
Tophus, 72
Topographic areas, of the hand, 21
Top 100 Secrets, **1–8**
Total wrist arthroplasty
  complications of, 193
  contraindications to, 193
  indications for, 193
Touch, as primary function of the hand, 15
Tourniquets, for intravenous regional anesthesia, 36
Toxicity, of local anesthetics, 33
Toxic substances, high-pressure injection of, 92
"Trampoline test," 133
Transillumination evaluation, of ganglion cysts, 311
Trapezial ridge, fractures of, 172, 173
Trapeziometacarpal joint
  arthroscopy of, 57
  center of rotation for, 17
  osteoarthritis of, 50, 53
    arthrodesis treatment of, 56, 57
    diagnostic tests for, 54
    surgical treatment of, 54,
    as swan-neck deformity cause, 54
    in the thumb, 51
Trapezium bone
  fractures of, 171
    classification of, 171
    injuries associated with, 171
  ossification of, 12
Trapezoid bone, fractures of, 171
Traumatic brain injury, 117–120
  as cubital tunnel syndrome cause, 118
  as elbow flexion spasticity cause, 118
  as "functional" spastic clenched fist cause, 120
  as heterotopic ossification cause, 117
  as spasticity cause, 117
  surgical procedures in
    functional, 120
    nonsurgical, 120
  as thumb-in-palm deformity cause, 117, 120

Traumatic brain injury *(Continued)*
  as upper extremity motor dysfunction cause, 118
  as wrist flexion deformity cause, 119
Triangular fibrocartilage complex (TFCC)
  calcification of, 41, 72
  components of, 132
  definition of, 150
  effect of radial fractures on, 154
  effect of ulnar styloid process fractures on, 154
  functions of, 133
  location of, 150
  relationship to the ulnar wrist palmar ligaments, 150
  tears of, 133
    central, 151
    classification of, 133, 151
    most common type of, 134
    peripheral, 153
    repair or debridement of, 133, 134
    treatment of, 151, 153
  ulnar attachment of, 150
  vascularity of, 151
  width of, relationship to ulnar variance, 154
Triangulation method, for midcarpal instability determination, 162
Triceps muscle paralysis, spinal cord injury-related, 278
Trigger finger, 76–78
  cause of, 76
  classification of, 77
  definition of, 76
  idiopathic, 78, 80
  open release of, 77
    as digital nerve injury cause, 78
  palmar pulleys in, 79
  secondary, 79
  treatment of, 77, 78
Triquetrohamate joint, in midcarpal joint contact, 18
Triquetrum bone
  fractures of
    as incidence of all carpal fractures, 171
    radiographic evaluation of, 40
    types of, 171
  ossification of, 12
  synostosis of, 326
Trisomy 21 (Down's syndrome), 325
Trunk nerve grafts, 270
Tumors
  benign, radiographic features of, 44
  bone tumors, **302–308**
    aggressive or malignant, 44, 302
    biopsy of, 308
    metastatic, 307, 308
  cutaneous malignant, 314
  radiographic features of, 44
  soft tissue, **310–314**
    magnetic resonance imaging of, 48
  vascular, 282

Tumor surgery, surgical margins in, 305
Two-point discrimination test, 21
Typing. *See* Keyboarding

U

Ulcers, vesicant extravasation-related, 85
Ulna bone
  distal
    dislocations of, 185
    dorsal subluxation of, 185
    excision or resection of, 196
    force distribution through, 16
    relationship to the palmar ligaments, 150
    shortening of, 129
      contraindication to, 154
Ulnar artery, 280
  occlusion and/or aneurysm of, 200, 281
Ulnar collateral ligament, injuries to
  avulsion fractures, 180
  examination of, 188
  as "gamekeeper's thumb," 188, 189
  as Stener's lesion, 188
  of the thumb metacarpophalangeal joint, 188
Ulnar deficiency, 318
Ulnar deviation, 15
  in force distribution, 16
  for maximal power grip, 17
  rheumatoid arthritis-related, 278
Ulnar dimelia, radiographic appearance of, 324
Ulnar finger extensor tendons, rheumatoid arthritis-related rupture of, 61
Ulnar head, role in distal radioulnar joint stability, 15, 16
Ulnar nerve, 23
  crossover in, 12
  as forearm innervation, 30
  muscles innervated by, 11
Ulnar nerve compression
  anatomic sites of, 261, 262, 263
  at the elbow, 27
Ulnar nerve palsy
  hand function deficits in, 275
  low
    claw deformity associated with, 276
    key-pinch function restoration in, 276, 278
    with low median nerve palsy, 278
    wrist and hand function restoration in, 278
Ulnar styloid process, fractures of, 154
Ulnar translocation, of the carpus, 185, 191
Ulnar variance
  definition of, 38, 150
  radiographic determination of, 38, 39
    posteroanterior views in, 127
  relationship to
    Kienböck's disease, 127
    the triangular fibrocartilage complex (TFCC) width, 154

Ulnocarpal abutment, 202
  differentiated from Kienböck's disease, 128
Ulnocarpal impaction syndrome, 134
  definition of, 154
  treatment of, 154
Ulnocarpal volar ligament, 158
Ultrasonography
  for carpal instability assessment, 160
  color duplex, advantages of, 280
  indications for, 48
Upper extremity
  developmental sequence of, 316
  occupational injuries and illnesses of, **328–332**
  overgrowth of, 324
  work-related (WRUED) disorders of, 330
Upper motor neuron lesions, 255

V

VACTERL syndrome, 317
Vascular disorders, **280–283**
Vascularized nerve grafts, 270
Vaughn-Jackson syndrome, 61
Venous insufficiency, in replanted digits, 293
Vesicants, as extravasation injury cause, 85
Vibration, workers' reduced exposure to, 330
Vincristine, 85
Vincula, 212
VISI (volar intercalated segmental instability), 44, 162
Volar beak ligament, 53
  reconstruction of, 56
Volar fixed-angle plates, 145
Volar intercalated segmental instability (VISI), 44, 162
Volar plate complex, injury to, 191
Volkmann's ischemic contracture, 103
von Recklinghausen's disease, 312

W

Wallerian degeneration, 255, 266
Wartenberg's sign, 23, 277
Watson scaphoid shift test, 159
Windblown hand, 323
Workers' compensation, 340, 342
  depositions in, 343
Worker's compensation claims, for upper extremity disorders, 328
Wound healing, 235
  effect of nicotine on, 233
  phases of, 235
    provisional wound matrix in, 235
    by secondary intention, 237
Wound infections, postoperative, 82
Wound management, 82
  after a fasciotomy, 104
Wounds
  contraction of, 236, 238
  free flap coverage for, 140
Wound strength, 236

Wrist
- arthrodesis (fusion) of
  - capitate-lunate-hamate-triquetral, 194
  - capitolunate, 194
  - contraindications to, 193
  - indications for, 193
  - intercarpal (limited), 194, 195
  - lunotriquetral, 194
  - nonunion rate in, 236
  - preferred position for, 193
  - scaphocapitate, 194
  - in SLAC pattern, 198
  - surgical goals for, 193
  - triscaphe, 194
  - unilateral, 194
- arthrography of, 45
  - triple-injection (three-compartment), 45
  - use in carpal instability assessment, 160
  - for wrist pain evaluation, 46
- arthroscopy of, **131–136**
  - complications incidence in, 135
  - indications for, 131
  - midcarpal, 132
  - portals for, 131, 132, 136
- biomechanics of, **15–19**
- carpal tunnel radiographic view of, 42
- center of rotation for, 15
- cerebral palsy-related absence of, 116
- "clenched fist view" of, 100
- compartments of
  - dorsal, tendons of, 11
  - separation of, 12
- degrees of freedom of, 15
- deviation of, proximal carpal row during, 16
- disarticulation of, 297
- extension of, 15, 16
  - for maximal power grip, 17
  - midcarpal joint in, 17
  - radiocarpal joint in, 17
- flexion deformity of, 116
- flexion of, 16
  - midcarpal joint in, 17
  - radiocarpal joint in, 17

Wrist *(Continued)*
- ganglion of
  - of the dorsal wrist, 310, 311
  - resection of, 135, 136
  - of the volar wrist, 282, 310
- instability of, radiographic views of, 42
- longitudinal axis of the scaphoid/longitudinal axis of the hand angle in, 13
- magnetic resonance arthrography of, 48
- midcarpal joint force distribution during, 18
- neutral position of, 18
- radiocarpal fusion-related loss of motion in, 15
- radiocarpal/ulnocarpal load transfer ratio in, 133
- role in activities of daily living (ADLs), 15
- scapholunate advanced collapse (SLAC), 55, 202
  - four-corner arthrodesis for, 198
- septic arthritis of, 70, 83
- stability of, 158 Carpal instability
- total arthroplasty of
  - complications of, 193
  - contraindications to, 193
  - indications for, 193
- ulnar deviation of, 15
  - for maximal power grip, 17
- ulnar neutral position of, 133
- weakness in, cerebral palsy-related, 116

Wrist pain, **200–209**
- arthrographic evaluation of, 46

WRUED (work-related upper extremity disorders), 330

X
Xenografts, 237
X-rays. *See* Radiographic evaluation

Z
Z-plasty
- indications for, 245
- 60-degree, 245
- technical errors associated with, 245
- technical points of, 245